Abstract Algebra with Applications

Abstract Algebra with Applications provides a friendly and concise introduction to algebra, with an emphasis on its uses in the modern world. The first part of this book covers groups, after some preliminaries on sets, functions, relations, and induction, and features applications such as public-key cryptography, Sudoku, the finite Fourier transform, and symmetry in chemistry and physics. The second part of this book covers rings and fields, and features applications such as random number generators, error-correcting codes, the Google page rank algorithm, communication networks, and elliptic curve cryptography.

The book's masterful use of colorful figures and images helps illustrate the applications and concepts in the text. Real-world examples and exercises will help students contextualize the information. Meant for a year-long undergraduate course in algebra for math, engineering, and computer science majors, the only prerequisites are calculus and a bit of courage when asked to do a short proof.

CAMBRIDGE MATHEMATICAL TEXTBOOKS

Cambridge Mathematical Textbooks is a program of undergraduate and beginning graduate-level textbooks for core courses, new courses, and interdisciplinary courses in pure and applied mathematics. These texts provide motivation with plenty of exercises of varying difficulty, interesting examples, modern applications, and unique approaches to the material.

A complete list of books in the series can be found at www.cambridge.org/mathematics
Recent titles include the following:

Chance, Strategy, and Choice: An Introduction to the Mathematics of Games and Elections, S. B. Smith
Set Theory: A First Course, D. W. Cunningham
Chaotic Dynamics: Fractals, Tilings, and Substitutions, G. R. Goodson
A Second Course in Linear Algebra, S. R. Garcia & R. A. Horn
Introduction to Experimental Mathematics, S. Eilers & R. Johansen
Exploring Mathematics: An Engaging Introduction to Proof, J. Meier & D. Smith
A First Course in Analysis, J. B. Conway
Introduction to Probability, D. F. Anderson, T. Seppäläinen & B. Valkó
Linear Algebra, E. S. Meckes & M. W. Meckes
A Short Course in Differential Topology, B. I. Dundas

Abstract Algebra with Applications

AUDREY TERRAS
University of California, San Diego, CA, USA

CAMBRIDGE
UNIVERSITY PRESS

University Printing House, Cambridge CB2 8BS, United Kingdom

One Liberty Plaza, 20th Floor, New York, NY 10006, USA

477 Williamstown Road, Port Melbourne, VIC 3207, Australia

314–321, 3rd Floor, Plot 3, Splendor Forum, Jasola District Centre, New Delhi – 110025, India

79 Anson Road, #06-04/06, Singapore 079906

Cambridge University Press is part of the University of Cambridge.

It furthers the University's mission by disseminating knowledge in the pursuit of education, learning, and research at the highest international levels of excellence.

www.cambridge.org
Information on this title: www.cambridge.org/9781107164079

First published 2019

Printed and bound in Great Britain by Clays Ltd, Elcograf S.p.A.

A catalogue record for this publication is available from the British Library.

ISBN 978-1-107-16407-9 Hardback

To the bears and koalas

Contents

List of Figures ix
Preface xiii

PART I GROUPS

1 Preliminaries 3

1.1 Introduction 3
1.2 Sets 6
1.3 The Integers 9
1.4 Mathematical Induction 14
1.5 Divisibility, Greatest Common Divisor, Primes, and Unique Factorization 19
1.6 Modular Arithmetic, Congruences 26
1.7 Relations 30
1.8 Functions, the Pigeonhole Principle, and Binary Operations 34

2 Groups: A Beginning 43

2.1 What is a Group? 43
2.2 Visualizing Groups 52
2.3 More Examples of Groups and Some Basic Facts 56
2.4 Subgroups 64
2.5 Cyclic Groups are Our Friends 72

3 Groups: There's More 81

3.1 Groups of Permutations 81
3.2 Isomorphisms and Cayley's Theorem 89
3.3 Cosets, Lagrange's Theorem, and Normal Subgroups 93
3.4 Building New Groups from Old, I: Quotient or Factor Groups G/H 98
3.5 Group Homomorphism 102
3.6 Building New Groups from Old, II: Direct Product of Groups 108
3.7 Group Actions 114

4 Applications and More Examples of Groups 124

4.1 Public-Key Cryptography 124
4.2 Chemistry and the Finite Fourier Transform 129
4.3 Groups and Conservation Laws in Physics 135
4.4 Puzzles 142
4.5 Small Groups 146

PART II RINGS

5 Rings: A Beginning 157

5.1 Introduction 157
5.2 What is a Ring? 158
5.3 Integral Domains and Fields are Nicer Rings 166
5.4 Building New Rings from Old: Quotients and Direct Sums of Rings 173
5.5 Polynomial Rings 180
5.6 Quotients of Polynomial Rings 185

6 Rings: There's More 189

6.1 Ring Homomorphisms 189
6.2 The Chinese Remainder Theorem 193
6.3 More Stories about $F[x]$ Including Comparisons with $\mathbb{Z}$ 198
6.4 Field of Fractions or Quotients 202

7 Vector Spaces and Finite Fields 206

7.1 Matrices and Vector Spaces over Arbitrary Fields and Rings like $\mathbb{Z}$ 206
7.2 Linear Functions or Mappings 218
7.3 Determinants 224
7.4 Extension Fields: Algebraic versus Transcendental 229
7.5 Subfields and Field Extensions of Finite Fields 233
7.6 Galois Theory for Finite Fields 239

8 Applications of Rings 244

8.1 Random Number Generators 244
8.2 Error-Correcting Codes 256
8.3 Finite Upper Half Planes and Ramanujan Graphs 265
8.4 Eigenvalues, Random Walks on Graphs, and Google 272
8.5 Elliptic Curve Cryptography 282

References 299
Index 305

Figures

0.1 Benzene C_6H_6 xiv
0.2 Photoshopped flower xiv
0.3 Hibiscus in Kauai xiv
0.4 Picture with symmetry coming from the action of 2×2 matrices with nonzero determinant and elements in a finite field with 11 elements xv
1.1 Intersection and union of square A and heart B 7
1.2 Cartesian product $[0, 1] \times \{2\}$ 8
1.3 $[0, 1]^3$ 8
1.4 Graph representing the hypercube $[0, 1]^4$ 9
1.5 Integers on the line – only even ones are labelled 13
1.6 The first principle of mathematical induction. A penguin surveys an infinite line of equally spaced dominos. If the nth domino is close enough to knock over the $(n+1)$th domino, then once the penguin knocks over the first domino, they should all fall over 14
1.7 Here is an attempt to picture the second mathematical induction principle in which we arrange dominos so that various numbers of dominos are needed to knock over the dominos to their left. In this picture step 1 would be for the penguin to knock over d_1 and d_2 17
1.8 A color is placed at the (m, n) entry of a 101×101 matrix according to the value of $\gcd(m, n)$. This is an ArrayPlot in Mathematica 25
1.9 Rolling up the integers modulo 3 27
1.10 Mathematica picture of the $x < y$ relation for the integers between 1 and 50 31
1.11 Mathematica picture of the $y|x$ relation for the integers between 1 and 50 31
1.12 Mathematica picture of the $x \equiv y \pmod 5$ relation for the integers between 1 and 50 32
1.13 Poset diagram of the positive divisors of 24 34
1.14 The pigeonhole principle 39
2.1 The symmetries of a regular triangle are pictured 44
2.2 Part of a design with translational symmetry which should be imagined to stretch out to ∞ and $-\infty$ 47
2.3 A figure with C_8 symmetry – not D_8 symmetry 48
2.4 Art from the Raja Ampat islands in the Indonesian part of New Guinea 48
2.5 Wallpaper from a Fourier series in two variables 50
2.6 Spherical wallpaper from spherical harmonics 51
2.7 Hyperbolic wallpaper from a modular form known as Δ on the upper half plane – a function with an invariance property under fractional linear transformation $(az+b)/(cz+d)$, where $a, b, c, d \in \mathbb{Z}$ and $ad - bc = 1$ 51

2.8 Group Explorer version of the multiplication table for C_6, a cyclic group of order 6 52
2.9 Group Explorer version of the multiplication table for D_3 (alias S_3) with our upper case R and F replaced by lower case letters 52
2.10 Cayley graph of cyclic group $G=\langle a\rangle$ of order 6 with generating set $S=\{a\}$ 53
2.11 Cayley graph of D_3 with generating set $\{R, F\}$. Our upper case letters are replaced by lower case in the diagram. If there are arrows in both directions on an edge, we omit the arrows 53
2.12 Undirected version of Cayley graph for $C_6=\langle a\rangle$, generating set $S=\{a, a^{-1}\}$ 53
2.13 Cayley graph of $C_6=\langle a\rangle$, generating set $S=\{a, a^3, a^5\}$ 53
2.14 Group Explorer version of the multiplication table for the Klein 4-group 54
2.15 Symmetrical designs 54
2.16 The Platonic solids 57
2.17 Poset diagram for subgroups of D_3 as defined in (2.8) 68
2.18 The Group Explorer version of the multiplication table for a cyclic group of order 10 73
2.19 Cayley graph $X(\langle a\rangle, \{a, a^{-1}\})$ for a cyclic group $\langle a\rangle$ of order 10 76
2.20 A less boring picture of a 10-cycle 76
2.21 Poset diagram of the subgroups of $\mathbb{Z}_{24}$ under addition 78
2.22 Cycle diagram in the multiplicative group $\mathbb{Z}_{15}^*$ 80
3.1 Tetrahedron 87
3.2 On the left is the Cayley graph for the Klein 4-group $K_4=\{e, h, v, hv\}$, with generating set $S=\{h, v\}$ using the notation of Figure 2.14. On the right is the Schreier graph for K_4/H, where $H=\{e, h\}$, with the same set $S=\{h, v\}$ 97
3.3 Roll up $\mathbb{Z}$ to get $\mathbb{Z}/n\mathbb{Z}$ 105
3.4 Roll up the real line to get a circle $\mathbb{T}\cong\mathbb{R}/\mathbb{Z}$ 106
3.5 Cayley graph for $\mathbb{Z}_2\oplus\mathbb{Z}_2$ with the generating set $\{(1,0),(0,1)\}$ and bears at the vertices 109
3.6 Cayley graph for $\mathbb{Z}_2\oplus\mathbb{Z}_2\oplus\mathbb{Z}_2$ with the generating set $\{(1,0,0),(0,1,0),(0,0,1)\}$ and koalas at the vertices 109
3.7 Cayley graph for $\mathbb{Z}_2\oplus\mathbb{Z}_2\oplus\mathbb{Z}_2\oplus\mathbb{Z}_2$ with the generating set $\{(1,0,0,0),(0,1,0,0),(0,0,1,0),(0,0,0,1)\}$ 110
3.8 A finite torus, which is the Cayley graph $X(\mathbb{Z}_{10}\oplus\mathbb{Z}_5, \{(\pm1,0),(0,\pm1)\})$ 111
3.9 The continuous torus (obtained from the plane modulo its integer points; i.e., $\mathbb{R}\oplus\mathbb{R}$ modulo $\mathbb{Z}\oplus\mathbb{Z}$) 111
3.10 Cayley graph for the quaternion group with generating set $\{\pm i, \pm j\}$ 113
3.11 The 13 necklaces with six beads of two colors 121
3.12 The dodecahedron graph drawn by Mathematica 122
3.13 The cuboctahedron drawn by Mathematica 123
4.1 Vibrating system of two masses 142
4.2 Group Explorer's multiplication table for the semi-direct product $C_3\rtimes C_4$ 149
4.3 Group Explorer draws the Cayley graph $X(C_3\rtimes C_4, \{a, b\})$ 149
4.4 The Cayley graph $X(\mathrm{Aff}(5), S_{1,2})$, with generating set defined by equation (4.16), has edges given by solid green lines while the dashed magenta lines are the edges of a dodecahedron 150

4.5 Butterfly from Cayley graph of Heis($\mathbb{Z}/169\mathbb{Z}$) 151
4.6 A spanning tree for the tetrahedron graph is indicated in solid fuchsia lines. Since the three dashed purple edges are left out, the fundamental group of the tetrahedron graph is the free group on three generators. The arrows show a closed path on the tetrahedron graph 153
4.7 The bouquet of three loops obtained by collapsing the tree in the tetrahedron graph of Figure 4.6 to point a 153
4.8 A passion flower 154
5.1 The color at point $(x, y) \in \mathbb{Z}_{163}^2$ indicates the value of $x^2 + y^2 \pmod{163}$ 157
5.2 Points (x, y), for $x, y \in \mathbb{Z}_{11^2}, y \neq 0$, have the same color if $z = x + y\sqrt{\delta}$ are equivalent under the action of non-singular 2×2 matrices $g = \begin{pmatrix} a & b \\ c & d \end{pmatrix}$ with entries in $\mathbb{Z}_{11}$. The action of g on z is by fractional linear transformation $z \to (az + b)/(cz + d) = gz$. Here δ is a fixed non-square in the field $\mathbb{F}_{121}$ with 121 elements 158
5.3 Poset diagram of the ideals in $\mathbb{Z}_{12}$ 178
5.4 A feedback shift register diagram corresponding to the finite field $\mathbb{Z}_2[x]/\langle x^3 + x + 1\rangle$ and the multiplication table given in the text 186
6.1 The Cayley graph $X(\mathbb{Z}_{15}, \{\pm 1 \pmod{15}\})$ 196
6.2 The Cayley graph $X(\mathbb{Z}_{15}, \{5, 6, 9, 10 \pmod{15}\})$ 196
7.1 The poset of subfields of $\mathbb{F}_{2^{24}}$ 234
8.1 The Cayley graph $X(\mathbb{Z}_{17}^*, \{3 \pmod{17}\})$ 245
8.2 The same graph as in Figure 8.1 except that now the vertices are given the usual ordering $1, 2, 3, 4, \ldots, 16$ 245
8.3 Plot of points $P_j = (j, v_j)$ whose second component is the real number $\frac{1}{499}$ times $7^j \pmod{499}$, identifying $7^j \pmod{499}$ as an integer between 1 and 498 248
8.4 Plot of points $P_j = (v_j, v_{j+1})$ whose first component is the real number $\frac{1}{499}$ times $7^j \pmod{499}$, identifying $7^j \pmod{499}$ as an integer between 1 and 498 248
8.5 Plot of points $P_j = (v_j, v_{j+1}, v_{j+2})$ whose first component is the real number $\frac{1}{499}$ times $7^j \pmod{499}$, identifying $7^j \pmod{499}$ as an integer between 1 and 498 249
8.6 Plot of points $P_j = (v_j, w_j)$ whose first component is the real number $\frac{1}{499}$ times $7^j \pmod{499}$, identifying $7^j \pmod{499}$ as an integer between 1 and 498 and whose second component is the analog with 499 replaced with 503 249
8.7 Points (v_i, w_i, z_i) from three vectors v, w, z formed from powers of generators of $\mathbb{F}_p^*$ for $p = 499, 503$, and 521, respectively 250
8.8 Feedback shift register corresponding to example 2 252
8.9 Sending a message of 0s and 1s to Professor Bolukxy on the planet Xotl 257
8.10 The matrix H_{32} where the 1s and -1s have become red and purple 263
8.11 Color at point $z = x + y\sqrt{\delta}$ in H_{163} is found by computing the Poincaré distance $d(z, \sqrt{\delta})$ 268
8.12 The graph on the left is $X_3(-1, 1)$, an octahedron, and that on the right is $X_5(2, 1)$ with the edges in green. The pink dashed lines on the right are the dodecahedron 269
8.13 Another version of Figure 5.2 270

8.14 A random walk on a pentagon. At time $t=0$, the big penguin is at vertex 1. At time $t=1$ the penguin has probability $\frac{1}{2}$ of being at vertex 2 and probability $\frac{1}{2}$ of being at vertex 5. So the penguins at these vertices are half size 275

8.15 Surfing a very small web 277

8.16 Real points (x, y) on the elliptic curve $y^2 = x^3 + x^2$ 282

8.17 Real points (x, y) on the elliptic curve $y^2 = x^3 - x + 1$ 283

8.18 Real points (x, y) on the elliptic curve $y^2 = x^3 - x$ 283

8.19 Addition $A + B = C$ on the elliptic curve $y^2 = x^3 - x$ over $\mathbb{R}$ 286

8.20 The rational points on the curve $y^2 + y = x^3 - x^2$ are a, b, c, d and the point at ∞ 287

8.21 The pink squares indicate the points (x, y) on the elliptic curve $y^2 = x^3 - x + 1$ mod 59. Points marked are: $A = (15, 36), B = (22, 40), C = (32, 46)$, with $A + B = -C = (32, 13)$ 289

8.22 Level "curves" of $y^2 - x^3 - x + 1 \pmod{29}$ 296

8.23 Smoothed level "curves" of $y^2 - x^3 - x + 1 \pmod{29}$ 297

8.24 A photoshopped version of the level curves of $(y + 2x)^4 + (x - 2y)^4 \pmod{101}$ 297

Preface

My goal for this book is to provide a friendly concise introduction to algebra with emphasis on its uses in the modern world – including a little history, concrete examples, and visualization. Beyond explaining the basics of the theory of groups, rings, and fields, I aim to give many answers to the question: What is it good for?" The standard undergraduate mathematics course in the 1960s (when I was an undergraduate) proceeded from Definition 1.1.1 to Corollary 14.5.59 with little room for motivation, examples, history, and applications. I plan to stay as far as possible from that old format, modeling my discussion on G. Strang's book [115], where the preface begins: "I believe that the teaching of linear algebra has become too abstract." My feeling is that the teaching of modern algebra (the non-linear part) has become even more abstract. I will attempt to follow Strang's lead and treat modern algebra in a way that will make sense to a large variety of students. On the other hand, the goal is to deal with some abstractions – groups, rings, and such things. Yes, it is abstraction and generalization that underlies the power of mathematics. Thus there will be some conflict between the applied and pure aspects of our subject.

The book is intended for a year-long undergraduate course in algebra. The intended audience is the less theoretically inclined undergraduates majoring in mathematics, the physical and social sciences, or engineering – including those in applied mathematics or those intending to get a teaching credential.

The prerequisites are minimal: comfort with the real numbers, the complex numbers, matrices, vector spaces at the level of calculus courses – and a bit of courage when asked to do a short proof.

In this age of computers, algebra may have replaced calculus (analysis) as the most important part of mathematics. For example:

1. Error-correcting codes are built into the DVD player and the computer. Who do you call to correct errors? The algebraist, that's who!
2. Digital signal processing (such as that involved in medical scanners, weather prediction, the search for oil) is dominated by the fast Fourier transform or FFT. What is this? The FFT is a finite sum whose computation has been sped up considerably by an algebra trick which goes back to Gauss in 1805. Once more algebra, not analysis, rules.
3. In chemistry and physics, one studies structures with symmetry such as the benzene molecule (C_6H_6) depicted in Figure 0.1. What does the 6-fold symmetry have to do with the properties of benzene? Group theory is the tool one needs for this.
4. The search for secret codes – cryptography. Much of the modern world – particularly that which lives on the internet – depends on these codes being secure. But are they? We will consider public-key codes. And who do you call to figure out these codes? An algebraist!

5. The quest for beauty in art and nature. I would argue that symmetry groups are necessary tools for this quest. See Figures 0.2–0.4.

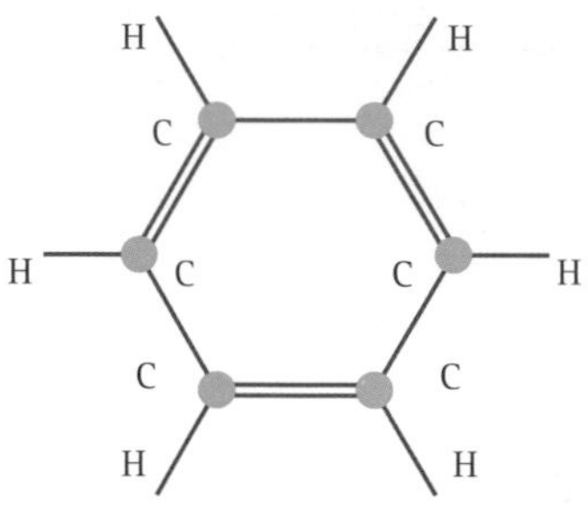

Figure 0.1 Benzene C_6H_6

Figure 0.2 Photoshopped flower

Figure 0.3 Hibiscus in Kauai

Our goal here is to figure out enough group and ring theory to understand many of these applications. And we should note that both algebra and analysis are necessary for the applications. In fact, we shall see some limits, derivatives, and integrals before the last pages of this book. You can skip all the applications if you just want to learn the basics

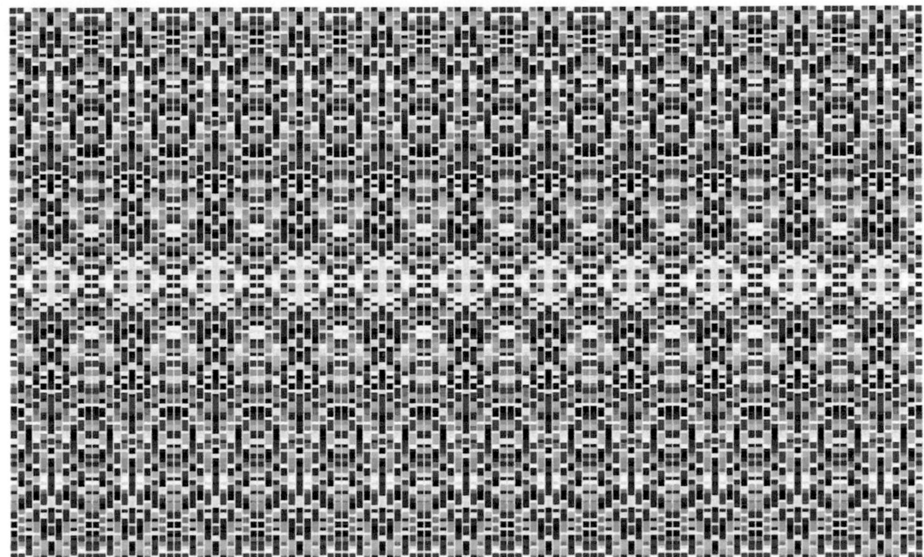

Figure 0.4 Picture with symmetry coming from the action of 2×2 matrices with nonzero determinant and elements in a finite field with 11 elements

of modern algebra. The non-applied sections are all independent of those on applications. However, you would be missing one of the big reasons that the subject is taught. And feel free to skip any applied sections you want to skip – or to add any missing application that you want to understand.

The first part of this book covers groups, after some preliminaries on sets, functions, relations, the integers, and mathematical induction. Of course every calculus student is familiar with the group of real numbers under addition and similarly with the group of nonzero real numbers under multiplication. We will consider many more examples – favorites being finite groups such as the group of symmetries of an equilateral triangle.

Much of our subject began with those favorite questions from high school algebra such as finding solutions to polynomial equations. It took methods of group theory to know when the solutions could not be found in terms only of nth roots. Galois, who died at age 21 in a duel in 1832, laid the foundations to answer such questions by looking at groups of permutations of the roots of a polynomial. These are now called Galois groups. See Edna E. Kramer [59, Chapter 16] or Ian Stewart [114] for some of the story of Galois and the history of algebra. Another reference for stories about Galois and the many people involved in the creation of this subject is *Men of Mathematics* by Eric Temple Bell [8]. This book is often criticized for lack of accuracy, but it is more exciting than most. I found it inspirational as an undergraduate – despite the title.

Another area that leads to our subject is number theory: the study of the ring of integers

$$\mathbb{Z} = \{0, \pm 1, \pm 2, \pm 3, \ldots\}.$$

The origins of this subject go back farther than Euclid's *Elements*. Euclid lived in Alexandria around 300 BC and his book covers more than the plane geometry we learned in high school. Much of the basic theory of the integers which we cover in Chapter 1 is to be found in Euclid's *Elements*. Why is it that the non-plane geometry part of Euclid's *Elements* does not seem to be taught in high school?

Polynomial equations with integer coefficients are often called Diophantine equations in honor of Diophantus who also lived in Alexandria, but much later (around AD 200). Yes, algebra is an old subject and one studied in many different countries. For example, the name "algebra" comes the word *al-jabr*, one of the two operations used to solve quadratic equations by the Persian mathematician and astronomer Mohamed ibn Musa al-Khwārizmī, who lived around AD 800.

A large part of this subject was created during many attempts to prove Fermat's last theorem. This was a conjecture of Pierre de Fermat in 1637 stating that the equation $x^n + y^n = z^n$ can have no integer solutions x, y, z with $xyz \neq 0$ and $n > 2$. Fermat claimed to have a proof that did not fit in the margin of the book in which he wrote this conjecture. People attempted to prove this theorem without success until A. Wiles with the help of R. Taylor in 1995. People still seek an "elementary" proof.

Groups are sets with one operation satisfying the axioms to be listed in Section 2.1. After the basic definitions, we consider examples of small groups. We will visualize groups using Cayley graphs and various other diagrams such as Hasse or poset diagrams as well as cycle diagrams. Other topics of study include subgroups, cyclic groups, permutation groups, functions between groups preserving the group operations (homomorphisms), cosets of subgroups, building new groups whose elements are cosets of normal subgroups, direct products of groups, actions of groups on sets. We will consider such applications as public-key cryptography, the finite Fourier transform, and the chemistry of benzene. Favorite examples of groups include cyclic groups, permutation groups, symmetry groups of the regular polygons, matrix groups such as the Heisenberg group of 3×3 upper triangular matrices with real entries and 1 on the diagonal, the group operation being matrix multiplication.

The second part of this book covers rings and fields. Rings have two operations satisfying the axioms listed in Section 5.2. We denote the two operations as addition $+$ and multiplication $*$ or $\cdot$. The identity for addition is denoted 0. It is NOT assumed that multiplication is commutative: that is, it is not assumed that $ab = ba$. If multiplication is commutative, then the ring is called commutative. A field F is a commutative ring with an identity for multiplication (called $1 \neq 0$) such that the nonzero elements of F form a multiplicative group. Most of the rings considered here will be commutative. We will be particularly interested in finite fields like the field of integers modulo p, $\mathbb{Z}/p\mathbb{Z}$ where p is prime. You must already be friends with the field $\mathbb{Q}$ of rational numbers – fractions with integer numerator and nonzero integer denominator. And you know the field $\mathbb{R}$ of real numbers from calculus: that is, limits of Cauchy sequences of rationals. We are not supposed to say the word "limit" as this is algebra. So we will not talk about constructing the field of real numbers $\mathbb{R}$. Ring theory topics include: definitions and basic properties of rings, fields, ideals, and functions between rings which preserve the ring operations (ring homomorphisms). We will also build new rings (quotient rings) whose elements are cosets $x + I$ of an ideal I in ring R, for x in ring R. Note that here R is an arbitrary ring, not necessarily the field of real numbers. We will look at rings of polynomials and their similarity to the ring of integers. We can do linear algebra for finite-dimensional vector spaces over arbitrary fields in a similar way to the linear algebra that is included in calculus sequences. Our favorite rings are the ring $\mathbb{Z}$ of integers and the quotient ring $\mathbb{Z}/n\mathbb{Z}$ of integers modulo n, in which x, is identified with all integers of the form $x + nk$, for integer k. Another favorite is the ring of Hamilton quaternions which is isomorphic to four-dimensional space over the real numbers with basis $1, i, j, k$ and with multiplication defined by $ij = k = -ji$, $i^2 = j^2 = k^2 = -1$.

Historically, much of our subject came out of number theory and the desire to prove Fermat's last theorem by knowing about factorization into irreducibles in rings like $\mathbb{Z}\left[\sqrt{m}\right]=\left\{a+b\sqrt{m}\mid a,b\in\mathbb{Z}\right\}$, where m is a non-square integer. For example, it turns out that, when $m=-5$, we have two different factorizations:

$$2\cdot 3=\left(1+\sqrt{-5}\right)\cdot\left(1-\sqrt{-5}\right).$$

So the fundamental theorem of arithmetic – true for $\mathbb{Z}$ as is shown in Section 1.5 – is false for $\mathbb{Z}\left[\sqrt{-5}\right]$.

Assuming that such factorizations were unique, Lamé thought that he had proved Fermat's last theorem in 1847. Dedekind fixed up arithmetic in such rings by developing the arithmetic of ideals, which are certain sets of elements of the ring to be considered in Section 5.4. One then had (at least in rings of integers in algebraic number fields) unique factorization of ideals as products of prime ideals, up to order. Of course, Lamé's proof of Fermat's last theorem was still invalid (lame – sorry for that).

The favorite field of the average human mathematics student is the field of real numbers $\mathbb{R}$. A favorite finite field for a computer is $\mathbb{F}_p=\mathbb{Z}/p\mathbb{Z}$, where $p=$ prime. Of course you can define $\mathbb{Z}/n\mathbb{Z}$, for any positive integer n, but you only get a ring and not a field if n is not a prime. We consider $\mathbb{Z}/n\mathbb{Z}$ as a group under addition in Section 1.6. In Chapter 5 we view it as a ring with two operations, addition and multiplication.

Finite rings and fields were really invented by Gauss (1801) and earlier Euler (1750). Galois and Abel worked on field theory to figure out whether nth degree polynomial equations are solvable with equations involving only radicals $\sqrt[m]{a}$. In fact, finite fields are often called "Galois fields."

The terminology of algebra was standardized by mathematicians such as Richard Dedekind and David Hilbert in the late 1800s. Much of abstract ring theory was developed in the 1920s by Emmy Noether. Discrimination against both women and Jews made it hard for her to publish. The work became well known thanks to B. L. Van der Waerden's two volumes [124] on modern algebra. Van der Waerden wrote these books after studying with Emmy Noether in 1924 in Göttingen. He had also heard lectures of Emil Artin in Hamburg earlier. See Edna E. Kramer [59] for more information on Noether and the other mathematicians who developed the view of algebra we are aiming to present.

The abstract theory of algebras (which are special sorts of rings) was applied to group representations by Emmy Noether in 1929. This has had a big impact on the way people do harmonic analysis, number theory, and physics. In particular, certain adelic group representations are central to the Wiles proof of Fermat's last theorem.

It would perhaps shock many pure mathematics students to learn how much algebra is part of the modern world of applied mathematics – both for good and ill. Google's motto: "Don't be evil," has not always been the motto of those using algebra. Of course, the Google search engine itself is a triumph of modern linear algebra, as we shall see.

We will consider many applications of rings in Chapter 8. Section 8.1 concerns random number generators from finite rings and fields. These are used in simulations of natural phenomena. In prehistoric times like the 1950s sequences of random numbers came from tables like that published by the Rand corporation. Random numbers are intrinsic to Monte Carlo methods. These methods were named for a casino in Monaco by J. von Neumann and S. Ulam in the 1940s while working on the atomic bomb. Monte Carlo methods are useful in computational physics and chemistry (e.g., modeling the behavior of galaxies, weather on earth), engineering (e.g., simulating the impact of pollution), biology (simulating the behavior of biological systems such as a cancer), statistics (hypothesis testing), game design, finance (e.g., to value options, analyze derivatives – the very thing that led to the horrible recession/depression of 2008), numerical mathematics (e.g., numerical integration, stochastic optimization), and the gerrymandering of voting districts.

In Section 8.2 we will show how the finite field with two elements and vector spaces over this field lead to error-correcting codes. These codes are used in DVDs and in the transmission of information between a Mars spacecraft and NASA on the earth. Section 8.3 concerns (among other things) the construction of Ramanujan graphs which can provide efficient communication networks.

Section 8.4 gives applications of the eigenvalues of matrices to Googling. Section 8.5 gives applications of elliptic curves over finite fields to cryptography.

The rush to abstraction of twentieth-century mathematics has had some odd consequences. One of the results of the abstract ring theory approach was to create such an abstract version of Fourier analysis that few can figure out what is going on. A similar thing has happened in number theory. On the other hand, modern algebra has often made it easier to see the forest for the trees by simplifying computations, removing subsubscripts, doing calculations once in the general case rather than many times, once for each example.

The height of abstraction was achieved in the algebra books of Nicolas Bourbaki (really a group of French mathematicians). I am using the Bourbaki notation for the fields of real numbers, complex numbers, rational numbers, and the ring of integers. But Bourbaki seems to have disliked pictures as well as applications. I do not remember seeing enough examples or history when I attempted to read Bourbaki's *Algebra* as an undergrad. In an interview, one of the members of Bourbaki, Pierre Cartier (see Marjorie Senechal [102]) said: "The Bourbaki were Puritans, and Puritans are strongly opposed to pictorial representations of truths of their faith." More information on Bourbaki as well as other fashions in mathematics can be found in Edna E. Kramer's history book [59]. She also includes a brief history of women in mathematics as well as the artificial separation between pure and applied mathematics.

As I said in my statement of goals, I will attempt to be as non-abstract as possible in this book and will seek to draw pictures in a subject where few pictures ever appear. I promise to give examples of every concept, but hope not to bury the reader in examples either, since I do aim for brevity. As I am a number theorist interested in matrix groups, there will be lots of numbers and matrices. Each chapter will have many exercises. It is important to do them – or as many of them as you can. Some exercises will be needed later in the book. The answers (mostly sketchy outlines) to odd-numbered exercises will be online hopefully. See my website. There may also be hints on others. No proof is intended to be very long. The computational problems might be slightly longer and sometimes impossible without the help of a computer. I will be using Mathematica, Scientific Workplace, and Group Explorer to help with computations.

Suggestions for Further Reading

A short list of possible references is: Garrett Birkhoff and Saunders Maclane [9], Larry L. Dornhoff and Franz E. Hohn [25], David S. Dummit and Richard M. Foote [28], Gertrude Ehrlich [29], John B. Fraleigh [32], Joseph A. Gallian [33], William J. Gilbert and W. Keith Nicholson [35], Israel N. Herstein [42], Audrey Terras [116]. There is also a free program: Group Explorer, which you can download and use to explore small groups. Another free but harder to use program is SAGE. I will be using Mathematica and Scientific Workplace. The Raspberry Pi computer ($35) comes with Mathematica (and not much else). There are many books on line as well. One example is Judson [50]. It includes computer exercises using SAGE. An on line group theory book making use of the Group Explorer program is that of Carter [12]. Wikipedia is often very useful – or just asking Google to answer a question. It is easier to be a student now than it was in my time – thanks to the multitude of resources to answer questions. On the other hand, it was nice just to have the one small book – in my case, Birkhoff and Maclane [9] – to deal with. And – perhaps needless to say – online sources can lie. Even the computer can lie – witness the arithmetic error in the Pentium chip that was revealed by number theorists' computations in the 1990s. But I have found that Wikipedia is usually very useful in its discussions of undergraduate mathematics, as is the mathematical software I have used.

It is often enlightening to look at more than one reference. Where something is mumbled about in one place, that same thing may be extremely clearly explained in another. Also feel free to read a book in a non-linear manner. If you are interested in a particular result or application, start there.

Acknowledgments

I should thank the Mathematical Sciences Research Institute, Berkeley, for support during the writing of some of this book, as well as the students in my applied algebra courses at the University of California, San Diego.

Part I

Groups

1

Preliminaries

1.1 Introduction

Notation. From now on, we will often use the abbreviations:

$\Longrightarrow$	implies
$\Longleftarrow$	is implied by
iff (or $\Longleftrightarrow$)	if and only if
$\forall$	for every
$\exists$	there exists
$\mathbb{Z}, \mathbb{Q}, \mathbb{R}, \mathbb{C}$	the integers, rationals, reals, complex numbers, respectively

We will not review the basics of proofs here. Hopefully you have figured out the basics, either from a high school plane geometry class or a college class introducing the subject of mathematical proof. See K. H. Rosen [93] for an introduction to proof. We will discuss proof by mathematical induction soon. There is an interesting book [60] by Steven Krantz on the subject of proof. Edna E. Kramer's history book [59] gives more perspective on the subject of proof. Another place to find a discussion of mathematical proof is Wikipedia. A cautionary tale concerns K. Gödel's incompleteness theorems from 1931, the first of which says that for any consistent formal system for the positive integers $\mathbb{Z}^+$, there is a statement about $\mathbb{Z}^+$ that is unprovable within this system.

There are those who argue against proofs. I have heard this at conferences with physicists. Nature will tell us the truth of a statement they argue. Ramanujan felt the goddess would inspire him to write true formulas. However, I have no such help myself and really need to see a proof to know what is true and what is false. This makes me very bad at real life, where there is rarely a proof of any statement. Thus I have grown to be happier writing an algebra book than a book on politics.

If you need more convincing about the need for proofs, look at the following two exercises, once you know what a prime is – an integer $p > 1$ such that $p = ab$, with positive integers a, b implies either a or $b = 1$. These exercises are silly if you can use your computer and Mathematica or some other similar program.

Exercise 1.1.1 *Show that $x^2 - x + 41$ is prime for all integers x such that $0 \leq x \leq 40$, but is not a prime when $x = 41$. Feel free to use a computer.*

Number theory has multitudes of statements like that in Exercise 1.1.1 that have been checked for a huge number of cases, but yet fail to be true in all cases. Of course, now computers can do much more than the puny 41 cases in the preceding exercise. For example, Mersenne primes are primes of the form $M_p = 2^p - 1$, where p is a prime. Mersenne compiled a list of Mersenne primes in the 1600s, but there were some mistakes after $p = 31$. Much computer time has been devoted to the search for these primes. Always bigger ones are found. In January, 2017 the biggest known prime was found to be $M_{77\,232\,917}$. It is conjectured that there are infinitely many Mersenne primes, but the proof has eluded mathematicians. See Wikipedia or Shanks [103] for more information on this subject and other unsolved problems in number theory. Wikipedia notes that these large primes have a cult following – moreover they have applications to random number generators and cryptography.

In the 1800s – before any computers existed – there was a conjecture by E. C. Catalan that M_{M_p} is prime, assuming that M_p is a Mersenne prime. Years passed before Catalan's conjecture was shown to be false. In 1953 the ILLIAC computer (after 100 hours of computing) showed that M_{M_p} is not prime when $p = 11$. M_{M_p} is prime for $p = 2, 3, 5, 7$. It was subsequently found that the conjecture is false for $p = 17, 19, 31$ as well. The next case is too large to test at the moment. Wikipedia conjectures that the four known M_{M_p} that are prime are the only ones. Anyway, hopefully, you get the point that you can find a large number of cases of some proposition that are true without the general proposition being valid. Stark gives many more examples in the introduction to [110].

Exercise 1.1.2 ***(Mersenne Primes).*** *Show that* $2^p - 1$ *is prime for* $p = 2, 3, 5, 7, 13$, *but not for* $p = 11$.

Hint. *The Mathematica command below will do the problem for the first 10 primes.*

```
Table[{Prime[n],FactorInteger[(2^Prime[n])-1]},{n,1,10}]
```

We assume that you can write down the **converse** of the statement "proposition A implies proposition B." Yes, it is "proposition B implies proposition A." Recall that $A \Longrightarrow B$ is not equivalent to $B \Longrightarrow A$. However $A \Longrightarrow B$ is equivalent to its **contrapositive:** (not B) $\Longrightarrow$ (not A).

We will sometimes use proof by contradiction. There are those who would object. In proof by contradiction of $A \Longrightarrow B$ we assume A and (not B) and deduce a contradiction of the form R and (not R). Those who would object to this and to any sort of "non-constructive" proof have a point, and so we will try to give constructive proofs when possible. See Krantz [60] for a bit of the history of constructive proofs in mathematics.

It is also possible that you can prove something that may at first be unbelievable. See the exercise below, which really belongs to an analysis course covering the geometric series – the formula for which follows from Exercise 1.4.7 below. If you accept the axioms of the system of real numbers, then you have to believe the formula.

Exercise 1.1.3 *Show that* $0.999\,999\ldots = 1$.

Hint. *See Exercise 1.4.7. The ... conceals an infinite series.*

A controversial method of proof is proof by computer. First you have to believe that the computer has been programmed correctly. This has not always been the case; e.g., the problem with the Pentium chip. Here I will choose to believe what my computer tells me when I use Mathematica to say whether an integer is a prime, or when used to compute eigenvalues of matrices, or graphs of functions, or to multiply elements of finite fields. There are more elaborate computer proofs that are hard to verify without even faster computers than the cheap laptop (vintage 2011) that I am using – for example, the proof of the four color problem in the 1970s or the recent proof of the Kepler conjecture on the densest packing of spheres in 3-space. See Krantz [60] for more information.

We are also going to assume that you view the following types of numbers as old friends:

the integers	$\mathbb{Z} = \{0, \pm 1, \pm 2, \pm 3, \ldots\}$,
the rationals	$\mathbb{Q} = \left\{ \frac{m}{n} \mid m, n \in \mathbb{Z}, n \neq 0 \right\}$,
the reals	$\mathbb{R} = \{\text{all decimals}\}$,
the complex numbers	$\mathbb{C} = \{x + iy \mid x, y \in \mathbb{R}\}$, for $i = \sqrt{-1}$.

We will list the axioms for $\mathbb{Z}$ in Chapter 1 and will construct $\mathbb{Q}$ from $\mathbb{Z}$ in Chapter 6. Of course, the construction of $\mathbb{Q}$ from $\mathbb{Z}$ just involves the algebra of fractions and could be done in Chapter 1 – minus the verbiage about fields and integral domains. We should define the real numbers as limits of Cauchy sequences of rationals rather then to say real numbers are represented by all possible decimals, but that would be calculus and we won't go there. Such a construction can be found for example in the book by Leon Cohen and Gertrude Ehrlich [17]. A serious student should really prove that $\mathbb{Z}$, $\mathbb{Q}$, $\mathbb{R}$, and $\mathbb{C}$ exist by constructing them from scratch, sort of like a serious chef makes a pie, but we will not do that here.

In contemplating the lower rows of our table of number systems, philosophers have found their hair standing on end. Around 500 BC the Pythagoreans were horribly shocked to find that irrational numbers like $\sqrt{2}$ existed. You will be asked to prove that $\sqrt{2} \notin \mathbb{Q}$ in Section 1.6. What was the problem for the Pythagoreans? You can read about it in Shanks [103, Chapter III]. What would they have thought about transcendental numbers like π? Later the complex numbers were so controversial that people called numbers like $i = \sqrt{-1}$ "imaginary." Non-Euclidean geometry was so upsetting that Gauss did not publish his work on the subject.

Warning. This course is like a language course. It is extremely important to memorize the vocabulary – the definitions. If you neglect to do this, after a week or so, the lectures – or the reading – will become meaningless. One confusing aspect of the vocabulary is the use of everyday words in a very different but precise way. Then one needs the axioms, the rules of constructing proofs. Those are our rules of grammar for the mathematical language. These too must be memorized. We should perhaps add that it is folly to argue with the definitions or axioms – unless you have found the equivalent of non-Euclidean geometry. To some our subject appears arcane. But they should remember that it is just a language – there is no mystery once you know the vocabulary and grammar rules.

Practice doing proofs. This means practice speaking or writing the language. One can begin by imitating the proofs in the text or other texts or those given by your professor. It is important to practice writing proofs daily. In particular, one must do as many exercises as possible. If your calculus class did not include proofs, this may be something of a shock. Mathematics seemed to be just calculations in those sad proof-less classes. And we will have a few calculations too. But the main goal is to be able to derive "everything" from a few basic definitions and axioms – thus to understand the subject. One can do this for calculus too. That is advanced calculus. If you do not practice conversations in a language, you are extremely unlikely to become fluent. The same goes with mathematics.

You should also be warned that sometimes when reading a proof you may doubt a statement and then be tempted to stop reading. Sadly, often the next sentence explains why that unbelievable statement is true. So always keep reading. This happens to "real" mathematicians all the time so do not feel bad. I have heard a story about a thesis advisor who told a student he did not understand the proof of a lemma in the student's thesis. The student almost had a heart attack worrying about that important lemma. But it turned out that the advisor had not turned the page to find the rest of thc proof.

Our second goal is to apply the algebra we derive so carefully. We will not be able to go too deeply into any one application, but hopefully we will give the reader a taste of each one.

1.2 Sets

We first review a bit of set theory. Georg Cantor (1845–1918) developed the theory of infinite sets. It was controversial. There are paradoxes for those who throw caution to the winds and consider sets whose elements are sets. For example, consider **Russell's paradox.** It was stated by B. Russell (1872–1970). We use the notation: $x \in S$ to mean that x is an element of the set S; $x \notin S$ means x is not an element of the set S. The notation $\{x|x \text{ has property } P\}$ is read as the set of x such that x has property P. Consider the set X defined by

$$X = \{\text{sets } S \mid S \notin S\}.$$

Then $X \in X$ implies $X \notin X$ and $X \notin X$ implies $X \in X$. This is a paradox. The set X can neither be a member of itself nor not a member of itself. There are similar paradoxes that sound less abstract. Consider the barber who must shave every man in town who does not shave himself. Does the barber shave himself? A mystery was written inspired by the paradox: *The Library Paradox* by Catharine Shaw. There is also a comic book about Russell, *Logicomix* by A. Doxiadis and C. Papadimitriou (see [26]). A nice reference for set theory illustrated by pictures and stories is the book by Vilenkin [122].

We will hopefully avoid paradoxes by restricting consideration to sets of numbers, vectors, and functions. This would not be enough for "constructionists" such as Errett Bishop who was on the faculty at the University of California San Diego. until his premature death. I am still haunted by his probing questions of colloquium speakers. Anyway, for applied mathematics, one can hope that paradoxical sets and barbers do not appear. Thus we will be using proof by contradiction, as we have already promised.

Most books on calculus do a little set theory. We assume you are familiar with the notation which we are about to review. We will draw pictures in the plane. We write $A \subset B$ (or $B \supset A$) if A is a **subset** of B: that is, $x \in A$ implies $x \in B$. We might also say B **contains** A. If $A \subset B$, the **complement** of A in B is $B - A = \{x \in B | x \notin A\}$.[1] The **empty set** is denoted $\emptyset$. It has no elements. The **intersection** of sets A and B is

$$A \cap B = \{x \mid x \in A \text{ and } x \in B\}.$$

The **union** of sets A and B is

$$A \cup B = \{x \mid x \in A \text{ or } x \in B\}.$$

[1] We will not use the other common notation B/A for set complement since it conflicts with our later notation for quotient groups.

Here – as is usual in mathematics – "or" means either or both. See Figure 1.1. Sets A and B are said to be **disjoint** iff $A \cap B = \emptyset$.

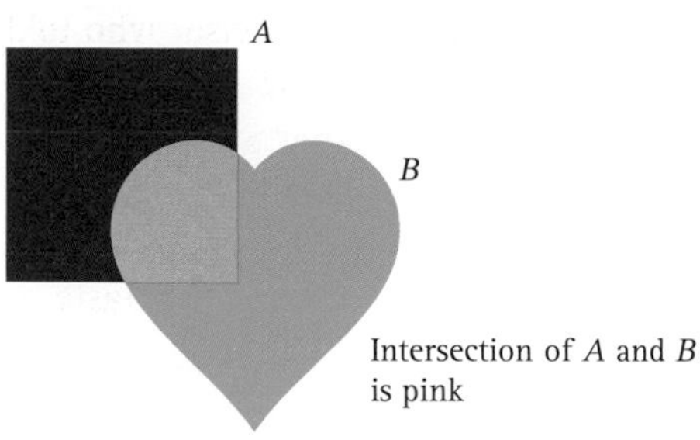

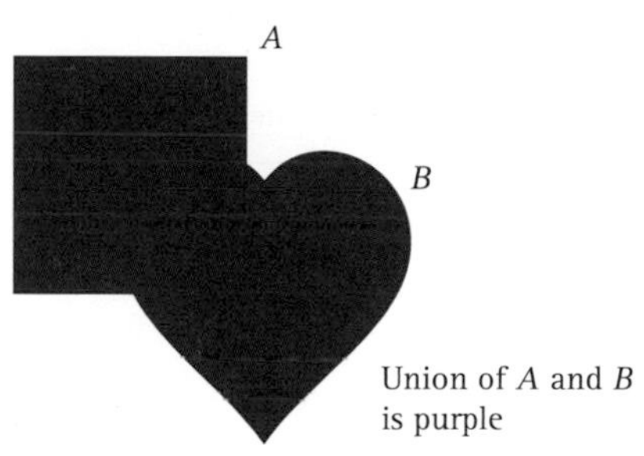

Figure 1.1 Intersection and union of square A and heart B

The easiest way to do the following exercises on the equality of various sets is to show first that the set on the left is contained in the set on the right and second that the set on the right is contained in the set on the left.

Exercise 1.2.1

(a) Prove that

$$A - (B \cup C) = (A - B) \cap (A - C).$$

(b) Prove that

$$A - (B \cap C) = (A - B) \cup (A - C).$$

Exercise 1.2.2 *Prove that* $A \cup (B \cup C) = (A \cup B) \cup C$. *Then prove the analogous equation with* $\cup$ *replaced by* $\cap$.

Exercise 1.2.3 *Prove that* $A \cap (B \cup C) = (A \cap B) \cup (A \cap C)$.

Definition 1.2.1 *If* A *and* B *are sets, the* ***Cartesian product*** *of* A *and* B *is the set of ordered pairs* (a, b) *with* $a \in A$ *and* $b \in B$*: that is,*

$$A \times B = \{(a, b) \mid a \in A, b \in B\}.$$

It is understood that we have equality of two ordered pairs $(a, b) = (c, d)$ iff $a = c$ and $b = d$.

Example 1. Suppose A and B are both equal to the set of all real numbers; $A=B=\mathbb{R}$. Then $A\times B=\mathbb{R}\times\mathbb{R}=\mathbb{R}^2$. That is, the Cartesian product of the real line with itself is the set of points in the plane. ▲

Example 2. Suppose C is the interval $[0,1]$ and D is the set consisting of the point $\{2\}$. Then $C\times D$ is the line segment of length 1 at height 2 in the plane. See Figure 1.2 below. ▲

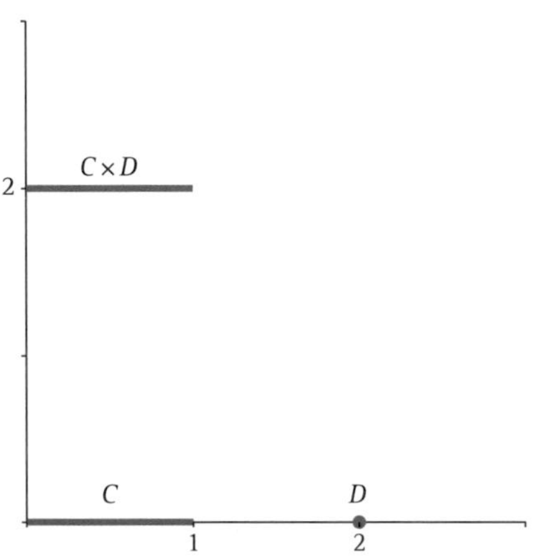

Figure 1.2 Cartesian product $[0,1]\times\{2\}$

Of course you can also define the Cartesian product of any number of sets – even an infinite number of sets. We mostly restrict ourselves to a finite number of sets here. Given n sets S_i, $i\in\{1,2,\dots,n\}$, define the Cartesian product $S_1\times S_2\times\cdots\times S_n$ to be the set of ordered n-tuples $(s_1,s_2,\dots,s_n)$ with $s_i\in S_i$, for all $i=1,2,\dots,n$.

Example 3. $[0,1]\times[0,1]\times[0,1]=[0,1]^3$ is the unit cube in 3-space. See Figure 1.3. ▲

Figure 1.3 $[0,1]^3$

Example 4. $[0,1]\times[0,1]\times[0,1]\times[0,1]=[0,1]^4$ is the four-dimensional cube or tesseract. Draw it by "pulling out" the three-dimensional cube. See T. Banchoff [6]. Figure 1.4 below shows the edges and vertices of the four-dimensional cube or tesseract (actually more of a 4-rectangular solid) as drawn by Mathematica. Of course both Figures 1.3 and 1.4 are really projections of the cube and hypercube onto the plane. ▲

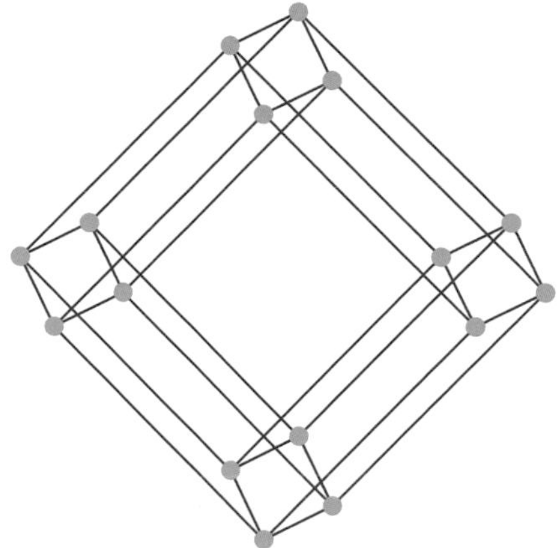

Figure 1.4 Graph representing the hypercube $[0, 1]^4$

Exercise 1.2.4 *Show that* $A \times (B \cap C) = (A \times B) \cap (A \times C)$. *Does the same equality hold when you replace* $\cap$ *with* $\cup$?

Exercise 1.2.5 *State whether the following set-theoretic equalities are true or false and give reasons for your answers.*

(a) $\{2, 5, 7\} = \{5, 2, 7\}$.
(b) $\{(2, 1), (2, 3)\} = \{(1, 2), (3, 2)\}$.
(c) $\emptyset = \{0\}$.

Exercise 1.2.6 *Prove the following set-theoretic identities:*

(a) $(A - C) \cap (B - C) = (A \cap B) - C$;
(b) $A \times (B - C) = (A \times B) - (A \times C)$.

1.3 The Integers

Notation.

$\mathbb{Z}^+$	$\{1, 2, 3, 4, \ldots\}$	the positive integers
$\mathbb{Z}$	$\{0, \pm 1, \pm 2, \pm 3, \ldots\}$	the integers

We assume that you have been familiar with the basic facts about the integers since childhood. Despite that familiarity, we must list the 10 basic axioms for $\mathbb{Z}$ in order to be able to prove anything about $\mathbb{Z}$. By an axiom, we mean a basic unproved assumption. We must deduce everything we say about $\mathbb{Z}$ from our 10 axioms – forgetting what we know from elementary school. In Section 5.3 we will find that much of what we do here – especially in the pure algebra part (R1 to R6) – works for any integral domain and not just $\mathbb{Z}$. Sometimes $\mathbb{Z}^+$ or $\mathbb{Z}^+ \cup \{0\}$ is referred to as the "**natural numbers.**" This seems somewhat prejudicial to the other types of numbers one may use and so we will try to avoid that terminology.

Algebra Axioms for $\mathbb{Z}$

For every $n, m \in \mathbb{Z}$ there is a unique integer $m+n$ and a unique integer $n \cdot m$ such that the following laws are valid for all $m, n, k \in \mathbb{Z}$. This says the set of integers is **closed** under addition and multiplication.

R1 **Commutative laws:** $m+n=n+m$ and $m \cdot n = n \cdot m$.
R2 **Associative laws:** $k+(m+n)=(k+m)+n$ and $k \cdot (m \cdot n)=(k \cdot m) \cdot n$.
R3 **Identities:** There are two special elements of $\mathbb{Z}$, namely
0 (identity for addition) and 1 (identity for multiplication) in $\mathbb{Z}$
such that $0+n=n$, $1 \cdot n=n$, for all $n \in \mathbb{Z}$, and $0 \neq 1$.
R4 **Inverse for addition:** For every $m \in \mathbb{Z}$ there exists an element
$x \in \mathbb{Z}$ such that $m+x=0$. Write $x=-m$, once you know x is unique.
R5 **Distributive law:** $k \cdot (m+n)=k \cdot m+k \cdot n$.
R6 **No zero divisors:** $m \cdot n=0$ implies either m or n is 0.

We sometimes write $n \cdot m=n * m=nm$. Thanks to the associative laws, we can leave out parentheses in sums like $k+m+n$ or in products like kmn. Of course we still need those parentheses in the distributive law.

As a result of axioms R1–R5, we say that $\mathbb{Z}$ is a "commutative ring with identity for multiplication." As a result of the additional axiom R6 we say that $\mathbb{Z}$ is an "integral domain." Rings will be the topic for the last half of this book – starting with Chapter 5.

Exercise 1.3.1

(a) Show that the identities 0 *and* 1 *in R3 are unique.*
(b) Show that the inverse x of the element m in R4 is unique once m is fixed.

Exercise 1.3.2 *Show that $a \cdot 0=0$ for any $a \in \mathbb{Z}$.*

Exercise 1.3.3 *In axiom R4, we can write $1+u=0$ and then define $u=-1$. Show that then, for any $m \in \mathbb{Z}$, if x is the integer such that $m+x=0$, we have $x=(-1) \cdot m$. Thus $x=-m=(-1) \cdot m$. Prove that $-(-m)=m$.*

Exercise 1.3.4 *(Cancellation Laws)*. *Show that if $a, b, c \in \mathbb{Z}$, then we have the following laws.*

(a) If $a+b=a+c$, then $b=c$.
(b) If $a \neq 0$ and $ab=ac$, then $b=c$.

Exercise 1.3.5 *Prove the other distributive law: $(m+n) \cdot k=m \cdot k+n \cdot k$.*

Exercise 1.3.6 *Prove that for any $a, b \in \mathbb{Z}$ we can solve the equation $a+x=b$ for $x \in \mathbb{Z}$.*

Additional axioms for $\mathbb{Z}$ involve the ordering $<$ of $\mathbb{Z}$ which behaves well with respect to addition and multiplication. The properties of inequalities can be derived from three simple axioms for the set $P=\mathbb{Z}^+$ of positive integers.

Order Axioms for $\mathbb{Z}$

O1 $\mathbb{Z}=P\cup\{0\}\cup(-P)$, where $-P=\{-x|x\in P\}$. Moreover this is a disjoint union. That is,

$$0\notin P,\quad 0\notin -P,\quad P\cap(-P)=\emptyset.$$

O2 $n, m\in P \Longrightarrow n+m\in P$.

O3 $n, m\in P \Longrightarrow n\cdot m\in P$.

As a result of the nine axioms R1–R6 and O1–O3, we say that $\mathbb{Z}$ is an **ordered integral domain.** There is still one more axiom needed to define $\mathbb{Z}$, but we discuss this below – after saying more about the order relation $a<b$.

> **Definition 1.3.1** *If $a, b\in\mathbb{Z}$ we say that $a<b$ (b is **greater than** a or a is **less than** b) iff $b-a\in P=\mathbb{Z}^+$. One can also write $b>a$ in this situation.*

Examples. By this definition, the set P consists of integers that are greater than 0. We can see that $0<1$ since otherwise, by O1, $0<-1$. But then, according to axiom O3 it follows that $(-1)(-1)=1\in P$. This contradicts $P\cap(-P)=\emptyset$.

It follows from our axioms that $P=\mathbb{Z}^+=\{1,2,3,4,\ldots\}$, using O2 and the last axiom (well-ordering) which we are about to state. This axiom will allow us to prove infinite lists of statements by checking two items (mathematical induction). See Exercise 1.3.13. ▲

We can use our axioms to prove the following facts about order.

Facts about Order. $\forall\, x, y, z, c\in\mathbb{Z}$

(1) **Transitivity.** $x<y$ and $y<z$ implies $x<z$.

(2) **Trichotomy.** For any $x, y, z\in\mathbb{Z}$ exactly one of the following inequalities is true:

$$x<y,\quad y<x,\quad \text{or}\quad x=y.$$

(3) **Addition.** $x<y$ implies $x+z<y+z$ for any $z\in\mathbb{Z}$.

(4) **Multiplication by a positive number.** If $0<c$ and $x<y$, then $cx<cy$.

(5) **Multiplication by a negative number.** If $c<0$ and $x<y$, then $cy<cx$.

Proof. We will leave most of these proofs to the reader as an exercise. But we will do (1) and (3).

Fact (1): $x<y$ means $y-x\in P$. $y<z$ means $z-y\in P$. Then by O2 and the axioms for arithmetic in $\mathbb{R}$, we have $y-x+z-y=z-x\in P$. This says $x<z$.

Fact (3): Since $x<y$ we know that $y-x\in P$. Then $(y+z)-(x+z)=y-x\in P$ which is what we needed to show. ▲

More Definitions. Of course we will write $a\leq b$ if either $a=b$ or $a<b$. We may also write $b\geq a$ in this case.

Exercise 1.3.7 *Prove the rest of the facts about order.*

Exercise 1.3.8 *Show that* $x^2 + 1 = 0$ *has no solution* $x \in \mathbb{Z}$ *(or in any ordered integral domain really). This says that if* $i = \sqrt{-1}$, *the domain of Gaussian integers* $\mathbb{Z}[i] = \{x + iy \mid x, y \in \mathbb{Z}\}$ *cannot be ordered – at least, not without dropping some properties of* $<$ *on* $\mathbb{Z}$.

Exercise 1.3.9 *By an ordered integral domain we mean a set satisfying all the axioms R1–R6 and O1–O3.*

(a) Show that there is no ordered integral domain D with 2 elements.
(b) Show that there is no finite ordered integral domain.

The set of real numbers $\mathbb{R}$ also satisfies axioms R1–R6 and O1–O3. In fact, it is what we call an ordered field. We will consider fields in Section 5.3.

Exercise 1.3.10 *Define the* ***absolute value*** $|x| = x$ *if* $x \in \mathbb{R}$ *and* $x \geq 0$, *and* $|x| = -x$, *if* $x \in \mathbb{R}$ *and* $x < 0$.

(a) Show that $|xy| = |x|\,|y|$ *for all* $x, y \in \mathbb{R}$.
(b) Prove the ***triangle inequality*** $|x + y| \leq |x| + |y|$, *for all* $x, y \in \mathbb{R}$.

Hint. *Use the fact that* $|a| = \sqrt{a^2}$.

Exercise 1.3.11 *Prove that the following inequality holds for all* $a, b \in \mathbb{R}$ *such that a and b are both positive,*

$$\frac{a+b}{2} \geq \sqrt{ab}.$$

This is called the ***arithmetic–geometric mean inequality****. The left-hand side is the arithmetic mean and the right-hand side is the geometric mean of a and b. The inequality can be generalized to an inequality involving n positive real numbers. We will consider that generalization in a later exercise.*

Because the creation of the real numbers involves limits – which reside within the domain of advanced calculus, we will not have much to say about this creation here. However we must still come to grips with the infinite, for example, infinite lists of theorems to prove. One thing that differentiates $\mathbb{Z}$ from the real numbers $\mathbb{R}$ or the rational numbers $\mathbb{Q}$ is the following well-ordering axiom. It really embodies the discreteness of $\mathbb{Z}$, as opposed to the continuity of $\mathbb{R}$.

The Well-Ordering Axiom. If $S \subset \mathbb{Z}^+$, and $S \neq \emptyset$, then S has a least element $a \in S$ such that $a \leq x$, $\forall x \in S$.

This axiom says that any non-empty set of positive integers has a least element. We usually call such a least element a minimum.

Note that you could state a similar axiom for $\mathbb{Z}^+ \cup \{0\}$ or $\mathbb{Z}^+ \cup \{0, -1\}$, or the union of $\mathbb{Z}^+$ and any finite set of integers.

Now we have our 10 axioms for $\mathbb{Z}$. We could do with fewer axioms. G. Peano (1858–1932) wrote down the five Peano postulates (or axioms) for the natural numbers $\mathbb{Z}^+ \cup \{0\}$. We will not list the Peano postulates here. See, for example, Birkhoff and MacLane [9]. Once one has these axioms, it would be nice to show that something exists satisfying the axioms.

We will not do that here – feeling pretty confident that you believe $\mathbb{Z}$ exists. See the comic book about Russell, *Logicomix* [26] by A. Doxiadis and C. Papadimitriou for a story of the writing of *Principia Mathematica* by Russell and Whitehead. One of the big events in the book is getting to the point to deduce that $1+1=2$.

It follows from all these axioms and the facts deduced from them that $\mathbb{Z}=\{0, \pm 1, \pm 2, \pm 3, \pm 4, \ldots\}$. Thus we can picture the integers as forming a line of equally spaced points stretching out to ∞ on the right and $-\infty$ on the left. See Figure 1.5, which shows part of the discrete set of integers embedded in the continuous real line.

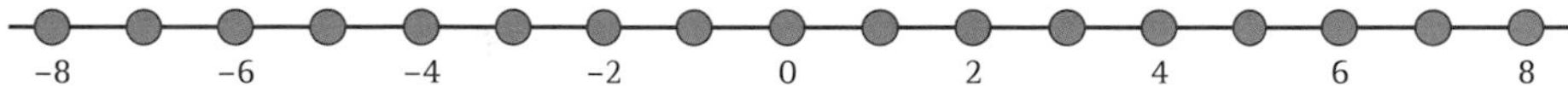

Figure 1.5 Integers on the line – only even ones are labelled

Now that we have listed all the axioms for $\mathbb{Z}$, we should at least mention that one could also list the defining axioms for $\mathbb{Q}$ and $\mathbb{R}$. Basically $\mathbb{Q}$ consists of fractions $\frac{m}{n}$, for $m, n \in \mathbb{Z}$, with $n \neq 0$. The rules for adding and multiplying fractions are the usual ones. See Definition 6.4.1 if you forget. We will assume in many exercises that you know these things. Of course $\mathbb{Q}$ has an ordering satisfying the same rules as that for $\mathbb{Z}$. It is what we will later call an ordered field. The real numbers $\mathbb{R}$ have an extra axiom that means that Cauchy sequences converge. A **Cauchy sequence** $\{x_n\}$ of real numbers has the property that $|x_m - x_n| \to 0$, as $m, n \to \infty$. One may view the real numbers as limits of Cauchy sequences of rationals. For concreteness, view the real numbers as decimals. You can of course add and multiply real numbers – also divide by nonzero ones. There is an ordering with the same rules as that for $\mathbb{Z}$. Most advanced calculus books discuss these things at length. Ordinary calculus books are based on these rules. The well-ordering axiom is false for subsets of the positive real numbers $\mathbb{R}^+$. There is no smallest element of the open interval $(0, 1)$, for example. Similarly $(0, 1) \cap \mathbb{Q}$ has no smallest element. We will say no more of this.

The following exercise may appear obvious from Figure 1.5, but we need to prove (using only our axioms) that the properties embodied in that figure are indeed true properties of $\mathbb{Z}$.

Exercise 1.3.12 *Show that there is no largest integer N such that* $\forall x \in \mathbb{Z}$, *we have* $x \leq N$.

Exercise 1.3.13

(a) Show that there is no integer a such that $0 < a < 1$.
(b) Deduce that then the set of positive integers

$$P = \mathbb{Z}^+ = \{1, 2, 3, 4, \ldots, n, n+1, \ldots\}.$$

Thus, in particular, no sum of 1*s can be* 0.

Hint.

(a) Consider the least such a and deduce a contradiction by considering the location of a^2 *with respect to* 0 *and a.*
(b) First explain why there is no integer a such that $1 < a < 2$. *Then explain why there is no integer a such that* $n < a < n+1$, *for any* $n = 2, 3, 4, \ldots$.

The most important fact about the well-ordering axiom is that it is equivalent to mathematical induction. We discuss that in the next section.

Exercise 1.3.14 *Show that if $a \in \mathbb{Z}$ and $a < 0$, then $a^2 > 0$.*

1.4 Mathematical Induction

Domino Version of Mathematical Induction. Given an infinite line of identical equally spaced dominos, we want to knock over all the dominos by just knocking over the first one in line. To be able to do this, we should make sure that the nth domino is so close to the $(n+1)$th domino (and so similar in weight) that when the nth domino falls over, it knocks over the $(n+1)$th domino. See Figure 1.6.

Figure 1.6 The first principle of mathematical induction. A penguin surveys an infinite line of equally spaced dominos. If the nth domino is close enough to knock over the $(n+1)$th domino, then once the penguin knocks over the first domino, they should all fall over

Translating this to theorems, we get the following.

Principle of Mathematical Induction I

Suppose you want to prove an infinite list of theorems T_n, $n = 1, 2, \ldots$ It suffices to do two things.

Step 1. Prove T_1.
Step 2. Prove that T_n true implies T_{n+1} true for all $n \geq 1$.

Note that this works by the well-ordering axiom. If $S = \{n \in \mathbb{Z}^+ | T_n \text{ is false}\}$, then either S is empty or S has a least element q. But we know $q > 1$ by the fact that we proved T_1 – which was Step 1. And we know that T_{q-1} is true since q is the least element of S. But then, by Step 2, we know T_{q-1} implies T_q, contradicting $q \in S$. It follows that S must be empty – meaning that all our T_n are true.

The assumption that T_n is true in Step 2 is often called the **induction hypothesis.**

Example: Formula for the Sum of an Arithmetic Progression. T_n is the formula used by Gauss as a youth to confound his teacher: The teacher had asked the students to sum the positive integers ≤ 50 or so.

$$T_n: \qquad 1 + 2 + \cdots + n = \frac{n(n+1)}{2}, \quad n = 1, 2, 3, \ldots.$$

To do this proof, we assume familiarity with the axioms for rational numbers. Sorry, we do not discuss fractions until Section 6.4 but surely you know how to add and multiply fractions already. If not, see Definition 6.4.1. ▲

Proof. We follow our procedure for Mathematical Induction I.

First, prove T_1. $1 = 1(2)/2$. Yes, that is certainly true by the rules for identifying fractions.

Second, assume T_n and use it to prove T_{n+1}, for $n = 1, 2, 3, \ldots$

$$T_n: \qquad 1 + 2 + \cdots + n = \frac{n(n+1)}{2} \tag{1.1}$$

Add the next term in the sum, namely, $n + 1$, to both sides of the equation and obtain

$$1 + 2 + \cdots + n + (n+1) = \frac{n(n+1)}{2} + (n+1).$$

Finish by simplifying the right-hand side of this last equality using our axioms for $\mathbb{Z}$ plus a bit of knowledge of the distributive law for $\mathbb{Q}$ and the rule for adding fractions. You obtain

$$\frac{n(n+1)}{2} + (n+1) = (n+1)\left(\frac{n}{2} + 1\right) = (n+1)\left(\frac{n+2}{2}\right),$$

which gives us equation (1.1) with n replaced by $n + 1$; that is, formula T_{n+1}. This completes the proof using Mathematical Induction I. ▲

One may find this proof a bit disappointing. Many students have complained at this point that they are not convinced of the truth of the formula for the sum of an arithmetic progression. Induction does not seem to reveal the underlying reason for the truth of such a formula. Some have even complained that this proof requires that one believe in mathematical induction. Well, yes, it is equivalent to an axiom and we have to believe it.

Of course, there are many other proofs of this sort of thing. For example, look at

$$\begin{array}{ccccccccc} 1 & + & 2 & + & \cdots & + & n \\ n & + & n-1 & + & \cdots & + & 1 \end{array}$$

When you add terms in the same column you always get $n + 1$. There are n such terms. Thus twice our sum is $n(n + 1)$. However, this proof still does not reveal any insights as to how to do the next exercise or to generalize the formulas to sums of arbitrary integer powers $1^k + 2^k + \cdots + n^k$. We will not go into such methods here – as our only interest in these formulas is to give us practice in the use of mathematical induction.

Exercise 1.4.1 *Use mathematical induction to show that*

$$1^2 + 2^2 + \cdots + n^2 = \frac{n(n+1)(2n+1)}{6}.$$

You will need to use the rules for adding fractions. See Definition 6.4.1.

Exercise 1.4.2 *Consider the matrix* $A = \begin{pmatrix} 1 & a \\ 0 & 1 \end{pmatrix}$, *for* $a \in \mathbb{R}$. *Use mathematical induction to show that* $A^n = \begin{pmatrix} 1 & na \\ 0 & 1 \end{pmatrix}$, *for all* $n \in \mathbb{Z}^+$.

We also have the concept of **inductive (or recursive) definition.** For example, to define n **factorial**, written $n!$, we define $0! = 1$ and, assuming that $n!$ is defined, then we define $(n+1)! = (n+1)n!$. Of course this means that $n! = n(n-1)(n-2)\cdots 2\cdot 1$ is the product of all integers between 1 and n. This number is the number of **permutations** or rearrangements of a set of n objects. To see how many ways there are to arrange n elements in a row, note that there are n choices for the first element in the row, $n-1$ choices for the second, and so on until you reach the last element in the row of n elements, for which you have one choice.

There are times when the first induction principle is not precisely what is needed. For that we need the second principle – also called strong induction or complete induction.

Principle of Mathematical Induction II

To prove an infinite list of theorems

$$d_1,\ d_2,\ d_3, \ldots, d_n,\ d_{n+1}, \ldots$$

you need to do two steps.

Step 1. Show that d_1 is true (or if necessary some finite number of d_i are true).
Step 2. Show that for every $n \geq 1$, the truth of $d_1, d_2, d_3, \ldots, d_n$ implies the truth of d_{n+1}.

Exercise 1.4.3 *Show that the well-ordering axiom implies the second principle of mathematical induction.*

The first and second principles of mathematical induction are both equivalent to the well-ordering principle. The fact that seemingly different principles are equivalent may appear to be surprising – welcome to the world of algebra.

Mathematical Induction II in Pictures. Step 2 says you have to organize the dominos so that if we knock over all the first n dominos, then the next domino must fall.

The moral is that, if your dominos are not all in a line, depending on the way the dominos are organized, it may take more than one domino to knock over the next one. In Figure 1.7 you are supposed to need a variable number of dominos to knock over the ones to their left – and again all dominos are supposed to be the same size and weight. Perhaps we need some real dominos to make a better picture – or a better artist.

In the examples, sometimes you know you need d_{n-1} and d_n to knock over d_{n+1}, but other times you may not know how many d_j with $j < n+1$ you need to knock over d_{n+1}, as in the proof of the fundamental theorem of arithmetic in the next section.

Figure 1.7 Here is an attempt to picture the second mathematical induction principle in which we arrange dominos so that various numbers of dominos are needed to knock over the dominos to their left. In this picture step 1 would be for the penguin to knock over d_1 and d_2

Example. **Fibonacci numbers** f_n are defined inductively by setting

$$f_1 = f_2 = 1 \quad \text{and} \quad f_{n+1} = f_{n-1} + f_n. \tag{1.2}$$

The first few Fibonacci numbers are

$$1, 1, 2, 3, 5, 8, 13, 21, 34, 55, 89, 144, 233, 377, 610, 987, 1597, 2584, 4181, 6765.$$

We could also have started with $f_0 = 0,\ f_1 = 1$. ▲

Exercise 1.4.4 *Show that* $f_n < 2^n$.

Hint. *You will need the second mathematical induction principle. Use the results for* $n-1$ *and* n *to prove the result for* $n+1$.

History. The numbers f_n in the preceding exercise are named for Fibonacci (1180–1228), aka Leonardo of Pisa, who used these numbers to model the number of pairs of rabbits on an island supposing one pair of baby rabbits is left on the island at the beginning and assuming no rabbit dies – ever. Each newborn pair takes two months to mature and produces a new pair in the third month and in every month thereafter. Mathematicians in India may have considered these numbers before Fibonacci. These numbers are so popular that there is even a journal devoted to them.

The Fibonacci numbers appear in many contexts – in pineapples, sunflowers, the family tree of a drone or male bee, the optics of light rays. For a discussion of such things, see the book on the golden ratio by Mario Livio [70]. We will look at the connection of the Fibonacci numbers and the golden ratio $\phi = \frac{1}{2}(1+\sqrt{5})$ in Section 8.1.

Exercise 1.4.5 *Use mathematical induction to prove that given n sets $B_1, \ldots, B_n$ and another set A, we have the generalized distributive law:*

$$A \cap \left(\bigcup_{i=1}^{n} B_i\right) = \bigcup_{i=1}^{n} (A \cap B_i).$$

Hint. *We are assuming Exercise 1.2.2 which says that the operation of union is associative as well as the extension to drop parentheses in unions of arbitrary numbers of sets. You will also need Exercise 1.2.3 in the case $n=2$. Then, of course, mathematical induction does the rest.*

Exercise 1.4.6 *What is wrong with the following "proof" by induction?*

We claim we can show that in any room of n people all have the same birthday.

This is clear when $n = 1$.

To show that the case $n-1$ implies the nth case, note that if a room has n people, we can send person A out. Those left are $n-1$ people, each having the same birthday by the induction assumption. Now bring person A back and send person B out. Then, by the induction assumption, A must have the same birthday as the rest of the people in the room. So all have the same birthday!

In the following exercises and throughout this text we use the summation symbol notation $\sum_{k=0}^{n} a_k = a_1 + \cdots + a_n$ and we replace n by ∞ to mean that we have taken the limit as $n \to \infty$.

Exercise 1.4.7 ***(Geometric Progression and Series).***

(a) Show using mathematical induction that if $x \in \mathbb{R}$ and $x \neq 1$, then

$$\sum_{k=0}^{n} x^k = \frac{x^{n+1}-1}{x-1}.$$

(b) Then show that if $|x| < 1$, assuming you believe in limits,

$$\sum_{k=0}^{\infty} x^k = \frac{1}{1-x}.$$

Exercise 1.4.8 *Use mathematical induction to prove that there are $n!$ ways to rearrange a row of n penguins.*

The following exercise implies the divergence of the harmonic series $\sum_{k=1}^{\infty} \frac{1}{k}$.

Exercise 1.4.9 *Prove that for $n = 1, 2, 3, \ldots$*

$$\sum_{k=1}^{2^n} \frac{1}{k} \geq 1 + \frac{n}{2}.$$

Exercise 1.4.10 *Prove that if f_n denotes the nth Fibonacci number defined in formula (1.2), with $f_0 = 0$, and $A = \begin{pmatrix} 1 & 1 \\ 1 & 0 \end{pmatrix}$, then, with the usual matrix multiplication as defined in formula (1.3) in Section 1.8 and the statement that follows it:*

$$\underbrace{A \cdot A \cdots A}_{n \text{ terms}} = A^n = \begin{pmatrix} f_{n+1} & f_n \\ f_n & f_{n-1} \end{pmatrix}, \quad \text{for } n \geq 1.$$

1.5 Divisibility, Greatest Common Divisor, Primes, and Unique Factorization

Number theory is an ancient subject which involves the study of various properties of the integers and generalizations thereof. One of the most fundamental properties is divisibility. This leads to the concept of prime number – a subject that is full of challenging questions that are easily stated but not easily answered. Here we present some of the basics of the subject. Some references for more information are: Ramanujachary Kumanduri and Cristina Romero [62], Steven J. Miller and Ramin Takloo-Bighash [78], Kenneth Rosen [91], Daniel Shanks [103], Joseph H. Silverman [105], and Harold Stark [110]. One of the charms of number theory is that one can easily do experiments, especially now that computers are ubiquitous. However, there are many cautionary tales of false conjectures that hold true for large numbers of cases. We gave some examples in the exercises of Section 1.1.

Definition 1.5.1 *Suppose that a and b are integers. We say* ***a divides b****, written $a|b$, if there is an integer c such that $b = ac$. We will also say that a is a* ***divisor*** *of b or b is a* ***multiple*** *of a. We could also say b is* ***divisible*** *by a.*

Examples. The set $\{\pm 1, \pm 2, \pm 3, \pm 4, \pm 6, \pm 12\}$ consists of all the divisors of 12. Every integer divides 0. But 0 divides none but 0. ▲

At this point, in order to find all the divisors of an integer n, we need to test all the integers m with $0 < m < |n|$ to see if m divides n. We will have a better way soon.

Exercise 1.5.1 *Show that if both $a|b$ and $b|a$, then $a = \pm b$.*

Definition 1.5.2 *An integer $p>1$ is **prime** iff $p=ab$ for integers a,b implies either a or b is ± 1.*

Thus a prime p is a positive integer greater than 1 with no positive divisors but itself and 1. Note that 1 is not a prime by definition. Yes, it has no non-trivial divisors but it is special in a different way. It is its own multiplicative inverse and is thus called a unit in $\mathbb{Z}$ – a concept that will be important in ring theory. See Definition 5.2.3.

Examples. The first few primes are $2,3,5,7,11,13,17,19,23,29,31$. Mathematica has a command that will tell you what is the nth prime. The size of n to plug into that command depends on your computer. ▲

Exercise 1.5.2 *Show that if $n\in\mathbb{Z}$ and $n>1$, then n has a prime divisor.*

Hint. *Use the second induction principle.*

Exercise 1.5.3 *Show that the preceding example of the primes ≤ 31 is correct. Do not use a computer.*

Exercise 1.5.4 *Show that there are infinitely many primes.*

Hint. *Do a proof by contradiction and assume that there are only finitely many primes. Call them $p_1,p_2,\ldots,p_n$. Consider the number $M=1+p_1p_2\cdots p_n$. Is M divisible by any p_j?*

Somewhere in the distant past you learned to do long division. So, for example, you divide 4 into 31 and get quotient 7 and remainder 3. This means $31=4*7+3$. This can be tabulated as:

$$\begin{array}{r|l} & 7 \\ \hline 4 & 31 \\ & 28 \\ \hline & 3 \end{array}$$

Note that if we define the **floor** of x to be $\lfloor x\rfloor$ = the largest integer $\leq x$, then the quotient here is $\left\lfloor \frac{31}{4}\right\rfloor=7$ and the remainder is then $3=31-4\left\lfloor\frac{31}{4}\right\rfloor$.

Exercise 1.5.5 *Use the well-ordering principle to explain why $\lfloor x\rfloor$ exists.*

Hint. *Look at $\{n\in\mathbb{Z}\mid n\geq -x\}$.*

Next we state a theorem whose proof justifies our knowledge that we can do long division. You might think us a bit crazy for proving this – especially if you have thought about the formula involving $\lfloor x\rfloor$ in the last paragraph and exercise. Later when we use a similar proof to show that we can divide polynomials, you might forgive us – or not.

Theorem 1.5.1 ***(The Division Algorithm).*** *Suppose that a, b are non-negative integers with $b > 0$. Then there are (unique) integers q (quotient) and r (remainder) such that $a = bq + r$ and $0 \le r < b$.*

Proof. We use the well-ordering principle. Look at the set S of non-negative integers of the form $a - bn$, for some $n \in \mathbb{Z}$. We know that S is not empty, since we can take $n = 0$, as we are assuming that $a \ge 0$. So now let r be the smallest element of S (which exists by the well-ordering axiom). We claim that r has the properties stated in the theorem. By the definition of S, we know $0 \le r = a - bn$, for some $n \in \mathbb{Z}$. If (by contradiction) $r \ge b$, then $r - b = a - bn - b = a - (n + 1)b \in S$, contradicting the minimality of r. We leave it as an exercise to show that q and r are unique. ▲

Exercise 1.5.6 *Prove the uniqueness of quotient and remainder in the division algorithm.*

Exercise 1.5.7 *Write an alternative proof of the division algorithm using the floor function to write $q = \left\lfloor \frac{a}{b} \right\rfloor$ and $r = a - bq$.*

Exercise 1.5.8 *Extend the division algorithm to negative integers a. For example: $-5 = 2 * (-3) + 1$. Here $a = -5, b = 2, q = -3, r = 1$.*

Note that b divides a means that when we use the division algorithm to divide a by b, the remainder is 0. This terminology may seem somewhat confusing and so you might want to say b divides a evenly, but we will not.

Definition 1.5.3 *A positive integer d is called the* ***greatest common divisor*** *(gcd) of two integers a, b, where a and b are not both 0, written $d = (a, b) = \gcd(a, b)$, if the following two properties hold:*

(1) d divides both a and b,
(2) if an integer c divides both a and b, then c must divide d.

Definition 1.5.4 *If $\gcd(a, b) = 1$, we say that a and b are* ***relatively prime.***

The existence of the greatest common divisor of two integers is not obvious until you have the Euclidean algorithm below. Once you have it, you can see that the greatest common divisor is just what the name says it is – the greatest of all common divisors of a and b. To see this, suppose that d' is the largest of all common divisors of a and b while $d = \gcd(a, b)$ from Definition 1.5.3. Certainly then $d \le d'$. But, by part (2) of the definition of d, since d' is a common divisor of a and b, we know that d' must divide d. This implies that $d' \le d$. Thus $d' = d$.

For very small numbers it is easy to find $\gcd(a, b)$ by factoring a, b. For example if a is prime, then $\gcd(a, b)$ must be either a or 1. It is a only if $a|b$. However, we should be careful at this point because we have not yet proved the fundamental theorem of arithmetic – Theorem 1.5.3 – which tells us we can factor integers as products of primes uniquely up to order. Thus when we say gcd(10,25) = gcd $(2\cdot 5, 5\cdot 5) = 5$, we are assuming the existence and uniqueness of the factorization. This will create what is called **circular reasoning** since we want to use the existence of the greatest common divisor to prove the fundamental theorem of arithmetic. Thus it is important that we devise a way to compute the greatest common divisor that does not require unique factorization. That is the Euclidean algorithm we are about to describe.

Examples. $1 = \gcd(0,1)$, $1 = \gcd(3,2)$, $5 = \gcd(10,25)$, $1 = \gcd(37,5)$. ▲

Exercise 1.5.9 *Compute* $\gcd(13,169)$ *and* $\gcd(11,1793)$.

The Euclidean Algorithm. To compute $\gcd(37,5)$ using the Euclidean algorithm, one compiles a list of divisions. The gcd will be the last nonzero remainder. We know that the list of remainders must end in 0 since it is strictly decreasing.

$$\begin{aligned} 37 &= 5\cdot 7 + 2 \\ 5 &= 2\cdot 2 + 1 \\ 2 &= 1\cdot 2 + 0 \end{aligned}$$

So $\gcd(37,5) = 1$.

It is not hard to check that the last nonzero remainder must divide all the preceding remainders and thus both 37 and 5. This is done by reading the preceding list from bottom to top. Moreover, if c is a common divisor of 37 and 5, we see that c must divide all the remainders by reading the list from top to bottom.

We can also use the Euclidean algorithm from bottom to top to write the $\gcd(a,b)$ as an integer linear combination of a and b:

$$\begin{aligned} 1 &= 5 - 2*2 && \text{from row 2} \\ 1 &= 5 - (37 - 5*7)*2 && \text{from row 1.} \end{aligned}$$

Thus

$$1 = 5*(1+14) - 37*2 = 5*15 - 37*2.$$

Exercise 1.5.10 *Compute* $d = \gcd(17,28)$ *using the Euclidean algorithm. Then find integers* m, n *such that* $d = 17m + 28n$.

Exercise 1.5.11 *Write out the general statement of the Euclidean algorithm for two positive integers* a *and* b, *and then prove it. How do we know that it ends after a finite number of steps?*

Hint. *Suppose that* $b \le a$. *To find* $\gcd(a,b)$ *perform the following divisions:*

$$\begin{aligned} a &= bq_1 + r_1, && \text{where } 0 \le r_1 < b \\ b &= r_1q_2 + r_2, && \text{where } 0 \le r_2 < r_1 \\ r_1 &= r_2q_3 + r_3, && \text{where } 0 \le r_3 < r_2 \\ &\vdots && \vdots \\ r_{n-2} &= r_{n-1}q_n + r_n, && \text{where } 0 \le r_n < r_{n-1} \\ r_{n-1} &= r_nq_{n+1} + 0, && \text{so that } r_n = \text{last nonzero remainder.} \end{aligned}$$

You need to show that $r_n = \gcd(a,b)$. *It is understood that if* $r_1 = 0$, *then* $\gcd(a,b) = b$.

The Euclidean algorithm leads to an identity named after the French mathematician É. Bézout, who was not really the first to prove it for the integers. Instead, he discovered the result for polynomials which we will describe in Section 5.5.

Theorem 1.5.2 ***(Bézout's Identity)*** *If a and b are integers, then $d = \gcd(a, b) = na + mb$ for some integers m, n. Moreover, d is the smallest positive integer which is an integer linear combination, $na + mb$, of a and b.*

Proof. We could prove this theorem constructively by reading the Euclidean algorithm backwards. But let us instead give a non-constructive proof which does not mention the Euclidean algorithm.

We might as well assume a, b are both positive. Look at the set of integers

$$S = \{k > 0 \mid k = na + mb, \quad n, m \in \mathbb{Z}\}.$$

Note that S is not empty and thus has a least element q by the well-ordering principle. In fact, q is the $\gcd(a, b)$. To see this, note that since q is an element of S, it is clear that any common divisor of a and b must divide q. To see that q divides a, use the division algorithm to write $a = qc + r$, with $0 \leq r < q$. But then $r = a - qc$ is in S. Thus r must be 0 by the minimality of q. So $q|a$. Similarly $q|b$. Thus $q = \gcd(a, b)$. ▲

Exercise 1.5.12 *Fill in the details in the proof of the preceding theorem. For example, explain why S is non-empty and $r \in S$.*

Exercise 1.5.13 *Use the Euclidean algorithm to compute* $\gcd(163, 1001)$ *and* $\gcd(163, 1141)$.

Lemma 1.5.1 ***(Euclid's Lemma)*** *Suppose that a and b are integers and let p be a prime. If p divides ab then either p divides a or p divides b.*

Proof. Suppose that p does not divide a. Then $1 = \gcd(a, p)$. Why? It follows from Theorem 1.5.2 that $1 = na + mp$ for some integers n, m. Multiply this equality by b. That gives $b = nab + mbp$. Since, by hypothesis, $ab = pc$, for some integer c, this means $b = p(nc + mb)$ and p divides b. ▲

We will need part (a) of the following exercise (proved by induction) in the proof of the next theorem.

Exercise 1.5.14

(a) Prove that if prime p divides a product $a_1 a_2 \cdots a_r$ then p must divide a_j for some j.
(b) Prove that if p is a prime and p does not divide the integer n, then $\gcd(p, n) = 1$.

Theorem 1.5.3 ***(The Fundamental Theorem of Arithmetic).*** *Every positive integer $n > 1$ factors uniquely (up to order) as a product of primes.*

Proof. Here we sketch only **the uniqueness** and leave the existence of the factorization as an exercise for the reader. Both parts of the proof require the 2nd principle of induction.
Step 1. We start with $n=2$. In this case, the factorization is unique as 2 is a prime.
Step 2. Our induction assumption says: each integer m with $2 \leq m < n$ has a unique prime factorization. We must use this to show that then so does $n+1$.

So suppose that $n+1$ has two prime factorizations

$$n+1 = p_1 p_2 \cdots p_u = q_1 q_2 \cdots q_v.$$

Here all the p_i and q_j are primes (not necessarily distinct). By part (a) of Exercise 1.5.14, if a prime divides a product, then it must divide one of the factors. So we know that the prime p_1 must divide the prime q_j for some j. But since the only positive divisors of the prime q_j are 1 and q_j, it follows that $p_1 = q_j$. This means we can divide $p_1 = q_j$ out of both sides and obtain two distinct factorizations for the smaller number $(n+1)/p_1$. But that is not possible by the induction assumption: that is, all the remaining primes must also coincide, and $n+1$ has a unique factorization. ▲

Exercise 1.5.15 *Complete the proof of the preceding theorem by proving the existence of the factorization.*

Euclid proved the existence part of the fundamental theorem. The uniqueness was not proved until Carl Friedrich Gauss (1777–1855) wrote his book *Disquisitiones Arithmeticae* in 1798 when he was 21. The uniqueness was perhaps considered obvious. However, when people attempted to prove Fermat's last theorem using arithmetic in more general rings like $\mathbb{Z}[e^{2\pi i/n}]$ whose elements are polynomials with integer coefficients in the nth root of unity $\zeta_n = e^{2\pi i/n}$, it was soon learned that unique factorization can and does fail for large enough values of n: for example, $n=23$. We call $\zeta_n = e^{2\pi i/n}$ an nth root of unity because it is a root of the polynomial $x^n - 1$.

Exercise 1.5.16 *State whether the following statements are true or false and give a brief explanation of your answer.*

(1) If $a, b, r \in \mathbb{Z}^+$ and r divides ab, then either r divides a or r divides b.
(2) For $a, b \in \mathbb{Z}^+$ if a divides b and b divides a, then $a = b$.

Exercise 1.5.17 *Find all the positive divisors of* 24, 36, *and* 81.

Exercise 1.5.18 *Suppose the prime factorization of the integer $n \geq 2$ is $n = p_1^{e_1} \cdots p_r^{e_r}$, with exponents $e_i > 0$ and pairwise distinct primes p_i. Write down an expression for any positive divisor of n. Then give a formula for the number of positive divisors of n.*

The following exercise would upset any Pythagoreans – should there be any existing today. The Pythagoreans formed a secret society around 550 BC. They believed: "Natural numbers and their ratios rule the universe." See Edna E. Kramer [59, Chapter 2] for more information on them. Thus the Pythagoreans were OK with fractions but they drew the line at the next step – creating $\sqrt{2}$ – even though they loved the right triangle with sides of length 1 and hypotenuse of length $\sqrt{2}$ – using the Pythagorean theorem. Supposedly they would kill you for revealing the irrationality of $\sqrt{2}$ to someone. Mercifully they disappeared thousands of years ago.

Exercise 1.5.19 *Prove that $\sqrt{2}$ is irrational: that is, not of the form $\frac{m}{n}$, where $m, n \in \mathbb{Z}$, with $n \neq 0$ and $\gcd(m, n) = 1$. The object is to derive a contradiction to the assumption that m and n are relatively prime using the equation $\sqrt{2} = \frac{m}{n}$. You need to square both sides and note that squares of odd numbers are odd while squares of even numbers are even. Once you see how to do $\sqrt{2}$, go on to prove $\sqrt[k]{2}$ is also irrational, for all $k = 2, 3, 4, 5, 6, \ldots$*

Exercise 1.5.20 *Factor the following numbers. Feel free to use Mathematica or your favorite computer program.*

(a) 31 415
(b) 314 159 265
(c) 314 159 265 358 979

In the 1970s I went to number theory conferences that held factoring competitions using programmable calculators. So I had to include the last problem. Now many types of computer software will do this factorization. Mathematica works for me. I admit that I did not take part in these calculator races of the 1970s. Very clever methods must be used to factor large numbers – especially if they are products of two large primes. The principle that underlies some cryptographic systems is the difficulty of factorization. See Section 4.1.

Figure 1.8 is an ArrayPlot in Mathematica of the matrix of values $\gcd(m, n)$ for $-50 \leq m, n \leq 50$. A different color is associated to each value with the color range pretty evident along the main diagonal.

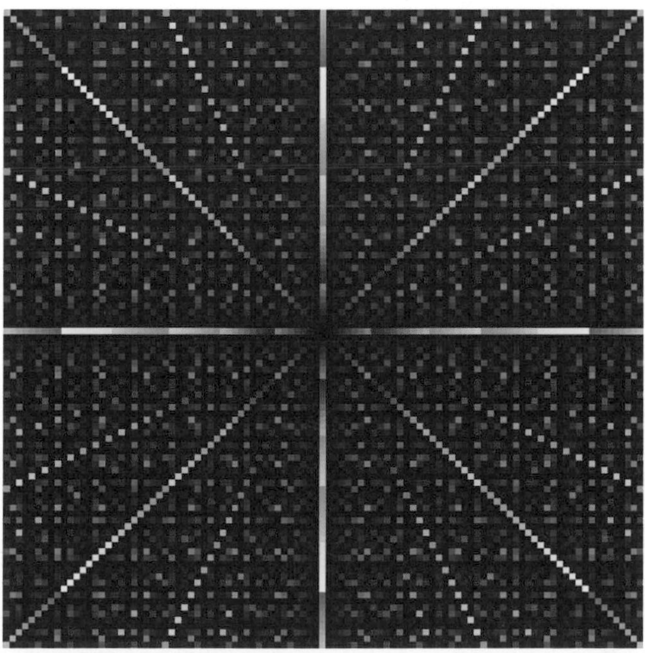

Figure 1.8 A color is placed at the (m, n) entry of a 101×101 matrix according to the value of $\gcd(m, n)$. This is an ArrayPlot in Mathematica

Exercise 1.5.21 *Explain the lines seen in Figure 1.8. The command in Mathematica used to generate the figure was:*

```
ArrayPlot[Table[GCD[m,n],{m,-50,50},{n,-50,50}],ColorFunction
->ColorData["BrightBands"],
Frame->False,DataReversed->{True,False}].
```

Hint. *The diagonal* $m=n$ *clearly comes from* $\gcd(m,m)=|m|$. *The color scheme chosen runs from red through blue, green, yellow, to orange. What about the other lines? For example, what happens to* $\gcd(m,n)$ *if* $m=2n$? *When* m *is a rational multiple of* n, *say* $m=\frac{p}{q}n$, *what can you say about* $\gcd(m,n)$?

Exercise 1.5.22 *Suppose that* $a,b,c\in\mathbb{Z}^+$ *such that* $\gcd(a,b)=1$ *and* $ab=c^n$, *for some* $n>1$. *Show that then there are* $r,s\in\mathbb{Z}^+$ *such that* $a=r^n$ *and* $b=s^n$.

In the following exercise we use the notation

$$\prod_{i=1}^{k} a_i = a_1a_2\cdots a_k.$$

Exercise 1.5.23 *Show that if the* p_i *are pairwise distinct primes, for* $i=1,\dots,k$, *then*

$$\gcd\left(\prod_{i=1}^{k}p_i^{e_i},\prod_{i=1}^{k}p_i^{f_i}\right)=\prod_{i=1}^{k}p_i^{g_i},\quad \text{where } g_i=\min\{e_i,f_i\}.$$

1.6 Modular Arithmetic, Congruences

The standard way to begin the story of modular arithmetic – a story which goes back to the 1700s – is to think about clocks.

Example: Clock Arithmetic

Question. If it is 3 o'clock now, what time is it after 163 hours?

Method of solution. Divide 163 by 12. Obtain the quotient 13 and the remainder 7. That is $163=13\times 12+7$.

Answer. It will be $10=(7+3)$ o'clock.

What if you want to know if it is a.m. or p.m.? Then you should divide by 24 rather than 12.

We say that $3+163$ is congruent to 10 modulo 12 and write

$$3+163\equiv 3+7\equiv 10 \pmod{12}.$$ ▲

C. F. Gauss invented this notation. L. Euler (1707–1783) had introduced the idea earlier (around 1750). It gives us a new way to do arithmetic. This is **modular arithmetic** and is the foundation of many of our applications.

> **Definition 1.6.1** *Fix a modulus m which is a positive integer. If a, b are integers, define* $a \equiv b \pmod{m}$ *(read as "a is* ***congruent*** *to b* ***modulo*** *m") iff m divides* $(a-b)$*. The relation* $a \equiv b \pmod{m}$ *is called a* ***congruence****. We will say that a and b are* ***in the same congruence class*** $[a]$ mod m.

When considering integers $a, b \pmod{m}$, we want to identify a and b when m divides $(a-b)$. That is, we are going to glomp together all integers that are congruent to a mod m and call the result $[a]$ a congruence class. This is the first example of a construction that we will see quite often in the rest of this book. See Definition 3.4.1 for example.

Exercise 1.6.1 *Suppose it is now 7 p.m. What time will it be after 101 hours? Is it a.m. or p.m.?*

Exercise 1.6.2 *Suppose it is 5 p.m. at the airport (Lindbergh field) in San Diego now. What time is it when you get off your plane at Kennedy airport after you take a 5 hour flight to New York?*

Exercise 1.6.3 *Prove that* $a \equiv b$ *(mod m) iff a and b have the same remainder upon division by m.*

Example. Let $m = 3$. When we create the integers mod 3, we are taking the infinite line of integers and rolling it up into a triangle. A picture of this is given in Figure 1.9.

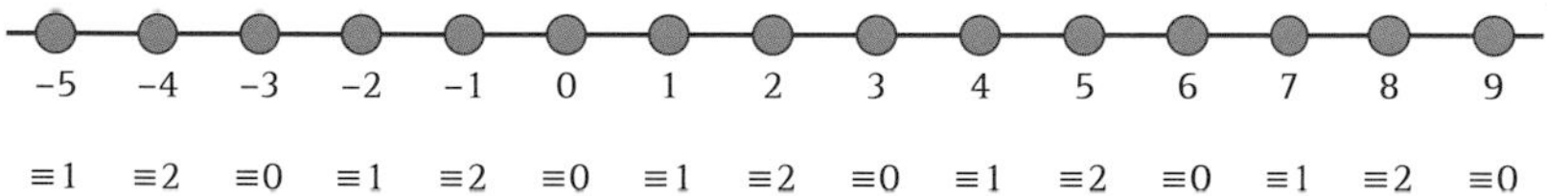

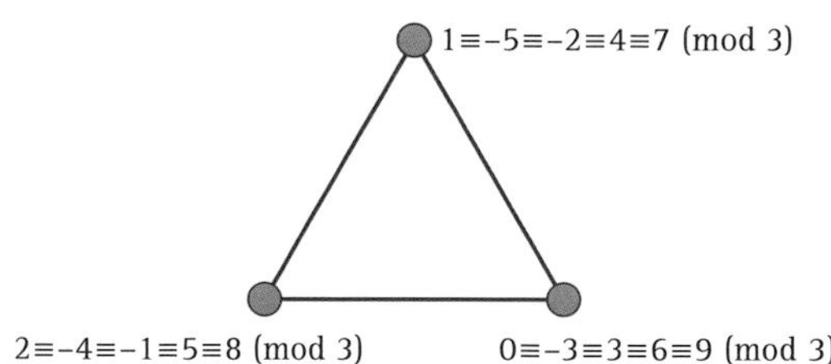

Figure 1.9 Rolling up the integers modulo 3

There are three congruence classes of integers that we have glomped together mod 3: namely $[0], [1], [2]$. Here $[a] = \{x \mid x \equiv a \pmod{3}\}$. The set of these congruence classes is denoted $\mathbb{Z}_3$. We will normally identify $\mathbb{Z}_3$ with $\{0, 1, 2\}$ or – equivalently – with $\{-1, 0, 1\}$. We can then use ordinary addition and multiplication of integers to define a sum and product on $\mathbb{Z}_3$.

Taking the modulus $m = 12$, you get a clock. Take the modulus $m = 1$, and you get one number; that is, $\mathbb{Z}_1$ can be identified with $\{0\}$. ▲

Example: Addition and Multiplication mod 3. We add mod 3 by adding representatives of the congruence classes: $[a]+[b]=[a+b]$. Thus

$$\begin{aligned}0+1&\equiv 1 \pmod 3,\\ 1+2&\equiv 3\equiv 0 \pmod 3,\\ 2+2&\equiv 4\equiv 1 \pmod 3.\end{aligned}$$

Multiplication is defined similarly: $[a]\cdot[b]=[a\cdot b]$. This gives, for example:

$$2\cdot 2\equiv 4\equiv 1 \pmod 3.$$

▲

With these definitions, you get addition and multiplication tables for $\mathbb{Z}_3$, which are also called Cayley tables, named for Arthur Cayley (1821–1895). Such tables first appeared in an 1854 paper of Cayley. In the first table, the top row and left column give the elements of $\mathbb{Z}_3$. Then the entry at row a, column b, for $a,b\in\mathbb{Z}_3$ is $a+b \pmod 3$. The second table is analogous with $+$ replaced by $\times$.

Addition table for addition in $\mathbb{Z}_3$

$+\pmod 3$	0	1	2
0	0	1	2
1	1	2	$0\equiv 1+2$
2	2	0	$1\equiv 2+2$

Multiplication table for multiplication in $\mathbb{Z}_3$

$\times\pmod 3$	0	1	2
0	0	0	0
1	0	1	2
2	0	2	$1\equiv 2\cdot 2$

In the multiplication table, it would make sense to leave out the first row and first column since they only contain 0s. Later – in Chapter 5 – we will learn that $\mathbb{Z}_3$ is a ring, in fact, a field. But before Chapter 5, we will just say $\mathbb{Z}_3$ is a group under addition, while $\mathbb{Z}_3-\{0\}$ is a group under multiplication.

More generally, we can define the **integers modulo** n, $\mathbb{Z}_n$ to consist of the elements of $\mathbb{Z}$ identified under congruence mod n. Thus the elements are **congruence classes** $[a]=\{a+nm|m\in\mathbb{Z}\}$. And we define $[a]+[b]=[a+b]$, $[a]\,[b]=[ab]$. One needs to check that these definitions give **well-defined** operations. This means that the sum or product is unambiguous. For example, in $\mathbb{Z}_3$, $[1]+[2]=[3]=[4]+[5]=[9]=[0]$. Yes, both sums are $[0]$. See the exercises below or Section 2.3 for more on this subject. With this notation, $\mathbb{Z}_3=\{[0],[1],[2]\}$. Usually we leave out the [ ].

Exercise 1.6.4 *Create the analogous addition and multiplication tables when $m=7$ and 8.*

We will have much more to say about these congruence groups. They are important for most of the applications we will discuss: for example, error-correcting codes and cryptography.

Exercise 1.6.5

(a) For $n \in \mathbb{Z}^+$, *and* $a, b, c \in \mathbb{Z}$, *show that* $a \equiv b \pmod{n}$ *implies* $a + c \equiv b + c \pmod{n}$.

(b) If a, b, c *and* n *are as in a) and* $d \in \mathbb{Z}$, *show that* $a \equiv b \pmod{n}$ *and* $c \equiv d \pmod{n}$ *imply that* $a + c \equiv b + d \pmod{n}$. *This means that addition in* $\mathbb{Z}_n$ *is well defined (i.e., unambiguous).*

Exercise 1.6.6

(a) For $n \in \mathbb{Z}^+$, *and* $a, b, c \in \mathbb{Z}$, *show that* $a \equiv b \pmod{n}$ *implies* $ac \equiv bc \pmod{n}$.

(b) If a, b, c *and* n *are as in (a) and* $d \in \mathbb{Z}$, *show that* $a \equiv b \pmod{n}$ *and* $c \equiv d \pmod{n}$ *imply that* $ac \equiv bd \pmod{n}$. *This means that multiplication in* $\mathbb{Z}_n$ *is well defined (i.e., unambiguous).*

Exercise 1.6.7 *Compute* $5^{100\,007} \pmod{4}$. *Then compute* $3^{100\,007} \pmod{4}$. *You should get an element of the set* $\{0, 1, 2, 3\}$. *Note that you should not compute* $5^{100\,007}$ *or* $3^{100\,007}$ *which are humongous numbers.*

Exercise 1.6.8

(a) Compute $\gcd(83, 38) = d$ *using the Euclidean algorithm from the preceding section.*

(b) Use the result of part (a) to write $d = 83m + 38n$, *with integers* m, n.

(c) Then use part (b) to solve $38x = 1 \pmod{83}$.

The following exercise is motivated by the preceding one and will be used in Section 2.5.

Exercise 1.6.9 *(Solving Linear Congruences).* *Suppose* $n \in \mathbb{Z}^+$, *and* $a, b \in \mathbb{Z}$. *Consider the linear congruence*

$$ax \equiv b \pmod{n}.$$

The question is: when can this congruence be solved for $x \pmod{n}$? *Prove that the answer is: when* $d = \gcd(a, n)$ *divides* b.

Hint. *Use the Bézout identity (Theorem 1.5.2).*

Exercise 1.6.10 *Suppose* $n \in \mathbb{Z}^+$ *and* n *is odd. Show that*

$$1 + 2 + 3 + \cdots + (n - 1) \equiv 0 \pmod{n}.$$

Is this congruence still true for even n?

Exercise 1.6.11 *Solve* $5x \equiv 1 \pmod{163}$.

One of the morals of this section is that you can often replace real numbers with elements of $\mathbb{Z}_n$. In particular, you can do linear algebra with $\mathbb{R}$ replaced by $\mathbb{Z}_n$. It is even better when n is a prime.

Exercise 1.6.12 *Solve the following pair of linear congruences simultaneously for* $x, y \pmod{5}$

$$2x + 2y \equiv 1 \pmod{5},$$
$$3x + 2y \equiv 2 \pmod{5}.$$

Then rewrite the congruences as a matrix equation of the form $Mv = b$, where M is a 2×2 matrix with entries in $\mathbb{Z}_5$, while b and v are column vectors with entries in $\mathbb{Z}_5$. We multiply matrix times vector as in formula (1.3).

Exercise 1.6.13 *Find the inverse* mod 7 *of the matrices* $\begin{pmatrix} 1 & 2 \\ 0 & 1 \end{pmatrix}$ *and* $\begin{pmatrix} 1 & 2 \\ 3 & 4 \end{pmatrix}$, *where the inverse of a matrix is defined as usual in linear algebra. See Example 3 in Section 2.3.*

Exercise 1.6.14 *Show that if f_n denotes the nth Fibonacci number from Section 1.4, then* $\gcd(f_n, f_{n+1}) = 1$.

Gallian [33, Chapter 1] gives many applications of modular arithmetic in everyday life: for example, in the assignment of check digits in Universal Product Codes read by optical scanners in large stores.

Exercise 1.6.15 *Solve the simultaneous congruences*

$$2x \equiv 1 \pmod{5},$$
$$3x \equiv 2 \pmod{7}$$

with $x \in \mathbb{Z}_{35}$.

The preceding sort of problem will be considered more generally in Section 3.6 under the heading of the Chinese remainder theorem – a result with many applications.

1.7 Relations

Many of the ideas that we have already discussed are examples of relations on the set $\mathbb{Z}$: for example, $a < b$ or $a \equiv b \pmod{m}$. The modern way to think of such things is as a subset of the Cartesian product $\mathbb{Z} \times \mathbb{Z}$.

Definition 1.7.1 *A* ***(binary) relation*** *R on a set A is a subset of the set*

$$A \times A = \{(a, a') \mid a, a' \in A\}.$$

We may write aRa' if $(a, a') \in R$. It is also possible to have a relation R from set A to set B, which is a subset of $A \times B$.

Examples: Relations

1. A relation on $\mathbb{Z}$ is $<$. That is, the relation is the set of pairs (a, b) in $\mathbb{Z} \times \mathbb{Z}$ with $a < b$. Figure 1.10 shows the relation using Mathematica to plot a 50×50 grid whose (i, j) square is turquoise if $i < j$ and purple otherwise. We created the figure with the Mathematica command:

```
ListDensityPlot[Table[i-j,{j,1,50},{i,1,50}],
   ColorFunction->(If[#>0,Purple,White]&),
   ColorFunctionScaling->False,InterpolationOrder->0]
```

2. Another relation on $\mathbb{Z}$ is divisibility: that is, the pairs (a, b) in $\mathbb{Z} \times \mathbb{Z}$ with $a|b$ (i.e., a divides b). Figure 1.11 shows the relation using Mathematica to plot a 50×50 grid whose (i, j) square is turquoise if $j|i$ and purple otherwise.

3. Congruence (mod m) gives yet another relation on $\mathbb{Z}$: that is, $a \equiv b \pmod{m}$. Figure 1.12 shows the relation using Mathematica to plot a 50×50 grid whose (i, j) square is turquoise if $i \equiv j \pmod 5$ and purple otherwise. ▲

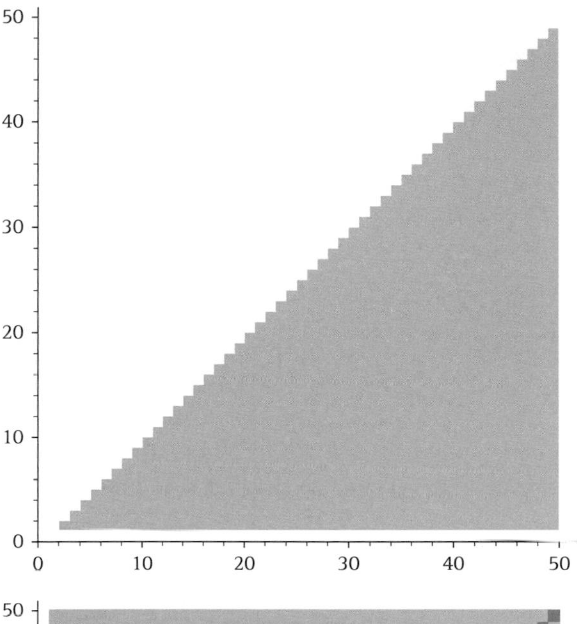

Figure 1.10 Mathematica picture of the $x < y$ relation for the integers between 1 and 50

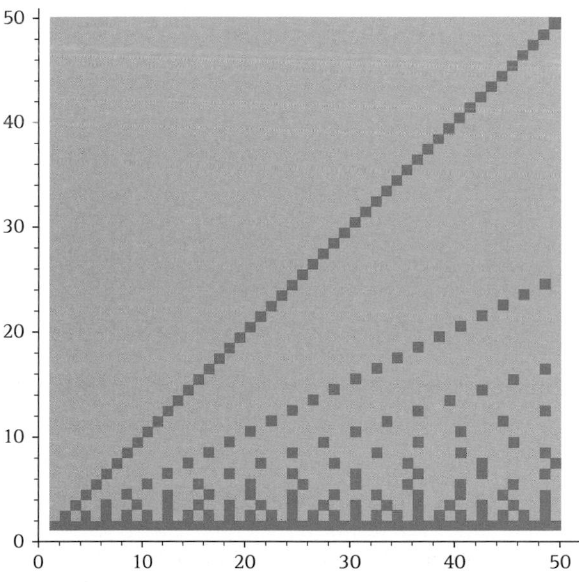

Figure 1.11 Mathematica picture of the $y|x$ relation for the integers between 1 and 50

Exercise 1.7.1 *Explain the lines in Figures 1.10–1.12.*

Definition 1.7.2 *A relation R on a set S is an* **equivalence relation** *iff it has the following three properties. We will write $a \sim b$ instead of $(a, b) \in R$.*

1. *$a \sim a$ for all $a \in S$* ***(reflexivity)****.*
2. *$a \sim b \Longleftrightarrow b \sim a$* ***(symmetry)****.*
3. *$a \sim b$ and $b \sim c \Longrightarrow a \sim c$* ***(transitivity)****.*

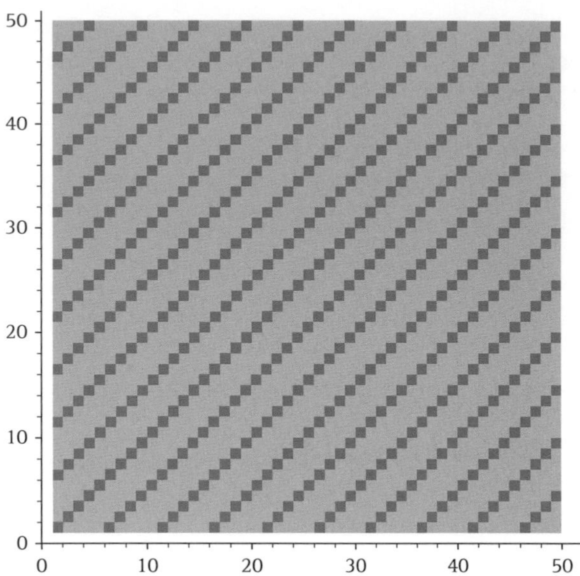

Figure 1.12 Mathematica picture of the $x \equiv y \pmod{5}$ relation for the integers between 1 and 50

Next we want to decide whether the three relations in the preceding examples are equivalence relations on $\mathbb{Z}$. The first two relations are considered in the exercises below. What about the third relation – congruence?

Example. The congruence relation $a \equiv b \pmod{m}$ is an equivalence relation on $a, b \in \mathbb{Z}$.

To see this, we check the three properties of an equivalence relation.

1. **Reflexivity.** Certainly $a \equiv a \pmod{m}$ since m divides $0 = a - a$.
2. **Symmetry.** If $a \equiv b \pmod{m}$ then $a - b = km$ for some integer k,and then $b - a = -km$, which implies that $b \equiv a \pmod{m}$.
3. **Transitivity.** Assume $a \equiv b \pmod{m}$ and $b \equiv c \pmod{m}$. Then $a - b = km$ and $b - c = k'm$. It follows that

$$a - c = a - b + b - c = km + k'm = (k + k')m.$$

Therefore $a \equiv c \pmod{m}$.

Thus $a \equiv b \pmod{m}$ is indeed an equivalence relation on $\mathbb{Z}$. We will see many generalizations of this equivalence relation when we envision quotient groups and quotient rings in Sections 3.4 and 5.4. ▲

Exercise 1.7.2 *Show that $a < b$ is not an equivalence relation on $\mathbb{Z}$. Can you see this from Figure 1.10?*

Exercise 1.7.3 *Show that $a|b$ is not an equivalence relation on $\mathbb{Z}$. Can you see this from Figure 1.11?*

Definition 1.7.3 *Suppose $a \sim b$ denotes an equivalence relation on a set S. Define the **equivalence class** of $a \in S$ to be*

$$[a] = \{b \in S | b \sim a\}.$$

Definition 1.7.4 *A **partition** of a set S is a collection of non-empty pairwise disjoint subsets $S_j \subset S$ whose union is S. More precisely, we mean that*

$$S = \bigcup_j S_j, \quad \text{with } S_i \cap S_j = \emptyset, \ \forall i \text{ and } j \text{ such that } i \neq j, \text{ and } S_j \neq \emptyset, \ \forall j.$$

Equivalence relations are closely connected with partitions of sets as we shall see in the next theorem.

Examples. Consider congruence modulo 3 as an equivalence relation on $\mathbb{Z}$. There are only three equivalence classes

$$\begin{aligned}
[0] &= \{\text{all integers which are divisible by } 3\} = \{3n \mid n \in \mathbb{Z}\} \\
&= \{a \in \mathbb{Z} \mid a \equiv 0 \pmod 3\}; \\
[1] &= \{\text{all integers with remainder 1 when divided by } 3\} \\
&= \{1 + 3n \mid n \in \mathbb{Z}\} = \{a \in \mathbb{Z} \mid a \equiv 1 \pmod 3\}; \\
[2] &= \{\text{all integers with remainder 2 when divided by } 3\} \\
&= \{2 + 3n \mid n \in \mathbb{Z}\} = \{a \in \mathbb{Z} \mid a \equiv 2 \pmod 3\}.
\end{aligned}$$

Note that we have a partition of $\mathbb{Z}$ into the three equivalence classes mod 3:

$$\mathbb{Z} = [0] \cup [1] \cup [2], \quad [0] \cap [1] = \emptyset, \quad [0] \cap [2] = \emptyset, \quad [1] \cap [2] = \emptyset.$$

▲

Theorem 1.7.1 *Equivalence classes from an equivalence relation $\sim$ on a set S give a partition of S. Conversely, given a partition of a set S as a union of pairwise disjoint non-empty subsets S_j:*

$$S = \bigcup_j S_j, \quad \text{with } S_i \cap S_j = \emptyset, \text{ when } i \neq j,$$

we can define an equivalence relation on $x, y \in S$ by saying $x \sim y$ iff both x and y lie in the same subset S_j, for some j.

Proof. $\Longrightarrow$ As usual, we write $[a] = \{x \in S \mid x \sim a\}$ for the equivalence class of $a \in S$.

Why are the equivalence classes non-empty? By reflexivity we know $a \in [a]$.

Why is S a union of the equivalence classes? Every $a \in S$ is in the equivalence class $[a]$ by reflexivity.

Why are the classes pairwise disjoint? If $c \in [a] \cap [b]$, then $c \sim a$ and $c \sim b$. By symmetry, then $a \sim c$ and $c \sim b$. So by transitivity, $a \sim b$. This implies if $x \in [a]$ then $x \sim a$ and $a \sim b$, so $x \sim b$ and thus $x \in [b]$. Thus $[a] \subset [b]$. Similarly $[b] \subset [a]$. Therefore $[a] = [b]$ if $[a] \cap [b] \neq \emptyset$.

$\Longleftarrow$ We leave the proof of the converse as an exercise. ▲

Exercise 1.7.4 *Prove the converse part of the preceding theorem.*

Exercise 1.7.5 *Define a relation on $a, b \in \mathbb{R}$ by $a \sim b \iff a - b \in \mathbb{Z}$. Show that this is an equivalence relation on $\mathbb{R}$. Find a nice set of representatives for the equivalence classes.*

Now we consider another sort of relation which will appear repeatedly in our discussions.

Definition 1.7.5 *A* ***partial order*** *denoted* $\leq$ *on a set* S *is a relation that has the following three properties, for elements* $a, b, c \in S$:

1. ***reflexive:*** $a \leq a$;
2. ***antisymmetric:*** $a \leq b$ *and* $b \leq a$ *implies* $a = b$;
3. ***transitive:*** $a \leq b$ *and* $b \leq c$ *implies* $a \leq c$.

Examples

1. Consider the set of subsets of a given set. Then set-theoretic inclusion is a partial order.
2. If the set $S = \mathbb{Z}$ or $\mathbb{R}$, then $a \leq b$ is a partial order.
3. If the set $S = \mathbb{Z}^+$, divisibility $a|b$ is a partial order.

▲

Given a finite partially ordered set S, or **poset** one can make a diagram called a **poset or Hasse diagram.** The diagram consists of a set of vertices such that each vertex corresponds to an element of S. Then we draw a rising line between vertices a and b if $a \leq b$ and there is no $c \in S$ with $a \leq c \leq b$.

Example. Consider the poset diagram for the divisibility on the set of positive divisors of 24. This is shown in Figure 1.13. ▲

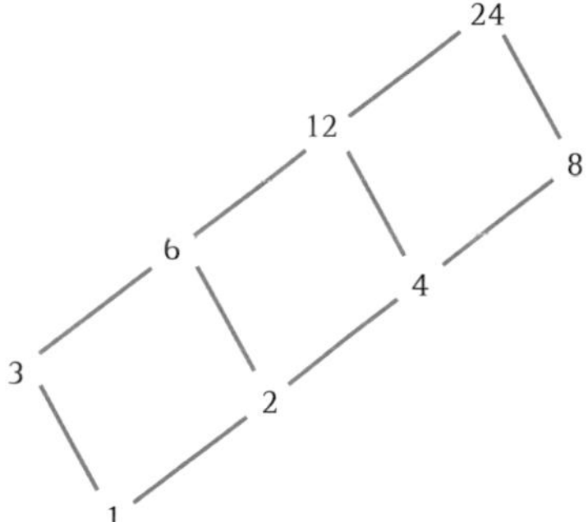

Figure 1.13 Poset diagram of the positive divisors of 24

Exercise 1.7.6 *Draw the poset diagram for the set of all subsets of* $\{1, 2, 3, 4\}$ *under the relation* $\subset$.

Exercise 1.7.7 *Draw the poset diagram for the set of positive divisors of* 30.

Exercise 1.7.8 *Draw the poset diagram for the set of positive integers* ≤ 20 *under the relation* $\leq$.

1.8 Functions, the Pigeonhole Principle, and Binary Operations

We now have a new way to think about functions. Identifying a function with its graph, we see that a function is a special kind of relation.

Definition 1.8.1 *Suppose A and B are sets. A **function (or mapping)** $f\colon A \longrightarrow B$ is a relation from A to B such that every $a \in A$ is the first entry of precisely one $(a, b) \in f$; that is,*

(i) if $a \in A$, $\exists\, b \in B$ such that $(a, b) \in f$,
(ii) if $(a, b) \in f$ and $(a, c) \in f$, then $b = c$.

We write $b = f(a)$ instead of $(a, b) \in f$.

Condition (ii) means that the function f is **well defined:** that is, $f(a)$ is a unique element of the set B. No ambiguity is allowed. It is not possible to make a computation with ambiguously defined functions or group elements or sets or numbers. We will often call a function a "map," which is short for mapping.

Warning. Some algebra books (e.g., Herstein [42]) write af instead of $f(a)$. This is often called Reverse Polish Notation. That old Hewlett-Packard calculator that I bought in the early 1970s used it. We will always write $f(a)$ and not af.

Definition 1.8.2 *If $f\colon A \longrightarrow B$ is a function, then the set $f(A) = \{f(a) \mid a \in A\}$ is called the **image** of A under f.*

If $f\colon A \longrightarrow B$ is a function, the image set $f(A)$ is a subset of B. Later we will define a subset of A called the inverse image.

Exercise 1.8.1 *State whether each of the following equations is true or false and explain.*
(a) $f(A \cup B) = f(A) \cup f(B)$,
(b) $f(A \cap B) = f(A) \cap f(B)$.

Examples

(1) Favorite functions from $\mathbb{Z}$ to $\mathbb{Z}$:

$f(x) = x^2$, for all $x \in \mathbb{Z}$.
$g(x) = x + 1$, for all $x \in \mathbb{Z}$.

(2) A non-function:

$h(x) =$ either 1 or -1, for all $x \in \mathbb{Z}$. ▲

Definition 1.8.3 ***(Composition of Functions).*** *If $f: A \to B$ and $g: B \to C$, define the **composition** of f and g to be $(g \circ f)(x) = g(f(x))$, for all $x \in A$. Then $g \circ f$ is a function and $g \circ f: A \to C$.*

Note that $g \circ f$ is not usually the same as $f \circ g$. For example, assuming $A = B = C = \mathbb{R}$, if $f(x) = x^2$ and $g(x) = x + 1$, then

$$(g \circ f)(x) = x^2 + 1,$$

while

$$(f \circ g)(x) = (x+1)^2 = x^2 + 2x + 1.$$

Exercise 1.8.2 *Show that composition of functions is associative: that is, if $f: A \to B$, $g: B \to C$, and $h: C \to D$, then*

$$h \circ (g \circ f) = (h \circ g) \circ f.$$

Definition 1.8.4 *A function $f: A \to B$ is* ***one-to-one*** *(1–1 or* ***injective****) iff $f(u) = f(v)$ implies $u = v$.*

Examples

1. $f: \mathbb{Z} \to \mathbb{Z}$ defined by $f(x) = 5x$ is 1–1 since $\mathbb{Z}$ has no zero divisors. Thus $5u = 5v$ implies $5(u - v) = 0$ and therefore $u - v = 0$. This means that $u = v$. If we replace $\mathbb{Z}$ by $\mathbb{Z}_{10}$, then this function maps 2 (mod 10) to 0 (mod 10) and is thus not 1–1, since $f(5) = f(0)$.
2. $f: \mathbb{Z} \to \mathbb{Z}$ defined by $f(x) = x^2$ is not 1–1 since $f(x) = f(-x)$. If we replace $\mathbb{Z}$ by $\mathbb{Z}_2$, then this function maps x to $x^2 \equiv x \pmod 2$. In short, it is what we call the identity function on $\mathbb{Z}_2$, since $f(x) = x$.
3. This example is one of the most important in the following chapters. You should remember such functions from linear algebra. Consider the set $\mathbb{R}^{m \times n}$ consisting of $m \times n$ matrices A over $\mathbb{R}$. A matrix $A = (a_{ij})_{1 \le i,j \le n} \in \mathbb{R}^{m \times n}$ gives rise to a linear function $\mathbb{R}^n \to \mathbb{R}^m$ defined by $y = T_A(x) = Ax$, where we are thinking of the elements of $\mathbb{R}^n$ as column vectors and we define **matrix multiplication** as usual by writing

$$A = \begin{pmatrix} a_{11} & \cdots & a_{1n} \\ \vdots & \ddots & \vdots \\ a_{m1} & \cdots & a_{mn} \end{pmatrix}_{1 \le i,j \le n}, \quad x = \begin{pmatrix} x_1 \\ \vdots \\ x_n \end{pmatrix}, \quad y = Ax = \begin{pmatrix} y_1 \\ \vdots \\ y_m \end{pmatrix}, \tag{1.3}$$

$$y_j = \sum_{i=1}^{n} a_{ji} x_i, \quad \text{for } j = 1, \ldots, m.$$

You can similarly define the multiplication of matrix $A \in \mathbb{R}^{m \times n}$ by matrix $B \in \mathbb{R}^{n \times k}$, by writing $B = (b_1 \cdots b_k)$ where b_j denotes the jth column of B, and then $AB = (Ab_1 \cdots Ab_k)$, multiplying each column vector of B by A to get the corresponding column of AB. These formulas will work with $\mathbb{R}$ replaced by any ring such as $\mathbb{Z}$ or $\mathbb{Z}_n$. This definition of matrix multiplication is due to Arthur Cayley. ▲

Definition 1.8.5 *$f: A \to B$ is* ***onto*** *(or* ***surjective****) iff for every $b \in B$ there exists $a \in A$ such that $f(a) = b$ or, equivalently, $f(A) = \{f(x) \mid x \in A\} = B$.*

Examples

1. $f: \mathbb{Z} \to \mathbb{Z}$ defined by $f(x) = x + 5$ is onto. For given y, you can solve $y = x + 5$ for x. Answer: $x = y - 5$.

2. $f:\mathbb{Z}\to\mathbb{Z}$ defined by $f(x)=5x$ is not onto, since not every integer is divisible by 5. Of course, if you replace $\mathbb{Z}$ with $\mathbb{R}$, with the same definition of the function, namely $f(x)=5x$, then $f:\mathbb{R}\to\mathbb{R}$ is onto. ▲

A function that is **1–1 and onto** is also called **bijective** or a **bijection**. We will not usually use these words – preferring to say 1–1 and onto.

Example. The function $g:\mathbb{Z}\to\mathbb{Z}$ defined by $g(x)=x+1$, for all $x\in\mathbb{Z}$, is both 1–1 and onto. ▲

There is a **right identity** for the operation of composition of functions. If $f:A\to B$ and $I_A(x)=x$, $\forall x\in A$, then $f\circ I_A=f$. Similarly I_B is a **left identity** for f: that is, $I_B\circ f=f$.

Definition 1.8.6 *If $f:A\to B$ is 1–1 and onto, it has an **inverse function** $f^{-1}:B\to A$ defined by requiring $f\circ f^{-1}=I_B$ and $f^{-1}\circ f=I_A$. If $f(a)=b$, then $f^{-1}(b)=a$.*

When the sets A and B are subsets of $\mathbb{R}$ and the function $f:A\to B$ is 1–1 and onto, then one gets the graph of the inverse function from that of the function by interchanging the x- and y-axes.

Exercise 1.8.3 *Suppose that $f:A\to B$ is 1–1 and onto and $g:B\to C$ is 1–1 and onto. Show that $g\circ f:A\to C$ is 1–1 and onto.*

Exercise 1.8.4

*(a) Prove that if $f:A\to B$ is 1–1 and onto, it has an **inverse function** f^{-1}.*
(b) Conversely show that if f has an inverse function, then f must be 1–1 and onto.
(c) Show that if $f:A\to B$ is 1–1 and onto, then $f^{-1}:B\to A$ is also 1–1 and onto.

Exercise 1.8.5 *State which of the following functions are 1–1 then state which are onto.*

(a) $f:\mathbb{Z}\to\mathbb{Z}$ given by $f(x)=x^6$;
(b) $f:\mathbb{Z}\to\mathbb{Z}$ given by $f(x)=3x$;
(c) $f:\mathbb{Z}\to\mathbb{Z}$ given by $f(x)=x-3$.

Maybe we should have introduced the following definition earlier, but we are now more capable of dealing with it precisely.

Definition 1.8.7 *We say that the empty set has 0 elements. If $n\in\mathbb{Z}^+$, we say that a set S **has n elements** iff there is a 1–1, onto function*

$$f:\{1,2,\ldots,n\}\to S.$$

*Then we write $n=|S|$. If a set cannot be said to have n elements for any $n\in\mathbb{Z}^+\cup\{0\}$, we call it **infinite**.*

To a mathematician, the infinite is nothing more than "not finite" – no need to philosophize. However, it is also possible to differentiate between different orders of magnitude of infinite sets. A **countable** or **denumerable** set S is in 1–1 correspondence with $\mathbb{Z}^+$. That is, there is a 1–1, onto function $f:\mathbb{Z}^+ \to S$. Cantor basically invented this subject and was not really popular for doing that. Many amazing things come to light – $\mathbb{Q}$ is countable, but $\mathbb{R}$ is not, for example. We do not really need to think about this much – mercifully. See Birkhoff and MacLane [9, Chapter XII] or Vilenkin [122]. The last reference explains the properties of countable sets with the story of the hotel with a countable number of rooms and its amazing abilities to accommodate guests. Sadly, however, this countable hotel is not able to produce a countable list telling which rooms are occupied.

We can use the function $f:\{1,2,\ldots,n\}\to S$ in the preceding definition to give a labeling of the elements of the finite set S with n elements:

$$S=\{f(1)=s_1,f(2)=s_2,\ldots,f(n)=s_n\}.$$

Similarly for a countably infinite set S, one can use the 1–1, onto function $f:\mathbb{Z}^+\to S$ to label the elements of S:

$$S=\{f(1)=s_1,f(2)=s_2,\ldots,f(n)=s_n,\ldots\}.$$

That is, one can view a countable infinite set as a sequence $S=\{s_n\}_{n\geq 1}$.

The following exercises give the basic principles of counting finite sets. The general theory of counting things is called **combinatorics.** A discussion of the basic principles can be found in the book by Clifford Stein, Robert L. Drysdale, and Kenneth P. Bogart [113] as well as those of Kenneth Rosen [92, 93].

Exercise 1.8.6 *Show that for finite sets S,T, if $S\cap T=\emptyset$, then $|S\cup T|=|S|+|T|$. Then extend the result to a union of a finite number of pairwise disjoint finite sets using mathematical induction.*

Exercise 1.8.7 *Show that for finite sets S,T, $|S\times T|=|S|\,|T|$.*

Hint. *You can derive this from the preceding exercise by writing $S\times T$ as a finite disjoint union of sets having the same order as S.*

Exercise 1.8.8 *Show that for finite sets S,T, $|\{f:S\to T\}|=|T|^{|S|}$.*

Hint. *This can be derived from the preceding exercise, once that exercise is extended to a product of n finite sets using mathematical induction.*

The following definition may appear silly but we will need it in later sections. Of course, you might be thinking that f and g in the definition are the same function, but that would be wrong thinking. The sets S and T in the notation $f:S\to T$ actually matter greatly to the notion of the function f and its properties.

Definition 1.8.8 *Suppose $f:S\to T$ and $R\subset S$. Define the* **restriction of f to R**, *written $f|_R=g$, to be the function $g:R\to T$ defined by setting $g(x)=f(x)$, for all $x\in R$.*

Example. Consider the function $f:\mathbb{Z}\to\mathbb{Z}$ defined by $f(x)=x^2$. We know that f is neither 1–1 nor onto. However $f|_{\mathbb{Z}^+}$ is 1–1, but still not onto. In fact $f|_{\mathbb{Z}^+}$ does not even map $\mathbb{Z}^+$ onto $\mathbb{Z}^+$. ▲

Exercise 1.8.9 ***(The Pigeonhole Principle).*** *If A and B are finite sets, each with the same number n of elements, then $f:A\to B$ is 1–1 iff f is onto.*

Hint. *Think of the elements of set A as pigeons and the elements of the set B as holes. Think of the $f(a)=b$ as putting pigeon a into hole b. So f is 1–1 means no 2 pigeons share a hole. And f is onto means every hole has a pigeon. See Figure 1.14.*

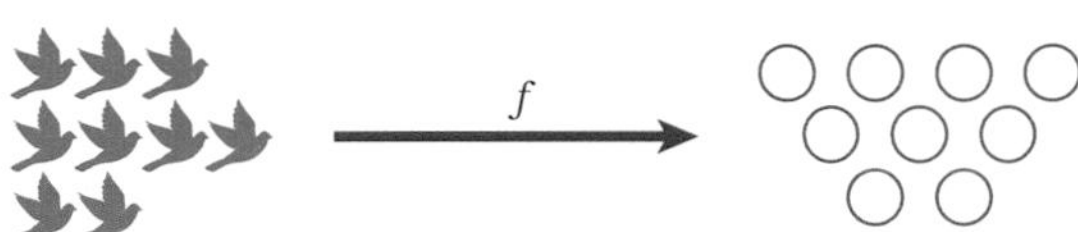

Figure 1.14 The pigeonhole principle

In 1834 Dirichlet formulated the pigeonhole principle. He called it the “Schubfachprinzip” which translates to “drawer principle.”

Definition 1.8.9 *For any function $f:A\to B$, we define the **inverse image** of a set $S\subset B$ to be the subset of A given by*

$$f^{-1}(S)=\{a\in A \mid f(a)\in S\}\,.$$

In the preceding definition *we do not assume* that the function f has an inverse function. For example, consider $f:\mathbb{Z}\to\mathbb{Z}$ defined by $f(x)=x^2$. Then

$$f^{-1}(\mathbb{Z})=\mathbb{Z},\quad f^{-1}(\mathbb{Z}^+)=\mathbb{Z}^+\cup(-\mathbb{Z}^+),\quad f^{-1}(\{0\})=\{0\}\,.$$

Exercise 1.8.10 *Show that the inverse image has the following properties for $f:S\to T$, and $A,B\subset T$.*

(a) $f^{-1}(A\cup B)=f^{-1}(A)\cup f^{-1}(B)$,
(b) $f^{-1}(A\cap B)=f^{-1}(A)\cap f^{-1}(B)$.

Now let us consider a few more counting problems. Recall that $n!=1\cdot 2\cdots(n-1)\cdot n$ and $0!=1$.

Definition 1.8.10 *The symbol $\binom{n}{k}$, read “n **choose** k” (the number of combinations of n things taken k at a time, not counting order) is defined to be*

$$\binom{n}{k}=\frac{n!}{k!(n-k)!}=\frac{n(n-1)\cdots(n-k+1)}{k(k-1)\cdots 1}, \tag{1.4}$$

for n, k non-negative integers.

The symbol $\binom{n}{k}$ represents the number of k-element subsets of a set with n elements. To see this, let us count the ways to create a k-element subset of an n-element set. First there are n ways to choose the first element of the subset. Then there are $n-1$ ways to choose the second element, since it must not equal the first element. Continue in this way until you reach the kth element. There will be $n-(k-1)$ ways to choose this element. The product of all these numbers is the numerator on the right in equation (1.4). This numerator is really the number of ordered k-tuples of elements from an n-element set. But two sets are the same if their elements are permuted or rearranged: for example, $\{1,2,3\}=\{2,3,1\}=\{3,1,2\}=\{1,3,2\}=\{3,2,1\}=\{2,1,3\}$. Thus there are $k!$ of the k-tuples corresponding to one k-element set. It follows that we must divide by $k!$.

The symbol $\binom{n}{k}$ is also a binomial coefficient. The **binomial theorem** says

$$\begin{aligned}(x+y)^n &= x^n + nx^{n-1}y + \frac{n(n-1)}{2}x^{n-2}y^2 + \cdots + \frac{n(n-1)}{2}x^2y^{n-2} + nxy^{n-1} + y^n \\ &= \sum_{k=0}^{n} \binom{n}{k} x^k y^{n-k}.\end{aligned}$$

Exercise 1.8.11 *Prove that*

$$(a)\ \binom{n}{k-1} + \binom{n}{k} = \binom{n+1}{k} \quad \text{and} \quad (b)\ \binom{n}{k} = \binom{n}{n-k}.$$

Hint. *For part (a), just use the definition and put everything on the left over a common denominator.*

Note that the equations in the preceding exercise are quite visible in **Pascal's triangle:**

$$\begin{array}{ccccccccccc} & & & & 1 & & 1 & & & & \\ & & & 1 & & 2 & & 1 & & & \\ & & 1 & & 3 & & 3 & & 1 & & \\ & 1 & & 4 & & 6 & & 4 & & 1 & \\ 1 & & 5 & & 10 & & 10 & & 5 & & 1 \end{array}$$

The nth row of Pascal's triangle gives the coefficients in the expansion of $(x+y)^n$. According to the equation in part (a) of the exercise, the kth coefficient in row n is the sum of the two coefficients nearest it in the row above (row $n-1$). Here we view the coefficients outside the triangle as 0.

We name this triangle for B. Pascal (1623–1662), but it was known earlier to Indian, Persian, Chinese and Italian mathematicians.

Exercise 1.8.12 *Use mathematical induction to prove the binomial theorem:*

$$(x+y)^n = \sum_{k=0}^{n} \binom{n}{k} x^k y^{n-k}.$$

There is also a short – and thus preferable – combinatorial proof of the binomial theorem. Look at the coefficient of $x^k y^{n-k}$ in the expansion of

$$\underbrace{(x+y)\cdots(x+y)}_{n \text{ terms}}.$$

This coefficient comes from choosing k of the xs and $(n-k)$ of the ys. This is the same as the number of k-element subsets of an n-element set, namely $\binom{n}{k}$.

Exercise 1.8.13 *Assume the standard facts from calculus for differentiable functions f and g that $(f+g)'=f'+g'$ and $(fg)'=f'g+fg'$. Prove the formula for the nth derivative of the product of two functions, assuming each function has an nth derivative. Of course binomial coefficients appear in this formula. Thus to prove the result by induction is somewhat similar to proving the binomial theorem by induction.*

Exercise 1.8.14 *Suppose that p is a prime and $1 \le k < p$. Show that p divides the binomial coefficient $\binom{p}{k}$.*

Exercise 1.8.15 *Prove the* ***arithmetic–geometric mean inequality****, which states that assuming $x_i \in \mathbb{R}^+$ for all $i=1,\ldots,n$:*

$$\frac{1}{n}(x_1+\cdots+x_n) \ge \sqrt[n]{x_1\cdots x_n}.$$

The left side is the arithmetic mean and the right side is the geometric mean of the numbers x_i.

Hint. *There are many proofs of this famous inequality. We are sort of hoping for a proof by mathematical induction here. That may be less revealing than the proof using Jensen's inequality (applied to the function $-\log(x)$) from the next exercise. You will find a huge number of proofs on the web.*

In order to do the next exercise, we must recall the definition of convex function from calculus and we should perhaps note that not all calculus books use the terminology of the exercise (for example, see Purcell [87] and Lang [63]).

Exercise 1.8.16 *Prove* ***Jensen's inequality*** *concerning convex functions $f: I \to \mathbb{R}$, where I is an interval on the real line. We call f* ***convex*** *if $f(\alpha_1 x_1+\alpha_2 x_2) \le \alpha_1 f(x_1)+\alpha_2 f(x_2)$ for all $x_1, x_2 \in I$ and for all $\alpha_1, \alpha_2 \in (0,1)$ such that $\alpha_1+\alpha_2=1$. This means that the part of the graph of $y=f(x)$ for x between any two points x_1 and x_2 in I lies below the line connecting the points $(x_i, f(x_i))$, $i=1,2$.* ***Jensen's inequality*** *says that if the weights $\alpha_i \in (0,1)$ are such that $\alpha_1+\cdots+\alpha_n=1$ and if $x_1,\ldots,x_n \in I$, then*

$$f(\alpha_1 x_1+\cdots+\alpha_n x_n) \le \alpha_1 f(x_1)+\cdots+\alpha_n f(x_n).$$

Setting all weights equal to $1/n$ and $f(x)=-\log x$ gives the preceding exercise. This inequality seems a little easier to prove by mathematical induction than the arithmetic-geometric mean inequality.

Exercise 1.8.17 ***(Principle of Inclusion–Exclusion).*** *Suppose that $S_1,\ldots,S_n$ are subsets of a finite set S. Show that the number of elements of $S_1 \cup S_2 \cup \cdots \cup S_n$ is*

$$\sum_{i=1}^{n}|S_i| - \sum_{1\le i<j\le n}|S_i \cap S_j| + \sum_{1\le i<j<k\le n}|S_i \cap S_j \cap S_k| + \cdots + (-1)^{n+1}|S_1 \cap \cdots \cap S_n|.$$

Hint. *Try the cases $n=1,2$, and 3 first. Then the easiest way is to proceed as follows. Consider a point p that is contained in exactly k of the sets S_i. Count the number of times that p is counted in each term of the formula. Then make use of the binomial theorem to see that these numbers add up to 1.*

Definition 1.8.11 *A **binary operation** on a set S is a function $\mu\colon S \times S \to S$. We will often write $\mu(a,b) = a \circ b$, for $a, b \in S$. Of course $\circ$ may be replaced by all sorts of symbols: for example, $+, \times, *$.*

We have already seen many examples of binary operations: for example, addition or multiplication on $\mathbb{Z}$ or $\mathbb{Z}_n$. The main property making for a binary operation is that it is well defined. In the next section we begin our discussions of groups – entities with one binary operation.

Exercise 1.8.18 *Which of the following are binary operations on the set $\mathbb{Z}$? Why?*

(a) $a \circ b = \frac{a}{b}$;
(b) $a \circ b = a^2 b^2$;
(c) $a \circ b = \sqrt{ab}$.

Exercise 1.8.19 *Consider the multiplication of matrices $A, B \in \mathbb{R}^{m\times m}$, defined by writing $B = (b_1 \cdots b_m)$, with columns $b_j \in \mathbb{R}^m$ and then $AB = (Ab_1 \cdots Ab_m)$, using the multiplication from formula (1.3). Show that this operation satisfies the associative law $A(BC) = (AB)C$, for all $A, B, C \in \mathbb{R}^{m\times m}$.*

2

Groups: A Beginning

2.1 What is a Group?

Cayley gave the first abstract definition of a group in 1854. We have already seen some examples of groups:

- the integers $\mathbb{Z}$ under addition;
- the integers mod n, $\mathbb{Z}_n$, under the operation $a + b \pmod{n}$.

The usefulness of the abstract concept is that it shortens the number of proofs we must produce. No longer must we prove the same thing over and over again for each special example. We pay for this by needing to imagine the abstract concept: a group with an arbitrary number of elements – maybe 20 billion, maybe 2, maybe infinitely many.

Definition 2.1.1 *A **group** G is a non-empty set of elements with one (binary) operation, which is a function from ordered pairs in $G \times G$ to G, taking $(a, b) \in G \times G$ to a unique element $a \cdot b \in G$, such that we have three laws:*

1. ***Associative law:***

$$a \cdot (b \cdot c) = (a \cdot b) \cdot c, \text{ for all } a, b, c \in G.$$

2. ***Identity** (call it e, or sometimes I or 0 or 1): there exists an element $e \in G$ such that*

$$a \cdot e = e \cdot a = a, \text{ for all } a \in G.$$

3. ***Inverses:** given $a \in G$, there is an element $a^{-1} \in G$ such that*

$$a \cdot a^{-1} = a^{-1} \cdot a = e.$$

*The group G is **Abelian or commutative** if for all $a, b \in G$, we have $a \cdot b = b \cdot a$. The fact that $a, b \in G \Longrightarrow a \cdot b \in G$ is called **closure**. We will usually write ab (or perhaps $a * b$ or $a + b$ or $a \circ b$) instead of $a \cdot b$.*

Example 1. $\mathbb{Z}$ under addition forms a commutative group. The identity is 0. The inverse of $x \in \mathbb{Z}$ is written $-x$. But $\mathbb{Z}$ under multiplication does not form a group, nor does $\mathbb{Z} - \{0\}$. ▲

Example 2. The dihedral group D_3 of symmetries of an equilateral triangle has six elements

$$D_3 = \{I, R, R^2, F, FR, FR^2\}.$$

The actions of these elements are pictured in Figure 2.1. We describe the elements in more detail in the following paragraphs – where we note that computing the product of elements of D_3 is made easier when one uses permutation notation. This group D_3 is the same as the group S_3 of all permutations of the three objects: Blue, Pink, Yellow. Write numbers instead. Write 1 for the blue position, 2 for pink, and 3 for yellow. A **permutation** ρ in S_3 means a function $\rho: \{1, 2, 3\} \to \{1, 2, 3\}$ which is 1–1 and onto.

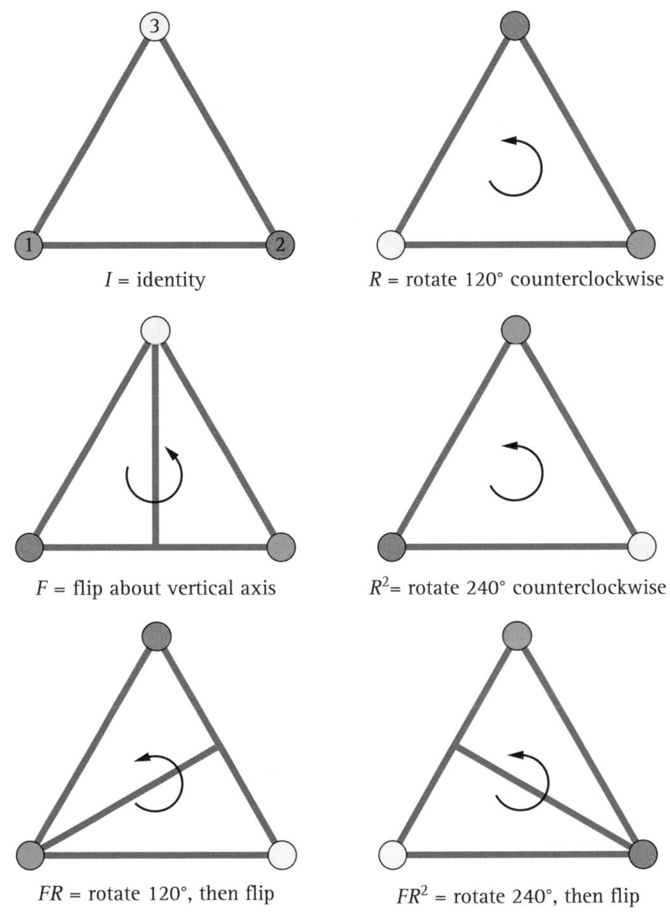

Figure 2.1 The symmetries of a regular triangle are pictured

To compute in D_3, it helps to use **permutation notation** because one may have trouble following what is going on when moving a triangle around in one's head. Moreover, one can easily come up with two different permutations for many elements of D_3. We will discuss the confusion in more detail in Section 3.7. Anyway, once you have translated the motions to permutations, the ambiguity is mostly gone – as long as you write permutations as functions in the usual way with the variable on the right, that is, $f(x)$, not xf. This forces you to compose functions in the usual way, which also forces you to multiply the group elements in the corresponding way.

To alleviate the confusion, one really needs two triangles – a fixed triangle of buckets and a moving triangle of balls. Then we can use the permutation notation. A permutation in S_3 is simply a 1–1, onto function f from $\{1,2,3\}$ to $\{1,2,3\}$:

$$f=\begin{pmatrix} 1 & 2 & 3 \\ f(1) & f(2) & f(3) \end{pmatrix}.$$

On the top we have the fixed buckets. On the bottom we have the moving balls. See Fraleigh [32, p. 36]. If you Google "permutation group action confusion," you will find many pages of attempts to clarify the situation. You might argue that instead of Figure 2.1, we should have drawn a figure with two triangles. We argue that the fixed triangle is easily imagined to be lying comatose underneath the figure.

Recall that we have identified the position of bucket 1 with the position of the blue ball in the top left triangle of Figure 2.1, the position of bucket 2 with the pink ball position, and bucket 3 with the yellow ball. Then we see that the flip F does not move the ball in bucket 3 and interchanges the balls in the other 2 buckets. Therefore F is given by the following permutation

$$F=\begin{pmatrix} 1 & 2 & 3 \\ 2 & 1 & 3 \end{pmatrix}.$$

The rotation R moves the blue ball at bucket 1 to bucket 2 and the pink ball at bucket 2 to bucket 3, finally the yellow ball at bucket 3 to bucket 1. Thus R is given by

$$R=\begin{pmatrix} 1 & 2 & 3 \\ 2 & 3 & 1 \end{pmatrix}.$$

So what is $R \cdot F$? Instead of looking at the figure, just compose the functions – first F, then R:

$$R \cdot F=\begin{pmatrix} 1 & 2 & 3 \\ 2 & 3 & 1 \end{pmatrix}\begin{pmatrix} 1 & 2 & 3 \\ 2 & 1 & 3 \end{pmatrix}.$$

This acts on a number in $\{1,2,3\}$ on the right so that first 1 goes to 2 and then 2 goes to 3 so that 1 goes to 3, for example. So we see that:

$$R \cdot F=F \cdot R^2=\begin{pmatrix} 1 & 2 & 3 \\ 3 & 2 & 1 \end{pmatrix}.$$

What is R^2?

$$R^2=\begin{pmatrix} 1 & 2 & 3 \\ 3 & 1 & 2 \end{pmatrix}.$$

What is R^3? That answer had better be the identity, I, which does not move any balls.

$$R^3=\begin{pmatrix} 1 & 2 & 3 \\ 1 & 2 & 3 \end{pmatrix}=I=\text{identity}.$$

We can create a multiplication table for the group once we have all the elements. The last is $R^2 \cdot F=F \cdot R$. Thus D_3 is not commutative. We see that

$$FR=R^2 \cdot F=\begin{pmatrix} 1 & 2 & 3 \\ 1 & 3 & 2 \end{pmatrix}.$$

With the above computations, we can create the multiplication table for D_3, which is incomplete until you do Exercise 2.1.3.

Multiplication table for D_3

$\cdot$	I	R	R^2	F	RF	R^2F
I	I	R	R^2	F	RF	R^2F
R	R	R^2	I	RF	R^2F	F
R^2	R^2	I	R	R^2F	F	RF
F	F	R^2F	RF			
RF	RF					
R^2F	R^2F					

▲

Exercise 2.1.1 *Prove that D_4, the group of symmetries of a square, is not the same as S_4, the group of permutations of four objects.*

Hint. *Note that S_4 has $4 \cdot 3 \cdot 2 = 24$ elements while D_4 has only eight elements. The group D_4 contains a 90° counterclockwise rotation R and two kinds of reflections: reflection F_1 across a diagonal and reflection F_2 across an axis connecting the midpoints of opposite sides.*

Definition 2.1.2 *For $n \in \mathbb{Z}^+$, the* **symmetric group** *S_n is the group of 1–1, onto functions from $\{1, 2, \ldots, n\}$ onto $\{1, 2, \ldots, n\}$, with the operation composition of functions. The elements of S_3 are called* **permutations** *of the integers $1, 2, \ldots, n$.*

Definition 2.1.3 *For $n \in \mathbb{Z}^+$, $n \geq 3$, the* **dihedral group** *D_n is the group of rigid motions of a regular n-gon. The group operation is composition.*

Neither S_n nor D_n is a commutative group when $n \geq 3$.

Exercise 2.1.2 *Show that D_n and S_n from Definitions 2.1.2 and 2.1.3 cannot have the same number of elements for $n \geq 4$.*

Hint. *Note that if you have two adjacent vertices of the regular n-gon, say 1 and 2, then any element $\sigma \in D_n$ maps them to adjacent vertices $\sigma(1)$ and $\sigma(2)$.*

Exercise 2.1.3 *Finish the multiplication table for D_3. What is the inverse of R? What is the inverse of RF? Explain why D_3 is a group.*

If you are given a multiplication table and asked to check that the set forms a group, it is easy to check properties 2 and 3 in the definition of a group, but hard to check the associative law (property 1). For a group of with six elements, there are 6^3 equations $a(bc) = (ab)c$ that must be checked. However, in our case this is automatic, since our group elements are functions mapping the set $\{1, 2, 3\}$ 1–1 onto itself and composition of functions is associative. See Exercise 1.8.2. Multiplication tables of groups always have every element of the group in each row and column, as we shall see in Corollary 2.3.1. See also Exercise 2.2.4.

When considering Figure 2.1, you could also imagine that R moves 1 to 3, moves 2 to 1, and moves 3 to 2. That would actually give you the inverse permutation to the one we wrote down above. This is the essence of the confusion. However, it does not really matter what you do – as long as you are consistent. Also, once you have the permutations, for R and F, you can stop looking at the triangle – always assuming that you are multiplying permutations correctly. See Section 3.7 for more information on permutations.

Symmetry groups arise in many areas such as chemistry, physics, statistics – even biology. You can replace the two-dimensional figures we just considered by three-dimensional or even higher-dimensional figures and look at their rigid motions.

How do you recognize the symmetry group of a figure in the plane? For the standard figures of classical plane geometry, it is the group of functions from the plane to itself that preserve the standard Euclidean distance and map the figure onto itself. There are three types of symmetries of figures in the plane:

(1) reflection across a line,
(2) rotation around the origin,
(3) translation by a fixed vector b, $x \to x + b$ for any vector $x \in \mathbb{R}^2$.

We have seen examples of figures with symmetries of types 1 and 2 for the dihedral group. To see examples of translational symmetries, look at Figure 2.2, which is supposed to stretch out to infinity to the left and to the right.

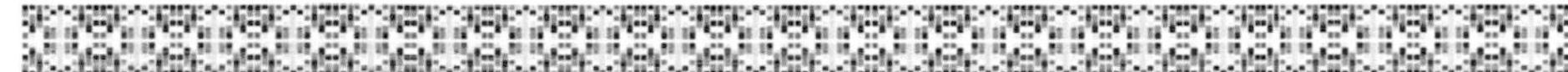

Figure 2.2 Part of a design with translational symmetry which should be imagined to stretch out to ∞ and $-\infty$

The rotations of a planar regular n-gon form a group with n elements called C_n, the **cyclic group of order** n. The group C_n is:

$$C_n = \{I, R, R^2, R^3, \ldots, R^{n-1}\} = \langle R \rangle, \tag{2.1}$$

where R is counterclockwise rotation by $2\pi/n$ radians. The group operation is again composition of functions. C_n has a particularly easy multiplication table. It is really the same as that for the integers modulo n under addition. Each row is obtained from the row above by shifting to the left and moving the first element to the end.

Multiplication table for C_n

$\cdot$	I	R	R^2	$\cdots$	R^{n-2}	R^{n-1}
I	I	R	R^2	$\cdots$	R^{n-2}	R^{n-1}
R	R	R^2	R^3	$\cdots$	R^{n-1}	I
R^2	R^2	R^3	R^4	$\cdots$	I	R
$\vdots$	$\vdots$	$\vdots$	$\vdots$	$\ddots$	$\vdots$	$\vdots$
R^{n-2}	R^{n-2}	R^{n-1}	I	$\cdots$	R^{n-4}	R^{n-3}
R^{n-1}	R^{n-1}	I	R	$\cdots$	R^{n-3}	R^{n-2}

The upper left quarter of the multiplication table for D_3 is the same as the multiplication table for C_3.

Example. The object in Figure 2.3 has rotational symmetry but not reflective symmetry. Thus its symmetry group is C_8 and not D_8. See Gallian [33] for many more examples of this sort. ▲

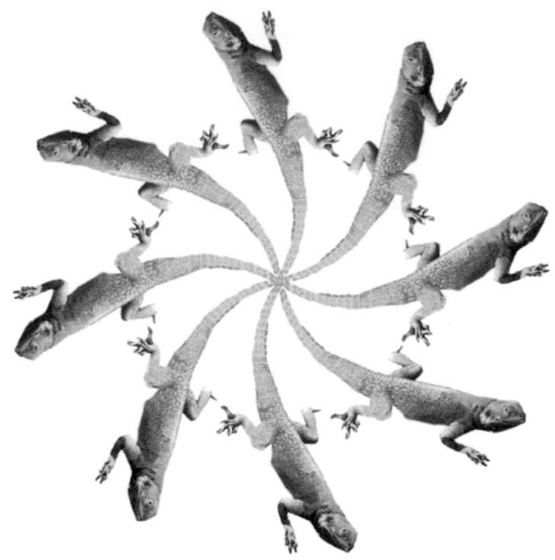

Figure 2.3 A figure with C_8 symmetry – not D_8 symmetry

Sometimes art is pleasing thanks to a mixture of symmetry and asymmetry. This is so with Figure 2.4 – a picture of a wall hanging from Raja Ampat in the Indonesian part of New Guinea. A good exercise would be to describe all the various symmetries and lack thereof in the figure.

Figure 2.4 Art from the Raja Ampat islands in the Indonesian part of New Guinea

Exercise 2.1.4 *Find the symmetry groups of Figures 0.1, 0.2, and 0.3, which are in the preface.*

Exercise 2.1.5 *Is the following table the multiplication table of a group $G = \{a, b, c, d\}$ of order 4?*

$*$	a	b	c	d
a	a	a	a	a
b	a	b	c	d
c	a	c	a	c
d	a	d	a	d

Exercise 2.1.6 *State which of the following are groups and why.*

(a) The integers $\mathbb{Z}$ under addition
(b) $\mathbb{Z}$ under multiplication
(c) The real numbers $\mathbb{R}$ under addition
(d) $\mathbb{R}$ under multiplication
(e) $\mathbb{R}-\{0\}$ under multiplication
(f) $\mathbb{Z}-\{0\}$ under multiplication

The following three exercises give a few basic facts about groups. We will deduce more such things in Section 2.3.

Exercise 2.1.7 *Show that the identity element of a group is unique. Then show that, for $a \in G$, the element a^{-1} is unique.*

Exercise 2.1.8 *Show that, in any group G, if $a, b \in G$, then $(ab)^{-1} = b^{-1}a^{-1}$.*

Exercise 2.1.9 *Show that in a group G, if $a \in G$, then $(a^{-1})^{-1} = a$.*

Exercise 2.1.10 *Show that in a group G, if $a, b \in G$ and $(ab)^2 = a^2b^2$, then $ab = ba$.*

Exercise 2.1.11 *Define the operation $\bigstar$ on $\mathbb{Z}$ by $a \bigstar b = a - b$. Does this operation make $\mathbb{Z}$ a group?*

Exercise 2.1.12 *State which of the following form groups:*

(a) the irrational numbers under multiplication;
(b) the rational numbers $\mathbb{Q}$ under addition;
(c) the rational numbers $\mathbb{Q}$ under multiplication;
(d) the nonzero rationals $\mathbb{Q} - 0$ under multiplication.

Exercise 2.1.13 *Show that S_n has $n!$ elements and D_n has $2n$ elements.*

Hint. *To count the elements of D_n, consider a regular n-gon. Label its vertices $1, 2, \ldots, n$. Define the rotation R to be a counterclockwise rotation through an angle of $\frac{2\pi}{n}$. Let F be a flip about the axis through vertex 1 and the center of the regular n-gon. Any element σ of D_n must fix the center of the n-gon and is determined by the image of two adjacent vertices such as 1 and 2. Vertex 1 can be taken by σ to any of the n vertices – say v. Vertex 2 must be taken to one of the two adjacent vertices to v. Then one can show that $\sigma = R^i$ or FR^i, for some i. Thus*

$$D_n = \langle R, F \rangle = \{I, R, R^2, \ldots, R^{n-1}, F, FR, FR^2, \ldots, FR^{n-1}\}.$$

In the preceding exercise, we use the notation $\langle R, F \rangle$ to denote the group of products of powers of R and F, which is called the **group generated by** R and F. Similarly we wrote $\langle R \rangle$ to denote the cyclic group generated by R in equation (2.1). We will say more about this idea when we discuss Cayley graphs of groups in the next section. It is also important to know the **relations** between the elements of $\langle R, F \rangle$ such as $R^n = I = F^2 = FRFR$. This is a different use of the word "relation" than that of Section 1.7.

There is much more that can be said about symmetry. Felix Klein (1849–1925) said "a geometry is a space with a group of transformations." Richard Brauer said in 1963: "Groups are the mathematical concept with which we describe symmetry." Hermann Weyl wrote an interesting book on the subject in which he gave an introduction to symmetry in physics and mathematics (see [125]). In a related shorter paper he said [126, pp. 592–610]: "By far the most fertile application of symmetry in the whole inorganic world has been made by quantum physics in studying the atomic molecular spectra." He also noted: "Relativity is nothing else than the problem of determining the group of automorphisms of space itself." Another reference for more general types of symmetry is the book by David Mumford, Caroline Series, and David Wright [80]. Non-Euclidean symmetries are used to produce some beautiful figures – including a few that go back to books of Klein and Fricke – figures produced by hand in the 1800s. The artist M. C. Escher created the beautiful circle limit designs using non-Euclidean symmetry. See Doris Schattschneider's book on Escher [100]. In the next section we will investigate ways to produce visions of symmetry which are infinitely more pedestrian than Escher's methods.

To close this section, we display three figures – one with Euclidean symmetry, one with spherical symmetry, and the other with non-Euclidean symmetry. Such symmetries are discussed in our book [118]. Figure 2.5 is a density plot of a two-term Fourier series in two variables and was produced with the Mathematica command below.

```
DensityPlot[Sin[4*Pi*x]*Sin[Pi*y]+Sqrt[2/3]*Sin[Pi*x]*Sin[4*Pi*y],
   {x,-4,4},{y,-4,4},ColorFunction->Hue,PlotPoints->500,Frame->False]
```

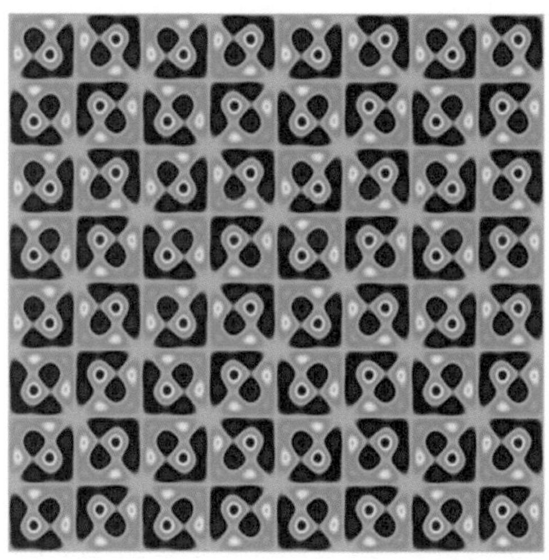

Figure 2.5 Wallpaper from a Fourier series in two variables

Figure 2.6 is a density plot of a spherical analog produced with the Mathematica command below.

```
ParametricPlot3D[{Cos[ϕ]*Sin[θ],Sin[ϕ]*Sin[θ],Cos[θ]},{ϕ,0,2π},{θ,0,π},
   PlotPoints->300,Mesh->None,
   ColorFunction->Function[{x,y,z,ϕ,θ},Hue[Re[SphericalHarmonicY[14,7,θ,ϕ]]]],
   ColorFunctionScaling->False,Axes->False,Boxed->False]
```

Figure 2.7 is a density plot of a non-Euclidean analog and was produced with the Mathematica command below.

```
DensityPlot[Abs[(y^6)*DedekindEta[x+I*y]^24],{x,-2.5,2.5},{y,0,1},
         ColorFunction->Hue,PlotPoints->500,AspectRatio->1/2,Frame->False]
```

Figure 2.6 Spherical wallpaper from spherical harmonics

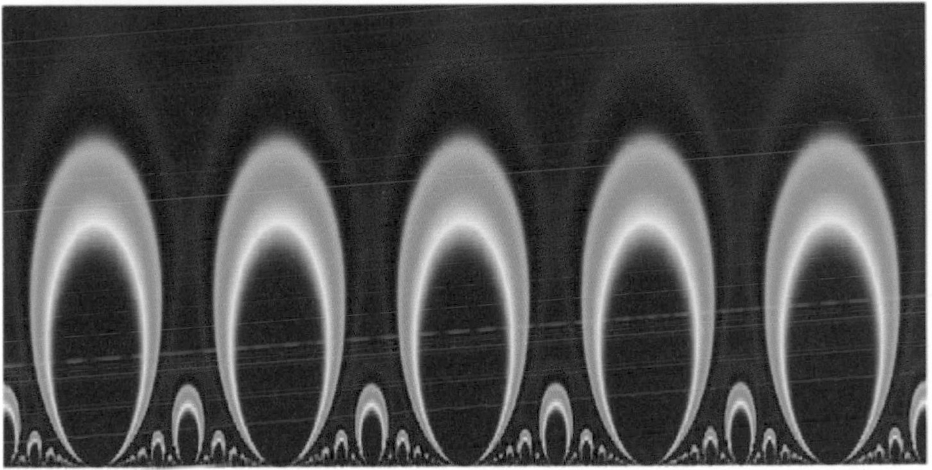

Figure 2.7 Hyperbolic wallpaper from a modular form known as Δ on the upper half plane – a function with an invariance property under fractional linear transformation $(az + b)/(cz + d)$, where $a, b, c, d \in \mathbb{Z}$ and $ad - bc = 1$

For Figure 2.5 the symmetry group is $\mathbb{Z}^2$ consisting of vectors (a, b) such that $a, b \in \mathbb{Z}$ with the operation vector addition. For Figure 2.6, the symmetry group is the group of rotations of the sphere with the operation composition. For Figure 2.7 the symmetry group is the **modular group** $SL(2, \mathbb{Z})$ of 2×2 matrices with integer entries and determinant 1 with the operation matrix multiplication. The functions being computed with the Mathematica commands above are those involved with Fourier analysis for the three groups. Such analysis can be applied to physical problems having symmetry described by these groups. In the plane, one might consider the vibration of a rectangular plate. That would involve Fourier series in two variables. In another example, earthquakes cause the earth to vibrate and such motions can be described using spherical harmonics. Sometimes the symmetry requires some torturous reasoning. Consider number theory problems such as Fermat's last theorem – which is surprisingly related to the modular group. A book that uses the same sort of methods that we used to create Figures 2.5–2.7 in order to create symmetric wallpaper is that of Farris [30].

2.2 Visualizing Groups

We have multiplication tables which help us to visualize groups. Using the program Group Explorer, for example, we get Figure 2.8 for the multiplication table of a cyclic group $G = \langle a \rangle$ of order 6 and Figure 2.9 for the multiplication table of the dihedral group D_3 (alias S_3).

e	a	a^2	a^3	a^4	a^5
a	a^2	a^3	a^4	a^5	e
a^2	a^3	a^4	a^5	e	a
a^3	a^4	a^5	e	a	a^2
a^4	a^5	e	a	a^2	a^3
a^5	e	a	a^2	a^3	a^4

Figure 2.8 Group Explorer version of the multiplication table for C_6, a cyclic group of order 6

e	r	r^2	f	rf	fr
r	r^2	e	rf	fr	f
r^2	e	r	fr	f	rf
f	fr	rf	e	r^2	r
rf	f	fr	r	e	r^2
fr	rf	f	r^2	r	e

Figure 2.9 Group Explorer version of the multiplication table for D_3 (alias S_3) with our upper case R and F replaced by lower case letters

But Cayley gave us another way to visualize these groups, using directed graphs (which are just sets of vertices with directed edges (arrows) connecting them). In fact, he even colored the edges. Every element of the cyclic group $\langle a \rangle$ is a power of a. We say that a is a **generator** of $\langle a \rangle$. Every element of D_3 is a finite product of powers of R and F, where R is a rotation by 120° counterclockwise and F is a flip about a vertical axis through the top vertex and the midpoint of the bottom edge. We say that $\{R, F\}$ generates D_3 and write $D_3 = \langle R, F \rangle$. In general, a subset S of a group G is said to be a set of **generators** of G iff all elements of G are finite products of elements of S. Associate a color to each element of S. The **Cayley graph** $X(G, S)$ has vertices corresponding (in 1–1 fashion) to the elements of the group G. Then for each element $s \in S$ and each vertex $g \in G$ draw an arrow with the color corresponding to s from vertex g to vertex gs. We get a directed graph with colored edges. Figures 2.10 and 2.11 show Cayley graphs for $G = C_6 = \langle a \rangle$ with $S = \{a\}$ and for D_3 with $S = \{R, F\}$. However, we have given the edges the same color in Figure 2.11. Because F has order 2, the edges corresponding to F are undirected.

We could also create Cayley graphs by drawing edges connecting vertex g to vertex sg, for $s \in S$. For commutative groups this would not matter but otherwise it could create a different graph – yet another right–left problem. Moreover there are many ways to draw a graph with more than two vertices.

Of course you can take different generating sets for the same group. In Figures 2.12 and 2.13 we choose $G = C_6 = \langle a \rangle$, with $S = \{a, a^5\}$ and $S = \{a, a^3, a^5\}$, respectively. Instead of putting arrows on the edges, since both directions are present on each edge of these two

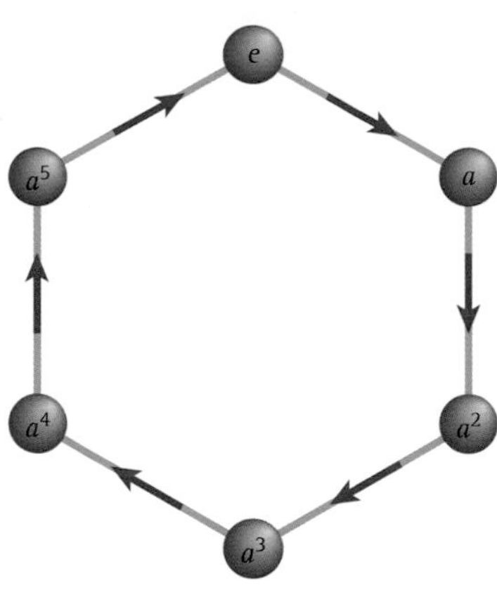

Figure 2.10 Cayley graph of cyclic group $G = \langle a \rangle$ of order 6 with generating set $S = \{a\}$

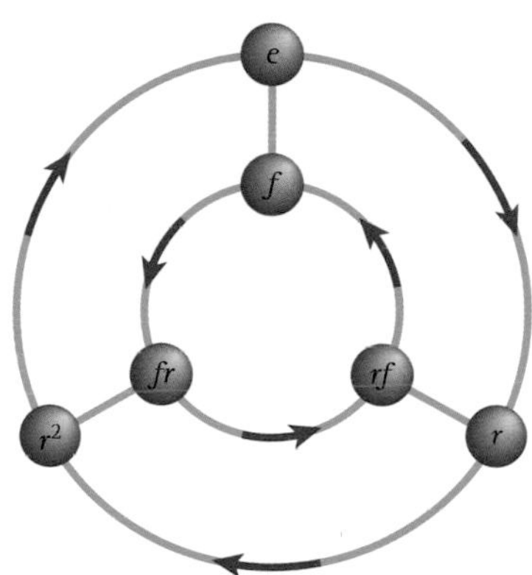

Figure 2.11 Cayley graph of D_3 with generating set $\{R, F\}$. Our upper case letters are replaced by lower case in the diagram. If there are arrows in both directions on an edge, we omit the arrows

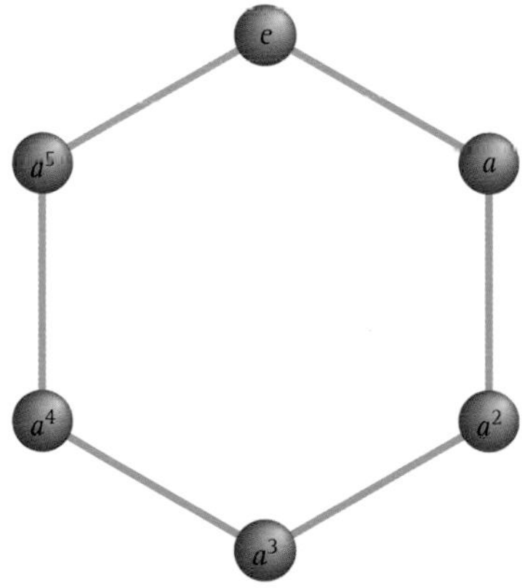

Figure 2.12 Undirected version of Cayley graph for $C_6 = \langle a \rangle$, generating set $S = \{a, a^{-1}\}$

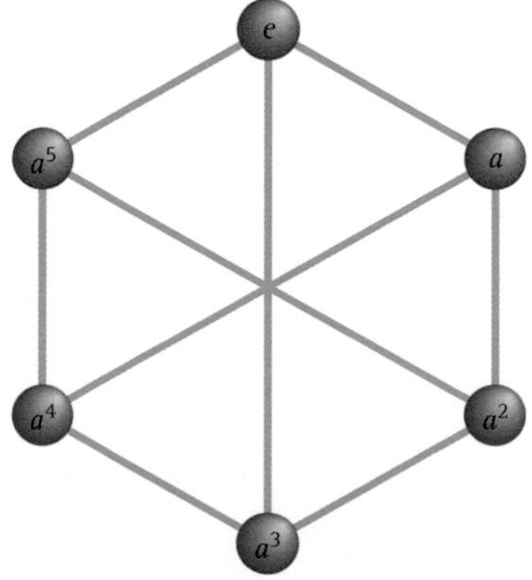

Figure 2.13 Cayley graph of $C_6 = \langle a \rangle$, generating set $S = \{a, a^3, a^5\}$

graphs, we just draw undirected graphs – that is, graphs with no arrows on the edges. Again, we have used the same color for the edges.

The fact that S is a generating set for the group G manifests itself in the Cayley graph $X(G,S)$ by insuring that any two vertices in the graph have a path connecting them consisting of a sequence of directed edges. In this situation, we say that the graph is **connected.**

Exercise 2.2.1 *Define the* ***Klein 4-group*** *as the group* $\mathbb{Z}_2^2$ *consisting of vectors* (a,b) *with* $a,b\in\mathbb{Z}_2$ *where the group operation in componentwise addition* mod 2. *Draw a Cayley graph for the Klein 4-group with generating set* $S=\{h,v\}$, $h=(1,0), v=(0,1)$. *The multiplication table is in Figure 2.14. Does this Cayley graph look like that for* $C_4=\langle a\rangle$ *with* $S=\{a,a^{-1}\}$? *What does that tell us about using Cayley graphs to understand groups?*

Hint. *Relations* $a^4=e$ *in* C_4 *and* $2h=2v=(0,0)$ *in* $\mathbb{Z}_2^2$ *correspond to closed paths in the graphs.*

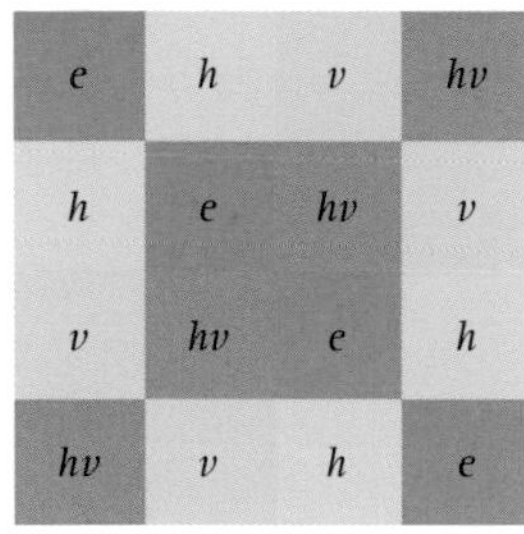

e	h	v	hv
h	e	hv	v
v	hv	e	h
hv	v	h	e

Figure 2.14 Group Explorer version of the multiplication table for the Klein 4-group

Exercise 2.2.2 *Find the symmetry group of a rectangle which is not a square.*

Exercise 2.2.3 *Find the symmetry group for each of the designs in Figure 2.15.*

Figure 2.15 Symmetrical designs

Exercise 2.2.4 *Prove that every multiplication table for a finite group is a* ***Latin square*** *– meaning that every row contains every element exactly once, the same for the columns.*

Exercise 2.2.5 *Draw the Cayley graph for D_4 with the generating set $\{R, F\}$, where R is a counterclockwise rotation of the square through 90° and F is a flip about an axis through vertex 1 and the center of the square.*

Exercise 2.2.6 *Consider the* ***affine group***

$$\mathrm{Aff}(3) = \left\{ \begin{pmatrix} a & b \\ 0 & 1 \end{pmatrix} \middle| \, a, b \in \mathbb{Z}_3, \ a \neq 0 \right\}.$$

Show that Aff(3) *is a group under matrix multiplication as defined in formula (1.3) of Section 1.8 and the statement that follows it. Draw a Cayley graph for this group with generating set* $S = \left\{ \begin{pmatrix} 2 & 0 \\ 0 & 1 \end{pmatrix}, \begin{pmatrix} 1 & 1 \\ 0 & 1 \end{pmatrix} \right\}$. *Compare with the Cayley graph $X(D_3, \{R, F\})$. Are these groups really the same in some sense (a sense to be known to us as isomorphic groups in Section 3.2)?*

We want to restrict ourselves mainly to **undirected graphs** here. This means that there are no arrows on the edges. A Cayley graph $X(G, S)$ is undirected if the edge set S is **symmetric** meaning that $s \in S$ implies $s^{-1} \in S$. Associated to an undirected graph with n vertices $V = \{v_1, \ldots, v_n\}$ is a symmetric $n \times n$ matrix of 0s and 1s called the **adjacency matrix** A. The i, j entry is 1 if there is an edge connecting vertex v_i with vertex v_j. The matrix A is **symmetric** because the i, j entry is the same as the j, i entry – which is true because the edge set is symmetric. The properties of the adjacency matrix are of great interest in graph theory, computer science, and chemistry. If the graph is a Cayley graph, the group has much to say about the adjacency matrix. We will say more about this in Section 4.2.

Example. Consider an adjacency matrix for the Cayley graph of $\mathbb{Z}_5$ under addition (mod 5) with generating set $\{1, -1 \pmod 5\}$; that is, $X(\mathbb{Z}_5, \{1, -1 \pmod 5\})$. We list the vertices in the usual order $\{0, 1, 2, 3, 4\}$ to obtain the matrix.

$$\begin{pmatrix} 0 & 1 & 0 & 0 & 1 \\ 1 & 0 & 1 & 0 & 0 \\ 0 & 1 & 0 & 1 & 0 \\ 0 & 0 & 1 & 0 & 1 \\ 1 & 0 & 0 & 1 & 0 \end{pmatrix}.$$

▲

The Cayley graph also tells us about **relations** in the group. This idea is not to be confused with the relations in Section 1.7. The relations we discuss here are equations involving "words" made up from the generators of our group and their inverses. For example, look at Figure 2.11. Follow the path going around the outside triangle clockwise starting at the top vertex and returning to it. This corresponds to the equation $RRR = R^3 = I$ which is a relation in D_3. Remember that, although we use capital letters for the group elements in the text, our figures have lower case letters. On the right side of Figure 2.11, the path which starts at e having four edges gives a second relation $RFRF = I$ in D_3. One way to define a group is to give its generators plus a set of relations. Cayley graphs do this in pictures. A **presentation** of a group G is a pair $\langle S : R \rangle$, where S is a set of generators of G and R is a set of words in these generators representing relations saying that the word = the identity and such that the relations in R generate all relations involving words in elements of G. We will say more about presentations of groups via generators and relations later. For example, a presentation of the dihedral group D_n is given by $D_n = \langle R, F : R^n = F^2 = RFRF = I\rangle$. Group Explorer lists groups this way. In one sense a group presentation tells you everything about the group. For a large group, however, not so much – just as a large Cayley graph is pretty hard to use to develop an understanding of the group.

The **free group** on a set S of generators means the set of all possible words in the generators and their inverses, with no relations. If S has one element, this group can be identified with $\mathbb{Z}$, but otherwise it is an infinite non-Abelian group.

Exercise 2.2.7 *Consider the set $SO(2)$ consisting of matrices*

$$m(\theta) = \begin{pmatrix} \cos\theta & -\sin\theta \\ \sin\theta & \cos\theta \end{pmatrix}, \quad \text{for } \theta \in \mathbb{R}.$$

Show that $SO(2)$ is a group under matrix multiplication as defined in formula (1.3) of Section 1.8 and the statement that follows it. This is called the ***special orthogonal group****. What is the effect of the group element $m(\theta)$ upon the vector $v = \begin{pmatrix} 1 \\ 0 \end{pmatrix}$, when we multiply $m(\theta)v$? Is the group $SO(2)$ commutative?*

Hint. *Identify vectors ${}^{\mathrm{T}}(x, y)$ in the plane with complex numbers $x + iy$, $i = \sqrt{-1}$. Here ${}^{\mathrm{T}}(x, y)$ denotes the transpose of (x, y): that is, the corresponding column vector. Then multiplication by $m(\theta)$ corresponds to multiplication of $x + iy$ with $e^{i\theta} = \cos\theta + i\sin\theta$.*

Exercise 2.2.8 *Suppose that G is a group with identity element e. Show that if $g^2 = g \cdot g = e$ for all $g \in G$, then G is Abelian.*

Exercise 2.2.9 *Write down an adjacency matrix for the Cayley graph of $\mathbb{Z}_6$ under addition $(\text{mod } 6)$ with generating set $\{1, -1 \ (\text{mod } 6)\}$; that is, $X(\mathbb{Z}_6, \{1, -1 \ (\text{mod } 6)\})$. List the vertices in the usual order $\{0, 1, 2, 3, 4, 5\}$.*

Exercise 2.2.10 *Write down an adjacency matrix for the Cayley graph $X(D_3, \{F, R, R^{-1}\})$, using the notation of the previous section and listing the vertices in the order $\{I, R, R^2, F, FR, FR^2\}$. Compare this result with that of the preceding exercise.*

2.3 More Examples of Groups and Some Basic Facts

For some more examples of groups, consider the symmetry groups of the five regular polyhedra (**the Platonic solids**): tetrahedron, cube, octahedron, dodecahedron, icosahedron; these are shown in Figure 2.16 – drawn by Mathematica. For more information on them, see Wikipedia under Platonic solids or groups or

http://www.dartmouth.edu/~matc/math5.geometry/unit6/unit6.html#Elements

or

http://www-history.mcs.st-and.ac.uk/~john/geometry/Lectures/L10.html.

The symmetry groups of the Platonic solids are quite interesting finite groups called the tetrahedral group (A_4) of proper symmetries of a tetrahedron, the octahedral group (S_4) of **proper symmetries** of an octahedron or cube, and the icosahedral group (A_5) of proper

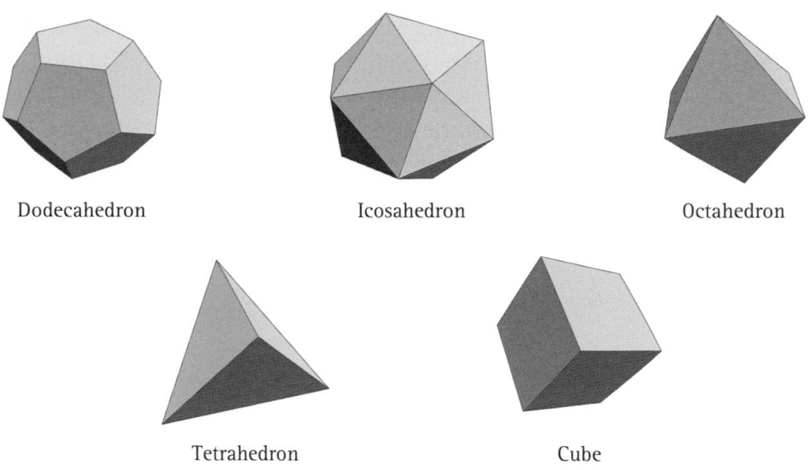

Figure 2.16 The Platonic solids

symmetries of an icosahedron or dodecahedron. Here "proper" means that the symmetry is obtained by an orientation-preserving rotation that one can actually perform in 3-space. They correspond to a 3×3 real matrix g in the **special orthogonal group** $SO(3, \mathbb{R})$, meaning that ${}^{\mathrm{T}}gg = I$ (the identity matrix, which is a diagonal matrix with 1s on the diagonal) and with $\det g = 1$. Here ${}^{\mathrm{T}}g$ denotes the transpose of the matrix g, which means the matrix whose rows are the corresponding columns of g. These groups can all be identified with permutation groups. We will define the alternating group A_n in Section 3.1. It is a subgroup of the symmetric group S_n which consists of half the elements of S_n (the "even"permutations).

We should note here that many authors throw in the improper rotations which cannot actually be performed in 3-space – just as the flips in D_n cannot be performed in the plane. The group corresponding to all proper and improper rotations of 3-space is the orthogonal group $O(3, \mathbb{R})$ of 3×3 real matrices g such that ${}^{\mathrm{T}}gg = I$. Allowing improper rotations produces groups of symmetries that are twice as large as those we spoke of above. Thus in Wikipedia, for example, if you read the entry on the tetrahedral group you will find that they are talking about S_4 rather than A_4. Moreover they will use the chemist's notation T_d instead of S_4.

The tetrahedral group is important to chemists because it is the symmetry group of various molecules such as methane CH_4. This was discovered in 1913 by W. H. Bragg and his son when they found that diamonds have tetrahedral symmetry. A diamond is a crystal of carbon atoms which are tetrahedrally bonded. The Braggs were pioneers in X-ray crystallography. You might ask why the molecule benzene (C_6H_6) – which is seemingly more complicated than methane – has a simpler symmetry group, the cyclic group of order 6. This was proved by Kathleen Lonsdale in 1929. She was a student of the senior Bragg and the first woman tenured at the University College London. Such things were proved using X-ray diffraction. This subject will be discussed a bit more in Section 4.2 but I fear that we will not really answer the question without doing a study of X-ray diffraction spectroscopy. This would require us to know more about representations of groups and quantum mechanics than is feasible for a short book.

Chemistry is not the only field that makes use of the groups of the Platonic solids. For example, the dodecahedral group is the symmetry group of many viruses – viruses such as those causing polio and herpes. There is hope that understanding the symmetries of viruses will help to find vaccines. Many papers on the subject have been written by Reidun Twarock at the Departments of Mathematics and Biology, York Centre for Complex Systems Analysis, University of York. Some beautiful pictures of viruses can be found in her papers and in Wikipedia articles on the subject. Or just Google images of icosahedral viruses.

As we said in Definition 2.1.2, the symmetric group S_n is the group of 1–1, onto functions

$$f:\{1,2,\ldots,n\}\to\{1,2,\ldots,n\}$$

with group operation given by composition of functions. We used the notation

$$\begin{pmatrix} 1 & 2 & \cdots & n \\ f(1) & f(2) & \cdots & f(n) \end{pmatrix}.$$

This allows us to see that S_n has $n!$ elements. For there are n ways to choose $f(1)$. Then, since f is 1–1, there are $n-1$ ways to choose $f(2)$, and $n-2$ ways to choose $f(3)$ as $f(3)$ cannot equal $f(1)$ or $f(2)$. Keep going in this way (inductively). At the end there is only one way to choose $f(n)$. Therefore we find that the number of elements in S_n is $n! = n(n-1)(n-2)\cdots 2\cdot 1$. Thus S_4 has $4\cdot 3\cdot 2\cdot 1 = 24$ elements The alternating group A_n consists of "even" permutations and always has half the number of elements of S_n, as we shall see in Section 3.1. Thus A_4 has 12 elements, and A_5 has 60 elements. We will study permutation groups in detail in Section 3.1 and in later sections. Let us return, for the rest of this section, to some more familiar groups.

Exercise 2.3.1 *Describe the elements of the group of proper symmetries of a regular tetrahedron as a subgroup of S_4. Show that – besides the identity – there are two sorts of rotations – one fixing only one vertex and the other fixing no vertex. We really need results from Sections 3.1 and 3.7 to give the whole story.*

The groups of symmetries of the Platonic solids are fairly complicated. Let us consider some simpler groups instead.

Example 1: The Group $\mathbb{Z}_n$ under Addition mod n. The elements of $\mathbb{Z}_n$ are really equivalence classes for the equivalence relation of congruence modulo n. Here we define the sum of equivalence classes by summing representative elements of the classes. That is, setting $[a] = \{b \mid b \equiv a \pmod{n}\}$, we have:

$$[a] + [b] = [a + b \pmod{n}].$$

Or one could just say $\mathbb{Z}_n = \{[0],[1],\ldots,[n-1]\}$ and the operation is addition mod n. One should really check that the operation is well defined. That means we should show that $a \equiv a' \pmod{m}$ and $b \equiv b' \pmod{m}$ implies $a + b \equiv a' + b' \pmod{m}$. This was Exercise 1.6.5. That the operation of addition mod n turns $\mathbb{Z}_n$ into a group requires us to find an identity which is $0 \pmod{n}$ and the inverse of $a \pmod{n}$ which is $-a \pmod{n}$. The associative law follows from that for $\mathbb{Z}$.

Compute the Cayley addition table for $\mathbb{Z}_6$. Each row of the table is obtained by shifting the row above it to the left (extending the row above by adding a copy of it to the right). This is quickly seen to be a colorless version of the Cayley table in Figure 2.8. ▲

Table for $\mathbb{Z}_6$ under addition

+ mod 6	0	1	2	3	4	5
0	0	1	2	3	4	5
1	1	2	3	4	5	0
2	2	3	4	5	0	1
3	3	4	5	0	1	2
4	4	5	0	1	2	3
5	5	0	1	2	3	4

The diagonals going from lower left to upper right are constant. This is a cyclic group generated by the element $[1]$, meaning that

$$\forall\,[x]\in\mathbb{Z}_6,\ \exists n\in\mathbb{Z}^+ \text{s.t. } [x]=n\,[1]=\underbrace{[1]+\cdots+[1]}_{n \text{ times}}.$$

These are the easiest groups to deal with. We study such groups in Section 2.5.

Example 2: The Unit Group $\mathbb{Z}_n^*$. What can we do about multiplication mod n? We make the analogous definition to that for addition. Thus, we multiply in $\mathbb{Z}_n$ by writing

$$[a]\,[b]=[ab].$$

This operation is also well defined: that is, $[a]=[a']$ and $[b]=[b']$ imply $[ab]=[a'b']$. This was proved in Exercise 1.6.6.

We need to take a subset of $\mathbb{Z}_n$ to get a group under multiplication. This subset is the unit group. ▲

Definition 2.3.1 *The* ***group of units mod*** *n is defined by*

$$\mathbb{Z}_n^*=\{[a]=a \pmod{n}\mid \gcd(a,n)=1\}, \tag{2.2}$$

with the operation $[a]\,[b]=[ab]$.

We call the elements of $\mathbb{Z}_n^*$ “units” because they are the invertible elements for multiplication in $\mathbb{Z}_n$. Thus $\mathbb{Z}_n^*$ is the set of $[a]=a \pmod{n}$ such that $\exists b$ with $ab\equiv 1 \pmod{n}$. To prove this, note that $a\in\mathbb{Z}_n^*$ implies that

$$1=\gcd(a,n)=ab+mn,\quad \text{for some integers } b, m. \tag{2.3}$$

Here we use Theorem 1.5.2 which says that $d=\gcd(a,n)$ is the smallest positive integer d of the form $d=ab+mn$, for integers b,m. But equation (2.3) implies that $ab\equiv 1 \pmod{n}$. Thus a is invertible for multiplication $\pmod{n}$.

Conversely if $ab\equiv 1 \pmod{n}$, then $ab-1=cn$, for some integer c, and then $1=ab-cn$. This implies that $1=\gcd(a,n)$ (again using Theorem 1.5.2).

Thus we can check that $\mathbb{Z}_n^*$ is a group under multiplication. To see that $\mathbb{Z}_n^*$ is closed under multiplication, note that $[a],[b]\in\mathbb{Z}_n^*$ means that there are integers c and d such that $ac\equiv 1 \pmod{n}$ and $bd\equiv 1 \pmod{n}$. This implies that $abcd\equiv 1 \pmod{n}$ and $[ab]\in\mathbb{Z}_n^*$.

The associative law follows from that for the integers; that is, $(ab)c = a(bc)$ for integers a, b, c implies $(ab)c \equiv a(bc) \pmod{n}$. The identity is $1 \pmod{n}$ using the fact that $\gcd(1, n) = 1$.

Inverses exist in $\mathbb{Z}_n^*$ as we proved above. For example: to find the inverse of $2 \pmod{5}$ you need to find b such that $2b \equiv 1 \pmod{5}$. The solution is $b \equiv 3 \pmod{5}$. You can find it by trial and error. Try all values of $b \pmod{5}$ and see which one works.

Another way to find b is to compute the $\gcd(2, 5) = 2b + 5c$, using the Euclidean algorithm. Note that $1/2$ does not make sense in $\mathbb{Z}_5^*$ except to write 2^{-1} is $3 \pmod{5}$.

Look at the case $n = 6$. $\mathbb{Z}_6^* = \{1, -1 \pmod{n}\}$. The multiplication table for $\mathbb{Z}_6^*$ below is essentially that of a cyclic group of order 2.

$\cdot \pmod{6}$	1	-1
1	1	-1
-1	-1	1

Exercise 2.3.2 *Show that there is no group G of integers* mod n *with the operation multiplication* mod n *such that $\mathbb{Z}_n^* \subsetneq G \subset \mathbb{Z}_n$ with the unit group $\mathbb{Z}_n^*$ defined by (2.2).*

How large is the unit group $\mathbb{Z}_n^*$? The answer depends on n. For example, if $n = 5$, we see that $\mathbb{Z}_5^*$ has all four nonzero elements of $\mathbb{Z}_5$. But $\mathbb{Z}_6^*$ has only the two elements, 1 and $4 \pmod{6}$. The number of elements of $\mathbb{Z}_n^*$ is defined to be the **Euler phi-function** $\phi(n)$. It is easy to check that

$$\phi(6) = 2, \quad \phi(8) = 4, \quad \text{and} \quad \phi(12) = 4.$$

Exercise 2.3.3

(a) Prove the last three results by listing all the elements of $\mathbb{Z}_n^*$, for $n = 6, 8,$ and 12.
(b) Show that $\phi(p^e) = p^e - p^{e-1}$, for any prime p, and exponent $e = 1, 2, 3, \ldots$.

Hint for part (b): *count the multiples of p between 1 and p^e.*

To compute $\phi(n)$ in general, you need to factor n into a product of pairwise distinct prime powers. Then for pairwise distinct primes p_i it can be proved (see Exercise 2.3.12) that

$$\begin{aligned}\phi(p_1^{e_1} p_2^{e_2} \cdots p_r^{e_r}) &= \phi\left(p_1^{e_1}\right)\phi\left(p_2^{e_2}\right)\cdots\phi\left(p_r^{e_r}\right) \\ &= \left(p_1^{e_1} - p_1^{e_1-1}\right)\left(p_2^{e_2} - p_2^{e_2-1}\right)\cdots\left(p_r^{e_r} - p_r^{e_r-1}\right). \end{aligned} \tag{2.4}$$

This fact really comes from the Chinese remainder theorem (see Section 6.2).

The multiplication table for $\mathbb{Z}_8^* = \{1, 3, -3, -1 \pmod{8}\}$ is a bit more interesting than that for $\mathbb{Z}_6^*$.

$\cdot \pmod{8}$	1	3	-1	-3
1	1	3	-1	-3
3	3	1	-3	-1
-1	-1	-3	1	3
-3	-3	-1	3	1

With a relabeling of entries this is the same multiplication table as that of the Klein 4-group. See Exercise 2.2.1 and Figure 2.14. Felix Klein named it the 4-group in 1884.

Note that every element a of this group has the property that $a \cdot a =$ identity. Also the group is commutative.

Next let us consider a group of 2×2 matrices under matrix multiplication.

Example 3: The General Linear Group $GL(2,\mathbb{Z}_p)$, where p is a prime. You could also replace $\mathbb{Z}_p$ with $\mathbb{R}$, the real numbers. The **general linear group** is defined by

$$GL(2,\mathbb{Z}_p) = \left\{ g = \begin{pmatrix} a & b \\ c & d \end{pmatrix} \middle| \, a,b,c,d \in \mathbb{Z}_p, \ \det g = ad - bc \in \mathbb{Z}_p^* \right\}, \tag{2.5}$$

where the operation is matrix multiplication

$$\begin{pmatrix} a & b \\ c & d \end{pmatrix} \begin{pmatrix} x & y \\ u & v \end{pmatrix} = \begin{pmatrix} ax+bu & ay+bv \\ cx+du & cy+dv \end{pmatrix}.$$

The definition of matrix multiplication is the same as for matrices of real numbers in formula (1.3) of Section 1.8, except that all computations are with integers modulo p. Note that (2.5) requires that the determinant be in the unit group $\mathbb{Z}_p^*$.

Why is $GL(2,\mathbb{Z}_p)$ closed under multiplication of matrices? You need to know that $\det(AB) = \det(A)\det(B)$. Since we are assuming $\det(A)$ and $\det(B)$ are in the unit group $\mathbb{Z}_p^*$, it follows that the product $\det(A)\det(B)$ is also in $\mathbb{Z}_p^*$. The reason is that $\mathbb{Z}_p^*$ is a group under multiplication – as we saw in the last example. We will have more to say about determinants in Section 7.3.

The identity in $GL(2,\mathbb{Z}_p)$ is the identity matrix $\begin{pmatrix} 1 & 0 \\ 0 & 1 \end{pmatrix}$.

The inverse of $\begin{pmatrix} a & b \\ c & d \end{pmatrix}$ is

$$\begin{pmatrix} a & b \\ c & d \end{pmatrix}^{-1} = (ad-bc)^{-1} \begin{pmatrix} d & -b \\ -c & a \end{pmatrix},$$

where all computations are done modulo p.

Here the reciprocal of the determinant is computed mod p, just as we compute inverses in $\mathbb{Z}_p^*$. We are not computing using fractions in the rational numbers. Thus $\frac{1}{a} = a^{-1} = b$ really means b is an integer such that $ab \equiv 1 \pmod{p}$. This is what happens in $\mathbb{Z}_p^*$. Note that the inverse is in $GL(2,\mathbb{Z}_p)$ since $\det(M^{-1}) = (\det(M))^{-1}$.

Matrix multiplication is associative because it comes from composition of functions. We saw this in Exercise 1.8.19. What is the function? For $F = \mathbb{Z}_p$, with prime p, let F^2 be the two-dimensional vector space over F:

$$F^2 = \left\{ \begin{pmatrix} x \\ y \end{pmatrix} \middle| \, x,y \in F \right\}.$$

Here we define the sum of vectors and product with scalar $\alpha \in F$ by

$$\begin{pmatrix} x \\ y \end{pmatrix} + \begin{pmatrix} u \\ v \end{pmatrix} = \begin{pmatrix} x+u \\ y+v \end{pmatrix}, \quad \alpha \begin{pmatrix} x \\ y \end{pmatrix} = \begin{pmatrix} \alpha x \\ \alpha y \end{pmatrix}. \tag{2.6}$$

where all operations are mod p.

Given a matrix $M=\begin{pmatrix} a & b \\ c & d \end{pmatrix}$, we have a linear function $T_M : F^2 \to F^2$ given by

$$T_M\begin{pmatrix} x \\ y \end{pmatrix} = \begin{pmatrix} a & b \\ c & d \end{pmatrix}\begin{pmatrix} x \\ y \end{pmatrix} = \begin{pmatrix} ax+by \\ cx+dy \end{pmatrix}. \tag{2.7}$$

The T_M function uniquely determines the matrix, since $M=\left(T_M\begin{pmatrix} 1 \\ 0 \end{pmatrix} \quad T_M\begin{pmatrix} 0 \\ 1 \end{pmatrix}\right)$. Composition of these functions corresponds to multiplying matrices. That is, $T_{MN}=T_M \circ T_N$. Thus matrix multiplication is associative. If you feel like filling up a page with computation, you could also check that multiplication of matrices is associative by multiplying out three matrices with parentheses arranged in two ways. See Section 7.1 for a discussion of vector spaces.

One can obtain an infinite general linear group $GL(2,\mathbb{R})$ by replacing $\mathbb{Z}_p$ with $\mathbb{R}$ and $\mathbb{Z}_p^* = \mathbb{Z}_p - \{[0]\}$ with $\mathbb{R}^* = \mathbb{R} - \{0\}$. The elements of $GL(2,\mathbb{R})$ are called non-singular matrices in linear algebra classes. A **non-singular square matrix** is a matrix with nonzero determinant or equivalently a matrix with a multiplicative inverse. See Exercise 7.3.13. ▲

Exercise 2.3.4 *For $R=\mathbb{Z}_n$, show that $R^2=\{(x,y) \mid x,y\in R\}$ is a group under addition using the componentwise definition of addition in (2.6).*

Take $p=5$ and look at the following example.

Special Case. In $GL(2,\mathbb{Z}_5)$, we get the following results. All computations are mod 5. Thus $1/(-2)$ is 2 (mod 5), since $2(-2)\equiv -4 \equiv 1$ (mod 5).

$$\begin{pmatrix} 1 & 2 \\ 3 & 4 \end{pmatrix}^{-1} = \frac{1}{-2}\begin{pmatrix} 4 & -2 \\ -3 & 1 \end{pmatrix} = 2\begin{pmatrix} 4 & 3 \\ 2 & 1 \end{pmatrix} = \begin{pmatrix} 3 & 1 \\ 4 & 2 \end{pmatrix}.$$

To check this, we multiply out the two matrices to see if we get the identity matrix.

$$\begin{pmatrix} 1 & 2 \\ 3 & 4 \end{pmatrix}\begin{pmatrix} 3 & 1 \\ 4 & 2 \end{pmatrix} = \begin{pmatrix} 3+8 & 1+4 \\ 9+16 & 3+8 \end{pmatrix} = \begin{pmatrix} 1 & 0 \\ 0 & 1 \end{pmatrix}.$$

Again the arithmetic was all mod 5. Similarly one checks that

$$\begin{pmatrix} 3 & 1 \\ 4 & 2 \end{pmatrix}\begin{pmatrix} 1 & 2 \\ 3 & 4 \end{pmatrix} = \begin{pmatrix} 1 & 0 \\ 0 & 1 \end{pmatrix}.$$

Question. How big is $GL(2,\mathbb{Z}_5)$?

Answer. 480 elements. To see this, note that the first row can be any vector (a,b) with $a,b\in\mathbb{Z}_5$, except $(0,0)$ since the determinant must be nonzero. Thus there are $5\cdot 5-1=24$ possible first rows of a matrix in $GL(2,\mathbb{Z}_5)$. Once the first row is given, the second row cannot be a scalar (in $\mathbb{Z}_5$) multiple of the first row (as the determinant must be nonzero). Thus there are $5^2-5=20$ possible second rows. This means that there are $24\cdot 20=480$ elements of this group. In 1832 Galois was the first to consider such groups.

Exercise 2.3.5 *Find the inverse of* $\begin{pmatrix} 1 & 2 \\ 3 & 4 \end{pmatrix}$ *in* $GL(2, \mathbb{Z}_7)$. *How big is* $GL(2, \mathbb{Z}_7)$?

Theorem 2.3.1 ***(Facts About Groups G).*** *The following hold for any group G:*

(1) The ***identity element*** *of G is* ***unique.***
(2) ***Inverses are unique.***
(3) ***Cancellation law.*** *For elements* a, x, y *in group G, we have* $ax = ay$ *implies* $x = y$.
(4) ***Solution of equations.*** *Given* a, b *in a group G, you can always solve* $ax = b$ *for* x *in G.*

Proof.
(1) Suppose e and f are both identities in G. Then $e = ef = f$, first using the fact that f is an identity and then using the fact that e is an identity.
(2) Suppose an element $a \in G$ has two inverses b and c. Then if e is the identity, $ab = e = ca$ which implies

$$b = eb = (ca)b = c(ab) = ce = c.$$

(3) Multiply both sides of the equality $ax = ay$ on the left by a^{-1}. You need to use the associative law.
(4) It is an exercise to prove this. ▲

Exercise 2.3.6 *Prove part (4) of the preceding theorem.*

Corollary 2.3.1 *Every row of the multiplication table of a finite group G is a permutation of the first row.*

Proof. The map from row 1 to row k is 1–1 by the cancellation law. This map is onto by part (4) of the preceding theorem or by the pigeonhole principle. ▲

The preceding corollary means that the multiplication table for a finite group G is a Latin square – as we saw already in Exercise 2.2.4.

Definition 2.3.2 *Assume G is a group. Define the* ***left multiplication function*** $L_a : G \to G$ *by* $L_a(x) = ax$, *for* $x \in G$.

Exercise 2.3.7 *Show that the function* L_a *is 1–1 and onto.*

If you would like to think about a really large finite group of symmetries, you might want to look at the symmetries of the icosahedron, or symmetries of various chemical structures, or the group of motions of Rubik's cube, or S_n for large n, or $GL(2, \mathbb{Z}_p)$ for large prime p. Infinite symmetry groups are as interesting – if not more so.

Exercise 2.3.8 *Which of the following groups are commutative? Give a reason for your answer. Assume that* $n \in \mathbb{Z}$ *and* $n \geq 2$.

(a) $\mathbb{Z}_n^{2\times 2}$, *the* 2×2 *matrices with entries in* $\mathbb{Z}_n$ *under addition;*
(b) the group of proper rotations of a cube;

(c) $\mathbb{Z}_n^2$, *the 2-vectors with entries in* $\mathbb{Z}_n$ *under addition;*

(d) $\mathbb{Z}_n^{2\times 2}$, *the* 2×2 *matrices* g *with entries in* $\mathbb{Z}_n$ *and such that* $\det(g)\in\mathbb{Z}_n^*$, *under matrix multiplication. Here* $\mathbb{Z}_n^*$ *denotes the unit group in Definition 2.3.1.*

Exercise 2.3.9 ***(Wilson's Theorem).*** *Show that if* p *is a prime, then* $(p-1)!\equiv -1(\text{mod } p)$.

Hint. *Pair the elements of the unit group* $\mathbb{Z}_p^*$ *with their inverses* a^{-1} *for multiplication. There are only two solutions of* $a\equiv a^{-1}$ $(\text{mod } p)$, *namely* 1 *and* -1 $(\text{mod } p)$. *Thus we can see that the product* $2\cdot 3\cdots(p-2)$ *is congruent to* 1 $(\text{mod } p)$.

Recall that Euler's phi function $\phi(n)$ is the number of elements of $\mathbb{Z}_n^*$. To compute it, you need to factor n into a product of distinct prime powers and use equation (2.4). The following exercises complete the proof of equation (2.4).

Exercise 2.3.10 *Show that if* p *and* q *are distinct primes then Euler's function satisfies* $\phi(pq)=(p-1)(q-1)$.

Exercise 2.3.11 *Show that if* m *and* n *satisfy* $\gcd(m,n)=1$, *then Euler's function satisfies* $\phi(mn)=\phi(m)\phi(n)$.

Hint. *First prove that the map sending* x $(\text{mod } mn)$ *such that* $\gcd(x,mn)=1$ *to the ordered pair* $(x\ (\text{mod } m), x\ (\text{mod } n))$ *maps* $\mathbb{Z}_{mn}^*$ *1–1, onto* $\mathbb{Z}_m^*\times\mathbb{Z}_n^*=\{(a,b)\ |a\in\mathbb{Z}_m^*, b\in\mathbb{Z}_n^*\}$. *This really comes from the Chinese remainder theorem (in Section 6.2).*

Exercise 2.3.12 *Use the preceding exercise (and Exercise 2.3.3 to prove equation (2.4) for* $\phi\left(p_1^{e_1}p_2^{e_2}\cdots p_r^{e_r}\right)$, *if the* p_i *are pairwise distinct primes.*

2.4 Subgroups

So let us talk about subgroups – that is, subsets of groups that are groups as well (using the same operations as in the big group). We have already seen examples, such as $\{I,F\}\subset D_3$. Throughout the definitions in this section we will be envisioning a group with the group operation being multiplication. Of course, for many examples, the operation is addition.

Before considering subgroups generated by a single element of a group, we should say a bit about **integer powers of an element** a in a group G. The rules are essentially the same as the rules for integer powers of real numbers. Define $a^0=e$, the identity in G, and, for a positive integer n, define

$$a^n=\underbrace{a\cdot a\cdot a\cdots a}_{n\text{ times}}.$$

Then define

$$a^{-n}=\left(a^{-1}\right)^n,\quad \text{for all } n=1,2,3,\ldots$$

One could also make the definition of a^n inductively.

Proposition 2.4.1 ***(Facts About Integer Exponents are the Usual).*** *Assuming a is an element of a group G and $r, s \in \mathbb{Z}$, we have the following two facts.*

(1) $a^r a^s = a^{r+s}$,
(2) $(a^r)^s = a^{rs}$.

Proof.

(1) If both exponents r and s are non-negative integers, this fact follows easily from the definition. You just have to think how many factors of a are on each side. If both r and s are negative integers, this fact is easily deduced from the definition in a similar way by counting how many factors of a^{-1} are on each side.

The only case that requires some effort is the case that one exponent is negative and one positive. If $r = -m$ with $m > 0$ and $s > 0$, then

$$a^r a^s = \underbrace{a^{-1} \cdots a^{-1}}_{m \text{ factors}} \underbrace{a \cdots a}_{s \text{ factors}} = a^{s-m} = a^{r+s},$$

since, by the associative law, we can cancel $a^{-1}a = e$. If $s \geq m$, what remains is $a^{s-m} = a^{r+s}$. If $s < m$, what is left is

$$a^r a^s = (a^{-1})^{m-s} = a^{-(m-s)} = a^{r+s}.$$

(2) The second fact about exponents can be shown as follows. If s is positive, then using the first fact (and mathematical induction), we have

$$(a^r)^s = \underbrace{a^r \cdots a^r}_{s \text{ factors}} = a^{\overbrace{r + \cdots + r}^{s \text{ terms}}} = a^{s \cdot r}.$$

If $s = 0$, the result is clear as both sides are the identity. If $s < 0$, we can argue in a similar way using the definition and the fact that

$$(a^r)^{-1} = a^{-r}, \text{ for all } r.$$

▲

Exercise 2.4.1 *Finish the proof of the second fact in Proposition 2.4.1 when* $s < 0$.

Example: The Unit Group $\mathbb{Z}_7^*$. The group $\mathbb{Z}_7^*$ consists of integers $a \pmod 7$ such that $\gcd(a, 7) = 1$. The operation is multiplication modulo 7. Consider the powers of 2. You get $2^0 \equiv 1 \pmod 7, 2^1 \equiv 2 \pmod 7, 2^2 \equiv 4 \pmod 7, 2^3 \equiv 8 \equiv 1 \pmod 7$. The powers will repeat after this. Hence $H = \{1, 2, 4 \pmod 7\}$ is a cyclic subgroup of $\mathbb{Z}_7^*$ consisting of all powers of $2 \pmod 7$. ▲

Exercise 2.4.2 *Find all the powers of 2 in $\mathbb{Z}_{11}^*$ and $\mathbb{Z}_{13}^*$.*

We have already been using the words "order of a group" in accordance with the following definition. But let us make it official.

Definition 2.4.1 *The* ***order of a finite group*** *G is the number of elements in G, denoted $|G|$ or $\#(G)$.*

Examples: Orders of Various Groups

1. The group of integers mod n under addition: $|\mathbb{Z}_n| = n$.
2. $\mathbb{Z}_n^* =$ the unit group of elements a of $\mathbb{Z}_n$ such that $\gcd(a, n) = 1$ under multiplication mod n:

   $$|\mathbb{Z}_n^*| = \phi(n) = \text{the Euler phi-function.}$$

3. $D_n =$ the dihedral group of motions of a regular n-gon: $|D_n| = 2n$ by Exercise 2.1.13.
4. The symmetric group of permutations of n objects: $|S_n| = n! = n(n-1)(n-2)\cdots 2\cdot 1$ by Exercise 2.1.13. ▲

There is yet another use of the word "order" – not to mention that of Section 1.3.

Definition 2.4.2 *The **order of an element** a in a group G is the smallest positive integer n such that $a^n = e =$ identity in G. If no such n exists, then we say that a has **infinite order**. We will usually write $|a| =$ the order of a.*

Example 1: The Unit Group $\mathbb{Z}_7^*$. The unit group $\mathbb{Z}_7^*$ consists of integers a (mod 7) such that $\gcd(a, 7) = 1$. The operation is multiplication modulo 7.

The order of 2 in $\mathbb{Z}_7^*$ was shown to be 3 in the first example of this section.

What about the order of 3 in $\mathbb{Z}_7^*$? The following computation shows that $|3| = 6$ in $\mathbb{Z}_7^*$.

$$3^2 \equiv 9 \equiv 2 \pmod 7,\ \ 3^3 \equiv 6 \equiv -1 \pmod 7,\ \ 3^4 \equiv -3 \equiv 4 \pmod 7,$$
$$3^5 \equiv -9 \equiv -2 \equiv 5 \pmod 7,\ \ 3^6 \equiv -6 \equiv 1 \pmod 7.$$

Moreover, this shows that $\mathbb{Z}_7^*$ is a cyclic group generated by 3 (mod 7):

$$\mathbb{Z}_7^* = \{3^n \pmod 7 \mid n \in \mathbb{Z}\} = \{1, 2, 3, 4, 5, 6 \pmod 7\} = \langle 3 \pmod 7\rangle.$$ ▲

Example 2: The Dihedral Group D_3. In the dihedral group D_3 of motions of an equilateral triangle, the order of R (which represents counterclockwise rotation by 120°) is 3 and the order of F (which is the reflection across an axis stretching from one vertex to the midpoint of the opposite side) is 2. ▲

Example 3: The Additive Group $\mathbb{Z}_6$. We investigate the orders of the elements of the group $\mathbb{Z}_6$ under addition mod 6. Here the powers

$$a^n = \underbrace{a \cdots a}_{n \text{ times}},$$

for integers $n > 0$ become multiples

$$n \cdot a = \underbrace{a \cdots a}_{n \text{ times}} \pmod 6, \quad \text{for } n > 0.$$

We list the elements and their orders in a table. The orders are all the divisors of the order of $\mathbb{Z}_6$. ▲

a = element of $\mathbb{Z}_6$ under + mod 6	$\lvert a \rvert$
0 (mod 6)	1
1 (mod 6)	6
2 (mod 6)	3
3 (mod 6)	2
4 (mod 6)	3
5 (mod 6)	6

Example 4: The Additive Group $\mathbb{Z}$. If x is a nonzero element of the group $\mathbb{Z}$ under addition, then the order of x is ∞. To see this, note that for any positive integer n, we know that $n \cdot x$ is not zero. There are no zero divisors in $\mathbb{Z}$. ▲

Exercise 2.4.3

(a) Find the orders of all elements of $\mathbb{Z}_8$ *under addition* mod 8.

(b) Do the same for the unit group $\mathbb{Z}_8^*$ *under multiplication* mod 8.

Exercise 2.4.4 *Find the orders of all elements of the unit group* $\mathbb{Z}_{11}^*$. *Then do the same for the unit group* $\mathbb{Z}_{13}^*$.

So finally we define subgroup.

Definition 2.4.3 *A* ***subgroup*** *H of a group G is a subset $H \subset G$ which is itself a group under the same operation as for G. If e denotes the identity of G, we will say that $H = \{e\}$ is the* ***trivial subgroup*** *of G. If the subgroup H of group G is such that $H \neq \{e\}$ and $H \neq G$, we will say that H is a* ***proper subgroup*** *of G.*

Example 1: Subgroups of the Additive Group $\mathbb{Z}_6$. Look at the subset $H = \{2n \pmod 6) | n \in \mathbb{Z}\}$ of $G = \mathbb{Z}_6$ under + mod 6 under the same operation as in G. This is a subgroup. To see this, note that if $2n \pmod 6$ and $2m \pmod 6$ are in H then $2n \pm 2m = 2(n \pm m) \pmod 6 \in H$, for all $n, m \in \mathbb{Z}$. It follows that $0 \in \mathbb{Z}$ and that H is closed under both addition and subtraction. Thus H is a subgroup as the associative law certainly holds. ▲

Example 2: Subgroups of the Dihedral Group D_3. As in Section 2.1, $D_3 = \{I, R, R^2, F, FR, FR^2\}$ is the dihedral group of motions of an equilateral triangle. Subgroups are

$$H_1 = \{I, F\}, \; H_2 = \{I, FR\}, \; H_3 = \{I, FR^2\}, \; H_4 = \{I, R, R^2\}. \tag{2.8}$$

There are also two other so-called **improper subgroups**

$$H_5 = \{I\} \quad \text{and} \quad H_6 = D_3 \text{ itself.}$$

Note that the orders of the subgroups are $1, 2, 3$, and 6. These are all the positive divisors of 6.

The collection of subgroups of D_3 forms a poset under the relation of $\subset$. We can draw the poset diagram of the subgroups of D_3 as in Section 1.7. See Figure 2.17. Here an ascending line means the lower group is a subgroup of the upper group and there is no subgroup between the lower and upper groups. See Dummit and Foote [28] for many more examples of poset diagrams for subgroups of various groups. ▲

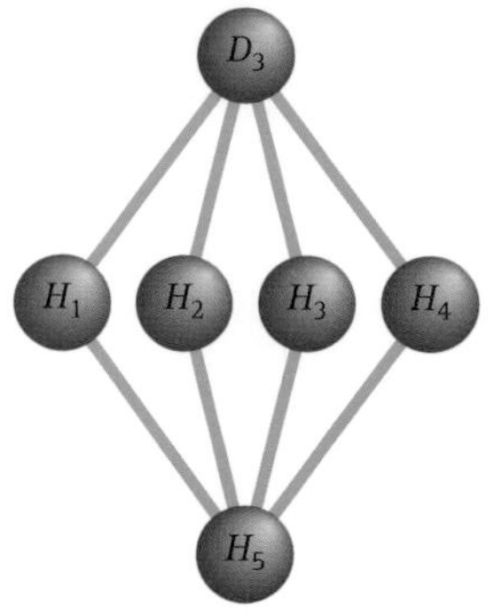

Figure 2.17 Poset diagram for subgroups of D_3 as defined in (2.8)

Exercise 2.4.5

(a) *Are there any proper subgroups of the dihedral group D_3 other than those listed in (2.8)? Why?*

(b) *Draw the poset diagram for subgroups of the Klein 4-group with the multiplication table in Figure 2.14.*

Exercise 2.4.6 *Draw the poset diagram for the subgroups of $\mathbb{Z}_{12}$ under addition* mod 12.

Exercise 2.4.7 *Draw the poset diagram for the subgroups of $\mathbb{Z}_{12}^*$ under multiplication* mod 12.

Exercise 2.4.8 *Draw the poset diagram for the dihedral group D_4.*

Example 3. The group $\mathbb{Z}_n$ of integers mod n under $+$ (mod n) is not a subgroup of the group $\mathbb{Z}$ of integers under $+$. Why? The two operations are different. For example $3+3 \equiv 0 \pmod 6$ in $\mathbb{Z}_6$ but $3+3=6$, which is not 0 in $\mathbb{Z}$. ▲

We have two tests for a non-empty subset H of G to be a subgroup of G under the operation of G.

Proposition 2.4.2 ***(Two-Step Subgroup Test).*** *Suppose that G is a group and $H \subset G$ such that $H \neq \emptyset$. Assume that $\forall\ a, b \in H$. It follows that $a \cdot b \in H$ and $a^{-1} \in H$. Then H is a subgroup of G.*

Proof. It is clear, by hypothesis, that H is closed under multiplication. Moreover, the associative law follows from that in G. Also, we have the existence of inverses in H, by hypothesis. But why does H have an identity? We know there is an $a \in H$ as $H \neq \emptyset$. We also know $a \in H$ implies $a^{-1} \in H$. Thus, by our hypothesis, with $b = a^{-1}$, the identity $e = aa^{-1} \in H$. ▲

There is also a shorter test.

Proposition 2.4.3 ***(One-Step Subgroup Test).*** *Suppose that G is a group and $H \subset G$ such that $H \neq \emptyset$. Assume that $\forall\ a, b \in H$ it follows that $a \cdot b^{-1} \in H$. Then H is a subgroup of G.*

Proof. Since H is non-empty, there is an element $a \in H$. Then $e = a \cdot a^{-1} \in H$, by hypothesis. Take $a = e$ to see that $b \in H$ implies $e \cdot b^{-1} = b^{-1} \in H$. Then $a, b \in H$ implies that $a \cdot (b^{-1})^{-1} = a \cdot b \in H$. So we are done by the two-step subgroup test. ▲

There is a third test for a non-empty finite subset H of G to be a subgroup of G under the operation of G.

Proposition 2.4.4 ***(Finite Subgroup Test).*** *Suppose that G is a group and $H \subset G$, H finite, such that $H \neq \emptyset$. Assume that $a, b \in H \Longrightarrow a \cdot b \in H$. Then H is a subgroup of G.*

Proof. We use the two-step test from Proposition 2.4.2. Thus we need only show that $a \in H$ implies $a^{-1} \in H$. If $a = e =$ identity, then $a^{-1} = e$ and we are done. Otherwise look at the subset $\{a, a^2, a^3, \ldots a^n, \ldots\} \subset H$. Since H is finite, we must have $a^i = a^j$ for some $i > j \geq 1$. Then $a^{i-j} = e$ and $i - j \geq 1$. Therefore $e \in H$ and $a \cdot a^{i-j-1} = e$ implies $a^{i-j-1} = a^{-1} \in H$. ▲

Definition 2.4.4 *If a is an element of a group G, the* **cyclic subgroup generated by** *a is*

$$\langle a \rangle = \{a^n \mid n \in \mathbb{Z}\}.$$

It is not hard to see that $\langle a \rangle$ is a subgroup using the one-step test from Proposition 2.4.3 and properties of exponents. Just look at the following calculation:

$$a^n (a^m)^{-1} = a^{n-m} \in \langle a \rangle.$$

Example 1: Cyclic Subgroups of the Additive Group $\mathbb{Z}_6$. In $\mathbb{Z}_6$ under addition mod 6, we find that all the subgroups are cyclic (which is a general fact about cyclic groups to be proved in Section 2.5). Remember that a^n becomes na for $a \in \mathbb{Z}_6$ and an integer n.

$$\langle 0 \rangle = \{n \cdot 0 \pmod 6 | n \in \mathbb{Z}\} = \{0 \pmod 6\}$$
$$\langle 1 \rangle = \{n \cdot 1 \pmod 6 | n \in \mathbb{Z}\} = \{0, 1, 2, 3, 4, 5 \pmod 6\} = \langle 5 \equiv -1 \pmod 6 \rangle = \mathbb{Z}_6$$
$$\langle 2 \rangle = \{n \cdot 2 \pmod 6 | n \in \mathbb{Z}\} = \{0, 2, 4 \pmod 6\} = \langle 4 \equiv -2 \pmod 6 \rangle$$
$$\langle 3 \rangle = \{n \cdot 3 \pmod 6 | n \in \mathbb{Z}\} = \{0, 3 \pmod 6\}$$

Note that the order of the element a in $\mathbb{Z}_6$, which was computed earlier in this section, is the same as the number of elements in $\langle a \rangle$. We prove this, in general, in Section 2.5. ▲

Exercise 2.4.9 *Are there any subgroups of the group $\mathbb{Z}_6$ other than those listed above? Draw the poset diagram for $\mathbb{Z}_6$.*

Example 2: Cyclic Subgroups of the Dihedral Group D_3. Recall that, as in Section 2.1, the dihedral group of symmetries of an equilateral triangle is

$$D_3 = \{I, R, R^2, F, FR, FR^2\}.$$

The cyclic subgroups of D_3 are the ones we found in Figure 2.17 except for D_3 itself which is not cyclic:

$$\langle F \rangle = \{I, F\}, \ \langle FR \rangle = \{I, FR\}, \ \langle FR^2 \rangle = \{I, FR^2\},$$
$$\langle I \rangle = \{I\}, \ \langle R \rangle = \{I, R, R^2\} = \langle R^2 \rangle.$$

▲

Next we look at the subgroup of elements that commute with everything in a group.

Definition 2.4.5 *The **center** of a group G is* $Z(G) = \{a \in G \mid ax = xa, \forall x \in G\}$.

Proposition 2.4.5 *The center of G is a commutative subgroup of G.*

Exercise 2.4.10 *Prove the preceding proposition using the one-step subgroup test from Proposition 2.4.3.*

Example 1: The Center of the Dihedral Group D_3. We claim that the center of the dihedral group D_3 is $Z(D_3) = \{I\}$. To see this, look at the group table in Figure 2.9 of Section 2.2. No row equals the corresponding column except the first row and column corresponding to the identity I. ▲

Example 2: The Center of the Dihedral Group D_4. We can show that the center of the dihedral group D_4 is $Z(D_4) = \{I, R^2\}$. Here R means rotate counterclockwise 90°. To generate D_4, we need R and F, a flip about an axis through a vertex and the point at the center of the square. To see that the center of D_4 is what we claim, we need to multiply a few elements of D_4. Using permutation notation, labelling the vertices of the square $1, 2, 3, 4$, we see that

$$R = \begin{pmatrix} 1 & 2 & 3 & 4 \\ 2 & 3 & 4 & 1 \end{pmatrix} \quad \text{and} \quad F = \begin{pmatrix} 1 & 2 & 3 & 4 \\ 1 & 4 & 3 & 2 \end{pmatrix}.$$

One finds then that $FR = R^{-1}F$. This is one of our defining relations for D_4. The others are $R^4 = I$ and $F^2 = I$. The center of D_4 is contained in the center of the subgroup $\langle R \rangle = \{I, R, R^2, R^3\}$. This means the center of D_4 is a subgroup of $\langle R \rangle$. The elements of D_4 are $\{I, R, R^2, R^3, F, FR, FR^2, FR^3\}$. Now $FR^i = R^{-i}F$. In order for $FR^i = R^iF$, for $i \in \{1, 2, 3, 4\}$, we need $R^i = R^{-i}$. That requires $2i \equiv 0 \pmod 4$ which implies $i \equiv 0 \pmod 2$. This means that $FR^2 = R^2F$ is the only non-trivial possibility. Thus the center of D_4 is as stated. ▲

Exercise 2.4.11 *Find the centers of the dihedral groups D_5 and D_6.*

Hint. *The center of any group is contained in the center of any of its subgroups. So the center of D_n is contained in the subgroup generated by R, the counterclockwise rotation through an angle of $2\pi/n$. But we can also prove that if F is the flip about an axis through a vertex and the center of the n-gon, then $FR = R^{-1}F$. It follows that $FR^i = R^iF$ can only happen if $2i \equiv 0 \pmod n$.*

Example 3: The Center of the General Linear Group $GL(2, \mathbb{R})$. We find the center of the general linear group $GL(2, F)$, where $F = \mathbb{R} =$ the (field of) real numbers. Here, as in equation (2.5) of Section 2.3, where $\mathbb{R}$ was replaced with $\mathbb{Z}_p$, for prime p

$$GL(2, \mathbb{R}) = \left\{ \begin{pmatrix} a & b \\ c & d \end{pmatrix} \middle| \, a, b, c, d \in \mathbb{R}, \ ad - bc \neq 0 \right\}.$$

The group operation is matrix multiplication. The center of $GL(2,\mathbb{R})$ is

$$Z(GL(2,\mathbb{R}))=\left\{ aI=\begin{pmatrix} a & 0\\ 0 & a\end{pmatrix} \middle|\, a\in\mathbb{R}, a\neq 0\right\}. \tag{2.9}$$

To prove this, you just have to see which matrices commute with all matrices in $GL(2,\mathbb{R})$. Let a, b, c, d be the entries of a matrix in the center of $GL(2,\mathbb{R})$. Then for all $u, v, w\in\mathbb{R}$ with $uw\neq 0$ we have

$$\begin{pmatrix} a & b\\ c & d\end{pmatrix}\begin{pmatrix} u & v\\ 0 & w\end{pmatrix}=\begin{pmatrix} u & v\\ 0 & w\end{pmatrix}\begin{pmatrix} a & b\\ c & d\end{pmatrix}.$$

Doing the multiplications we get

$$\begin{pmatrix} au & *\\ cu & *\end{pmatrix}=\begin{pmatrix} au+cv & *\\ wc & wd\end{pmatrix}.$$

It follows that $au=au+cv$ for all v in $\mathbb{R}$ – in particular, for $v=1$. Thus $c=0$. Similarly $b=0$ can be seen by looking at

$$\begin{pmatrix} a & b\\ 0 & d\end{pmatrix}\begin{pmatrix} u & 0\\ v & w\end{pmatrix}=\begin{pmatrix} u & 0\\ v & w\end{pmatrix}\begin{pmatrix} a & b\\ 0 & d\end{pmatrix}.$$

This implies

$$\begin{pmatrix} au+bv & bw\\ dv & dw\end{pmatrix}=\begin{pmatrix} ua & ub\\ va & *\end{pmatrix}.$$

So $au+bv=au$. Taking $v=1$ gives $b=0$.

Thus our matrix must be diagonal. To see that the diagonal entries must be equal, look at the preceding computations again, now with $b=c=0$. You see that $dv=va$. Taking $v=1$ implies $d=a$. So our matrix is aI, where I is the identity matrix.

Note that our arguments also work for matrices with entries in $F=\mathbb{Z}_p$, p a prime. ▲

Exercise 2.4.12 *For prime p, find the center of the* ***affine group***

$$\mathrm{Aff}(p)=\left\{\begin{pmatrix} a & b\\ 0 & 1\end{pmatrix}\middle|\, a,b\in\mathbb{Z}_p,\ a\in\mathbb{Z}_p^*\right\}, \tag{2.10}$$

where the group operation is matrix multiplication.

Exercise 2.4.13

(a) Suppose that H and K are subgroups of G. Show that $H\cap K$ is also a subgroup of G.
(b) Is the same true for the union of subgroups? Why?

If you try to write down a multiplication table for an abstract group G of order 4 that is not cyclic, you will be forced to write down a table like that below. What is the main property of the table? Every element has order 2 or 1. No element has order 3 or order 4. So the multiplication table looks like:

$\cdot$	e	a	b	c
e	e	a	b	c
a	a	e	c	
b	b		e	
c	c			e

Note that the main diagonal consists of e, the identity, since all elements $x \neq e$ have order 2. What is ab? There is only one choice $ab = c$. Why? If $ab = b = eb$, for example, then by cancellation $a = e$. But $a \neq e$. That is true since our group has order 4.

Exercise 2.4.14 *Consider an abstract group G of order 4 that is not cyclic. Explain why G cannot have elements of order 3 or 4. Complete the table above and show that the result gives a group. Do you recognize it? Later we will have Lagrange's theorem to make this problem easier. See Section 3.3.*

Exercise 2.4.15 *State whether each of the following is true or false and give a reason for your answer.*

(a) The unit group $\mathbb{Z}_n^*$ (with the operation of multiplication mod n*) is a subgroup of the group $\mathbb{R}^* = \mathbb{R} - \{0\}$ (under multiplication).*
(b) The affine group Aff(p) *for prime p defined in equation (2.10), is a subgroup of the general linear group $GL(2, \mathbb{Z}_p)$ defined in equation (2.5). Here the group operation is matrix multiplication for both* Aff(p) *and $GL(2, \mathbb{Z}_p)$.*
(c) It is possible to view the dihedral group D_n as a subgroup of the symmetric group S_n.

One of our goals should be to get to know the small groups really well. Ultimately, we should know how many different groups G with $|G| \leq 15$ exist – up to isomorphism, a word meaning mathematically identical, which will be defined in Section 3.2. And one should be able to write down Cayley tables for such groups. Group Explorer will help in this endeavor certainly. For prime orders, the only groups are cyclic. That is our next topic. However, if we were chemists, we would need to know much more about the 230 space groups (219 if you do not distinguish between mirror images). These are discrete subgroups of the group of Euclidean motions of 3-space: that is, the group generated by rotations and translations. Crystallography is based on the study of such groups. See Gallian [33, Chapter 28] for a discussion of the 17 crystallographic groups in the plane – also known as wallpaper groups. Wikipedia has a long article on space groups listing the numbers in dimensions up to 5.

2.5 Cyclic Groups are Our Friends

The cyclic groups are the simplest, thus our friends. They are actually old friends – as we met them a long time ago in Chapter 1. They are always Abelian (but the converse is not true). And they have many applications as we shall see in Chapter 4. Some applications come from thinking of large finite cyclic groups as good approximations for a circle – thus the name "cyclic." They might be called the groups of *déjà vu or* "what goes around comes around." If you keep multiplying a given element a of a finite multiplicative cyclic group by itself, you eventually get back to the identity and then the whole cycle repeats: $e, a, a^2, \ldots, a^n = e, a, a^2, \ldots$ I discuss many applications of cyclic groups in my book [116] and we shall see many before we reach the end of this story. The fast Fourier transform is one such application. It allows rapid signal and data processing.

Other groups are more complicated, especially the non-Abelian ones. The reason may be that – as J. H. Conway says – such groups "are adept at doing large numbers of impossible things before breakfast."

The following definition says what it means for a group G to be a cyclic group under multiplication.

Definition 2.5.1 *A group G is **cyclic** means $G = \langle a \rangle = \{a^m \mid m \in \mathbb{Z}\}$. We call a the **generator** of G. If the order of the cyclic group is n, we will call it $\mathbf{C}_n$.*

Another way to think about this is to say that the multiplication table for the group can be identified (by clever relabeling) with that for $\mathbb{Z}_n$ under addition mod n.

Example: The Multiplication Table for an Abstract Cyclic Group *G* of Order 10 under Multiplication. See Figure 2.18 for this table. The multiplication table for a cyclic group of order 10 generated by a is really the "same" as that of $\mathbb{Z}_{10}$ under addition mod 10. More precisely, this means the following. You identify e with 0 (mod 10), a^x with x (mod 10). Define a **finite logarithm function** f by

$$f(a^x) = x \pmod{10}.$$

e	a	a^2	a^3	a^4	a^5	a^6	a^7	a^8	a^9
a	a^2	a^3	a^4	a^5	a^6	a^7	a^8	a^9	e
a^2	a^3	a^4	a^5	a^6	a^7	a^8	a^9	e	a
a^3	a^4	a^5	a^6	a^7	a^8	a^9	e	a	a^2
a^4	a^5	a^6	a^7	a^8	a^9	e	a	a^2	a^3
a^5	a^6	a^7	a^8	a^9	e	a	a^2	a^3	a^4
a^6	a^7	a^8	a^9	e	a	a^2	a^3	a^4	a^5
a^7	a^8	a^9	e	a	a^2	a^3	a^4	a^5	a^6
a^8	a^9	e	a	a^2	a^3	a^4	a^5	a^6	a^7
a^9	e	a	a^2	a^3	a^4	a^5	a^6	a^7	a^8

Figure 2.18 The Group Explorer version of the multiplication table for a cyclic group of order 10

This is a well-defined 1–1 function since (as will be checked in the exercise below) $a^x = a^y$ iff $x \equiv y \pmod{10}$. The function also maps the order 10 cyclic group $\langle a \rangle$ onto $\mathbb{Z}_{10}$. Finally one can check that $f(c)+f(d) \equiv f(cd) \pmod{10}$. Later (see Definition 3.2.1) we will call such a function $f\colon G \to \mathbb{Z}_{10}$ a group isomorphism. ▲

Exercise 2.5.1 *Check that if G is a cyclic group of order* 10, $G=\langle a\rangle$, *then* $f(a^x)=x \pmod{10}$ *is a well-defined function and* $f(c)+f(d)\equiv f(cd) \pmod{10}$. *Show that f maps G 1–1 onto* $\mathbb{Z}_{10}$. *Then find the orders of* a^2 *and* a^3 *in G.*

Recall that the order of a finite group means the number of elements in the group. When we study groups of order n, we need only look at groups that are really different – that is, non-isomorphic groups. In particular, we should not distinguish between the two groups in Exercise 2.5.1. Thus (up to isomorphism) there is really only one group of order 1. There is only one group of order 2 and only one group of order 3. There are two groups of order 4, one group of order 5, and so on. We will be able to prove such things soon. See Section 4.5 for our table of groups of order ≤ 15. There are longer tables of small groups in books and on the internet. You can also use the Group Explorer program to see a list of small groups and to find multiplication tables, Cayley graphs and other information about them. In the old days I looked at a book by A. D. Thomas and G. V. Wood [120]. It lists the multiplication tables for all groups of order ≤ 32, except the cyclic groups. Another possibility is to use SAGE as explained in Robert A. Beezer's exercises in Thomas W. Judson's on line book [50].

Our goal in this section is to prove the main theorem about cyclic groups, which is Theorem 2.5.1 below. Before we can prove this theorem, we need more information about computation in cyclic groups.

Facts About Powers in Finite Cyclic Groups

Suppose that G is a finite group and $a\in G$. Suppose that a has order $n=|a|$, meaning that the positive integer n is the smallest such that $a^n=e=$ identity of G. Then we have the following facts.

1. $a^i=a^j \iff i\equiv j \pmod{n}$. Multiplication in G corresponds to addition of exponents mod n.
2. $\langle a\rangle=\{a,a^2,a^3,\ldots,a^{n-1},a^n=e\}$. The elements in the set are distinct and thus the two ways we use the word "order" are in agreement. The order n of a is the same as the order of the cyclic group $\langle a\rangle$. That is,

$$|\langle a\rangle|=|a|.$$

3. $a^k=e=\text{identity} \iff n \text{ divides } k \iff k\equiv 0 \pmod{n}$.

Proof of fact 3. This is just a special case of fact 1. So we will say no more about fact 3.
Proof of fact 1. $\Longleftarrow$. Suppose $i\equiv j \pmod{n}$. This means $i-j=n\cdot q$ for some integer q. Then, by Proposition 2.4.1,

$$a^{i-j}=a^{nq}=(a^n)^q=e^q=e.$$

It follows that $a^{i-j}=e$ and then that $a^i=a^j$ upon multiplying both sides by a^j, again using Proposition 2.4.1.

$\Longrightarrow$. Suppose $a^i = a^j$. Then multiply both sides by a^{-j}. Using Proposition 2.4.1, this gives $a^{i-j} = e$, the identity. We want to show that $n = |a|$ divides $i - j$. To do this, we apply the division algorithm to $i - j$ to get

$$i - j = nq + r, \quad \text{with } 0 \le r < n.$$

The proof will be complete if we can show the remainder $r = 0$. We know that $a^n = e$ and

$$e = a^{i-j} = a^{nq+r} = (a^n)^q a^r = a^r.$$

So $e = a^r$. But if r were not 0, this would contradict the definition of $|a| = n$, the smallest positive integer such that $a^n = e$. Thus $r = 0$ and $i \equiv j \pmod{n}$. We leave the statement about multiplication as an exercise.

Proof of fact 2. If n is the order of a, and we look at the powers a^j, $j \ge 1$, the first power to be the identity is the nth power. Moreover, the elements of the set

$$\langle a \rangle = \{a, a^2, a^3, \ldots, a^{n-1}, a^n = e\}$$

must be distinct by fact 1. For no two distinct elements in the set $\{1, 2, 3, \ldots, n\}$ can be congruent mod n. Why?

Exercise 2.5.2

(a) Show that if a is as in the preceding facts about powers in cyclic groups, then $a^i a^j = a^k$, where $i + j \equiv k \pmod{n}$.
(b) Answer the "why?" in the proof of fact 2 above.

Morals of the Three Facts About Powers

It is easier to compute in $\mathbb{Z}_n$ under addition mod n than it is to compute in the multiplicative group of powers of the element a. That is, one should take logs. In the old days (when I was a teenager) we learned to multiply real numbers using log tables which change multiplication to addition. The slide rules we used when I was a college student had the same effect. They were mechanical log tables and they made a horrible noise when hundreds of students were using them on a physics exam. Click click click ... Anyway, the same principle works in cyclic groups. Take logs and compute with the exponents $i \pmod{n}$ rather than a^i. Discrete logarithms in finite cyclic groups have applications in cryptography. The discrete log problem concerns the speed of computing discrete logs.

Another moral is that for a finite cyclic group $\langle a \rangle$ the Cayley graph $X(\langle a \rangle, \{a, a^{-1}\})$ is a finite circle. See Figures 2.19 and 2.20 for the case that the order of $\langle a \rangle$ is 10.

Example. Suppose the order of a is $|a| = n$. How many elements of

$$\langle a \rangle = \{a, a^2, \ldots, a^{n-1}, a^n = e\}$$

are squares? ▲

Answer. An element $b = a^k$ of $\langle a \rangle$ is a square if there exists an integer power y with $b = a^k = (a^y)^2$. By fact 1 about powers in cyclic groups, $a^k = (a^y)^2$ is solvable iff, given k, there exists a y solving the congruence

$$k \equiv 2y \pmod{n}. \tag{2.11}$$

Now we have a **linear** congruence to solve rather than an abstract **non-linear** equation in some multiplicative group. The congruence in (2.11) will have a solution y for a given k iff $\gcd(2, n)$ divides k. This follows from Exercise 1.6.9.

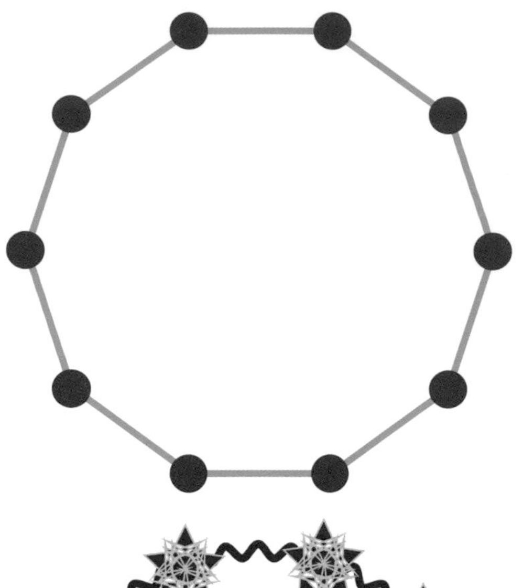

Figure 2.19 Cayley graph $X\left(\langle a\rangle, \left\{a, a^{-1}\right\}\right)$ for a cyclic group $\langle a\rangle$ of order 10

Figure 2.20 A less boring picture of a 10-cycle

So we have two cases.

Case 1. If n is even, we will only have a solution for even exponents k modulo n. That is, only half of the elements of $\langle a\rangle$ will be squares in this case.
Case 2. If n is odd, then everything in $\langle a\rangle$ is a square.

The following results involve generalizing this example from squares to kth powers.

Three More Facts About Powers in Cyclic Groups

Again assume $|a| = n$ in a finite group G. Then we have the following three facts.

4. $\langle a^k\rangle = \langle a\rangle$ iff $\gcd(k, n) = 1$.
5. $\langle a^k\rangle = \langle a^{\gcd(n,k)}\rangle$
6. $\left|a^k\right| = \dfrac{n}{\gcd(n, k)}$.

Proof of fact 4. This is just a special case of fact 5.
Proof of fact 5. To show the set equality, we use our logarithm argument mentioned earlier. So begin by noting that $x \in \left\langle a^k\right\rangle$ is equivalent to saying that $x = a^b$, where

$$b \equiv yk \pmod{n}, \tag{2.12}$$

for some $y \in \mathbb{Z}$. By Exercise 1.6.9, we know that we can solve (2.12) for y iff $\gcd(n, k)$ divides b. This says that $x = a^k \in \langle a^{\gcd(n,k)} \rangle$.

Proof of fact 6. We leave this as an exercise. Again use the logarithm method to switch the problem to one involving linear congruences.

Exercise 2.5.3 *Prove fact 6 above.*

Hint. *The multiplicative order of a^k in $\langle a \rangle$ is the same as the additive order of k in $\mathbb{Z}_n$. Set $d = \gcd(k, n)$. You need to show that $kx \equiv 0 \pmod{n}$ iff $\frac{n}{d}$ divides x.*

Now finally we can prove the main theorem about cyclic groups G in the case that G is finite. Part (1) also holds if G is an infinite cyclic group. See Exercise 2.5.8.

Theorem 2.5.1 ***(Main Theorem on Cyclic Groups and Their Subgroups).*** *Suppose G is a finite cyclic group of order $|G|$.*

(1) Any subgroup H of G is cyclic.
(2) The order $|H|$ of subgroup H of G must divide the order $|G|$.
(3) For every divisor d of $|G|$, there is a unique subgroup H of G such that $|H| = d$. If $G = \langle a \rangle$, then $H = \langle a^{n/d} \rangle$.

Proof.

(1) Suppose $G = \langle a \rangle$. Let m be the smallest positive integer such that $a^m \in H$. Then (1) is a consequence of the following claim.
Claim. $H = \langle a^m \rangle$.
Proof of Claim. By the division algorithm, if $a^t \in H$, we have $t = mq + r$ with $0 \leq r < m$. But then by Proposition 2.4.1, $a^r = a^{t-mq} = a^t (a^m)^{-q} \in H$, contradicting the definition of m unless $r = 0$. Thus H must equal $\langle a^m \rangle$. **QED Claim.**

(2) Using part (1) of this theorem, we know that $H = \langle a^m \rangle$. Facts 2 and 6 in our list of facts about powers in cyclic groups preceding this theorem say

$$\text{if } G = \langle a \rangle \text{ and } |a| = n, \text{ then } |H| = |a^m| = \frac{n}{\gcd(n, m)}.$$

Thus $|H|$ certainly divides n.

(3) Suppose d divides $n = |a|$, where $G = \langle a \rangle$. Then, by fact 6 in our list of facts about powers in cyclic groups preceding this theorem, a subgroup of G of order d is $A = \langle a^{n/d} \rangle$.
We should explain why this subgroup $A = \langle a^{n/d} \rangle$ is the only subgroup of G with order d. Suppose H is another subgroup of G with $|H| = d$. We know by (1) and (2) that $H = \langle a^m \rangle$ for some m. Then, by fact 5 in our list of facts about powers in cyclic groups preceding this theorem, we have $H = \langle a^{\gcd(m,n)} \rangle$. Let $\gcd(m, n) = k$. Then k divides n. Now $H = \langle a^k \rangle$ and $|H| = d$ imply (by fact 6) that

$$d = |a^k| = \frac{n}{\gcd(n, k)} = \frac{n}{k}$$

since k divides n. It follows that $\frac{n}{d} = k$. This proves $A = \langle a^{n/d} \rangle = \langle a^k \rangle = H$. ▲

We will see in Section 3.2 that part (2) of Theorem 2.5.1 holds for all finite groups – not just the cyclic ones.

Example. The poset diagram (as in Section 1.7) for the subgroups of a cyclic group of order 24 under the relation $H \subset K$ is now easy to draw. It is the same as the poset diagram of the subgroups of the group $\mathbb{Z}_{24}$, under addition (mod 24). See Figure 2.21. ▲

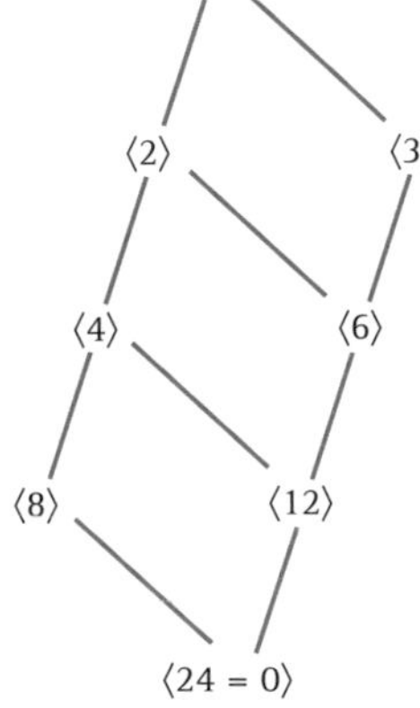

Figure 2.21 Poset diagram of the subgroups of $\mathbb{Z}_{24}$ under addition

If one wants to draw the poset diagram for a cyclic group of order n, then the more primes dividing n, the higher dimensional the diagram. Here there are only two primes and so the diagram is two-dimensional. The diagram is also the same as that for the divisors of 24 under the relation $a|b$, except that up and down are reversed. See Figure 1.13.

In these poset diagrams we only draw a line from the group above to the subgroup directly below. For example, $\langle 0 \rangle$ is a subgroup of $\langle 4 \rangle$, but we only draw a line from $\langle 8 \rangle$ down to $\langle 0 \rangle$. The line is drawn between subgroups $A \subset B$ iff there is no subgroup of G between A and B.

Exercise 2.5.4 *Draw the poset diagram for the subgroups of the cyclic group of order 60. There are three primes so the diagram is three-dimensional.*

We end this section with a formula counting the number of elements of order d in a cyclic group G if d divides the order of G.

Theorem 2.5.2 *Assume $G = \langle a \rangle$, where $|G| = |a| = n$ is finite. Then, if d divides n, the number of elements of order d in G is $\phi(d)$, the Euler phi-function (which is the order of $\mathbb{Z}_d^*$ or the number of integers a between 1 and d with $\gcd(a, d) = 1$).*

Proof. By part (3) of Theorem 2.5.1, there is a unique subgroup $H = \langle b \rangle$ of G with $|H| = d$. Then every element of order d in G must generate H. Fact 4 in our list of six facts about powers in cyclic groups says that $H = \langle b^k \rangle$ iff $\gcd(k, d) = 1$. There are $\phi(d)$ such numbers k modulo d. ▲

Example 1. Suppose p is a prime. Then $\phi(p)=p-1$, since all integers a between 1 and $p-1$ have $\gcd(a,p)=1$. The only positive divisors of p are 1 and p. Then the cyclic group $\mathbb{Z}_p$, under $+ \pmod p$, has $p-1$ generators $a \pmod p$, $a=1,2,\ldots,p-1$. ▲

Example 2. It turns out that when p is prime, $\mathbb{Z}_p^*=\{1,2,\ldots,p-1 \pmod p\}$, under multiplication mod p, is cyclic. This will be Exercise 6.3.10. However not all groups $\mathbb{Z}_n^*$ are cyclic. For example $\mathbb{Z}_8^*$ and $\mathbb{Z}_{12}^*$ are not cyclic. The theorem on the subject says that $\mathbb{Z}_n^*$ is cyclic iff $n=2,4,p^r$, or $2p^r$, where p is an odd prime. See Kenneth Rosen [91], Daniel Shanks [103, pp. 92ff], or Ramanujachary Kumanduri and Cristina Romero [62, p. 173]. ▲

The group $\mathbb{Z}_p^*$ has order $p-1$ and thus has $\phi(p-1)=\phi(\phi(p))$ generators. In the following examples, we list the smallest positive generator.

$$\mathbb{Z}_5^*=\langle 2 \pmod 5\rangle, \mathbb{Z}_7^*=\langle 3 \pmod 7\rangle, \mathbb{Z}_{11}^*=\langle 2 \pmod{11}\rangle, \mathbb{Z}_{13}^*=\langle 2 \pmod{13}\rangle,$$
$$\mathbb{Z}_{17}^*=\langle 3 \pmod{17}\rangle, \mathbb{Z}_{19}^*=\langle 2 \pmod{19}\rangle, \mathbb{Z}_{23}^*=\langle 5 \pmod{23}\rangle, \mathbb{Z}_{29}^*=\langle 2 \pmod{29}\rangle.$$

Finding generators of $\mathbb{Z}_p^*$ is so useful in number theory that many books on the subject include tables giving the generators g of $\mathbb{Z}_p^*$ for small values of p. These generators are called **"primitive roots"** mod p in number theory. Finding them by hand can be time consuming. Mathematica will find them for you. The number 2 works about 37% of the time according to a famous conjecture of E. Artin from 1927 which remains unproved though there is much evidence for the conjecture. See Shanks [103] for more information.

Exercise 2.5.5 *State whether* 2 *is a primitive root* mod p*: that is,* $\mathbb{Z}_p^*=\langle 2 \pmod p\rangle$ *for all the primes* $p\le 100$.

Exercise 2.5.6 *State whether each of the following statements is true or false and give a brief reason for your answer.*

(a) A cyclic group can have only one generator.
(b) Any Abelian group is cyclic.
(c) The group $G=\{a,b,c,d\}$ *of order* 4 *with the following multiplication table is cyclic.*

$*$	a	b	c	d
a	a	b	c	d
b	b	a	d	c
c	c	d	b	a
d	d	c	a	b

Exercise 2.5.7 *Find all the subgroups of the group* $\mathbb{Z}$ *of integers under addition.*

Hint. *Use well-ordering.*

Exercise 2.5.8 *Show that any subgroup of an infinite cyclic group is cyclic.*

Exercise 2.5.9 *Suppose* $G=\langle a\rangle$ *is an infinite cyclic group. Show that if two of its subgroups are equal, namely,* $\langle a^r\rangle=\langle a^s\rangle$ *for* $r,s\in\mathbb{Z}^+$*, then* $r=\pm s$ *and conversely.*

Exercise 2.5.10 *Let a be an element of a multiplicative group. Suppose the order of a is $|a| = n$. How many elements of $\langle a \rangle = \{a, a^2, \ldots, a^{n-1}, a^n = e\}$ are cubes?*

Exercise 2.5.11 *State whether each of the following statements is true or false and give a reason for your answer.*

(a) The multiplicative group $\mathbb{Z}_8^*$ is a subgroup of the multiplicative group $\mathbb{Q}^* = \mathbb{Q} - \{0\}$.
(b) $\mathbb{Q}$ is an additive subgroup of the additive group $\mathbb{R}$.
(c) The additive group $\mathbb{Z}_{175}$ has a subgroup of order 3.

There are some other interesting figures that can be associated to a group: **cycle diagrams.** These diagrams have as nodes every group element a and show all the cyclic subgroups $\langle a \rangle = \{e, a, a^2, \ldots, a^{n-1}\}$ as cycle graphs. The cycles $\langle a \rangle$ and $\langle b \rangle$ will be interconnected if one element is in the cyclic subgroup generated by the other. So, for example consider the cycle diagram for the multiplicative group $\mathbb{Z}_{15}^*$ in Figure 2.22 below. The cycles are $\{1, -1 \equiv 14\}, \{1, -4 \equiv 11\}, \{1, 2, 4, 8\}, \{1, -2 \equiv 13, 4, -8 \equiv 7\}$. See Shanks [103, pp. 87ff] for many more such figures. Group Explorer will also create these diagrams.

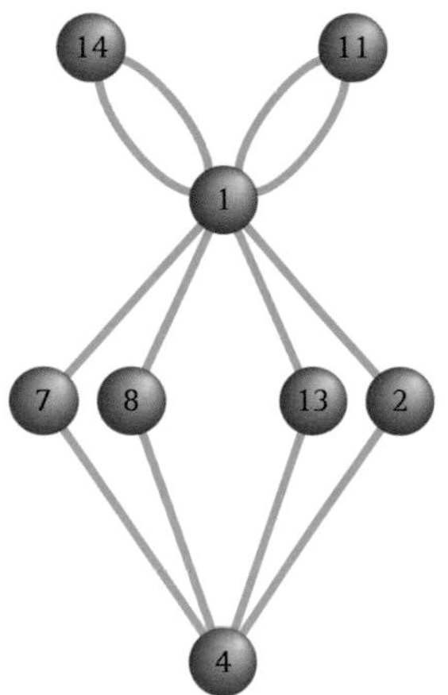

Figure 2.22 Cycle diagram in the multiplicative group $\mathbb{Z}_{15}^*$

Exercise 2.5.12 *Draw a cycle diagram for $\mathbb{Z}_{21}^*$. Then do the same for $\mathbb{Z}_{24}^*$.*

3

Groups: There's More

3.1 Groups of Permutations

We have already seen that permutation notation allows us to compute more easily in the symmetry groups of various geometric figures such as D_n. The symmetric group of permutations of n objects is an extremely important group for many reasons - not least of which being the applications in computer science. It also is an extremely large group for large n. We have already said many things about this group. Let us review them first.

Definition 3.1.1 *The symmetric group is*

$$S_n = \{\sigma : \{1,2,3,\ldots,n\} \to \{1,2,3,\ldots,n\} \mid \sigma \text{ is 1–1 and onto}\},$$

for $n = 2,3,4,5,\ldots$ *Multiplication is composition of functions.*

Throughout this section we will assume $n \geq 2$ for obvious reasons.

A **permutation** σ is uniquely determined by its list of values and we shall write

$$\sigma = \begin{pmatrix} 1 & 2 & \cdots & n \\ \sigma(1) & \sigma(2) & \cdots & \sigma(n) \end{pmatrix}.$$

It follows from this representation that there are $n! = n(n-1)(n-2)\cdots 2\cdot 1$ elements of S_n as we showed in Section 2.3.

In Section 3.2 we will see that every finite group G of order n can be viewed as a subgroup of S_n. This is Cayley's theorem - proved in 1854. Subgroups of S_n are known as permutation groups. Thus every finite group can be viewed as such a permutation group. We saw in Section 2.1 that the dihedral group D_3 of symmetries of an equilateral triangle can be viewed as S_3. The dihedral group D_4 is similarly considered as a proper subgroup of S_4. For the order of D_4 is 8 while the order of S_4 is 24. Similarly D_n is viewed as a proper subgroup of S_n for $n \geq 4$, since D_n permutes the vertices of a regular n-gon. In fact, we have shown in Exercise 2.1.13 that the order of D_n is $2n$. The order of S_n grows large quite rapidly. Some examples are $|S_5| = 120$, $|S_6| = 720$, $|S_{52}|$ is approximately 8 times 10^{67}.

The last group, S_{52}, is of interest for a card player. Statisticians study S_{52} to learn how many shuffles it takes to put a deck of cards in "random" order. See Diaconis [24] for example, which includes many other applications as well as a discussion of Fourier analysis on the symmetric group - a subject too intricate to include in our text, where we consider only the cyclic groups and the Abelian group $\mathbb{Z}_2^n$ in Section 4.2 and Exercises 8.2.15, 8.2.16 respectively. There are applications of the symmetric group S_n in a variety of fields, for example: computer science, economics, psychology, statistics, chemistry, physics. Mathematica will do computations in the symmetric group. So will SAGE. See Beezer's SAGE exercises in Judson [50].

Disjoint Cycle Notation for Permutations. Let us define the cycle notation by example. Consider the permutation

$$\sigma = \begin{pmatrix} 1 & 2 & 3 & 4 & 5 \\ 5 & 3 & 4 & 2 & 1 \end{pmatrix}.$$

This permutation sends 1 to 5 and 5 to 1. In addition, it sends 2 to 3, 3 to 4, and 4 to 2. So we write

$$\sigma = (15)(234) = (234)(15).$$

The cycles (15) and (234) are said to be **disjoint** since each acts on disjoint sets of numbers, namely $\{1,5\}$ and $\{2,3,4\}$. Since the cycles (15) and (234) act on disjoint sets of numbers, the cycles commute: that is, $(15)(234) = (234)(15)$.

Exercise 3.1.1 *Find the disjoint cycle decomposition for every permutation in S_3 and S_4.*

Exercise 3.1.2 *If $\sigma = (a_1 a_2 \cdots a_k)$ is a cycle in S_n, show that the order of σ is $|\sigma| = k$.*

Exercise 3.1.3 *Show, by example, that non-disjoint cycles need not commute.*

We will need the following definition to compute orders of permutations written in disjoint cycle notation.

Definition 3.1.2 *The **least common multiple** of two positive integers m, n, written $\mathrm{lcm}[m,n] = [m,n]$, is the smallest positive integer that is a multiple of both m and n.*

Examples. lcm[2,3] = 6, lcm[4,20] = 20. ▲

Note that one can give a definition of least common multiple which is similar to the one we gave for the greatest common divisor of two integers. That is, one can show that $\mathrm{lcm}[m,n] = c$ iff the following two properties hold:

(1) both m and n divide c;
(2) m divides c' and n divides $c' \Longrightarrow c$ divides c'.

Exercise 3.1.4 *Prove this last statement.*

Exercise 3.1.5 *Compute* $\mathrm{lcm}[13,169]$ *and* $\mathrm{lcm}[11,1793]$.

Exercise 3.1.6 *Show using unique factorization into primes that we can compute the* lcm *as follows, once we have factored the integers involved as a product of the pairwise distinct primes p_i, $i = 1, \ldots, k$:*

$$\mathrm{lcm}\left[\prod_{i=1}^{k} p_i^{e_i}, \prod_{i=1}^{k} p_i^{f_i}\right] = \prod_{i=1}^{k} p_i^{h_i}, \quad \text{where } h_i = \max\{e_i, f_i\}.$$

Exercise 3.1.7 *Prove that, for integers n, m, not both 0, we have*

$$\operatorname{lcm}[n, m] = \frac{nm}{\gcd(n, m)}.$$

Exercise 3.1.8 *Extend the definitions of* gcd *and* lcm *to sets of three or more integers. Do the formulas in Exercise 1.5.23 and the last two exercises extend?*

Proposition 3.1.1 *(Facts About Permutations). The following hold:*

(1) Every permutation in S_n can be written as a product of disjoint cycles. Here "disjoint" means that the numbers moved by each of the cycles form pairwise disjoint sets.
(2) Disjoint cycles commute.
(3) The order of a product of disjoint cycles is the least common multiple of the cycle lengths.

Proof.

(1) Given $\sigma \in S_n$, choose $a \in \{1, \ldots, n\}$ and look at the **orbit of** a under the cyclic subgroup $\langle\sigma\rangle \subset S_n$:

$$\operatorname{Orb}(a) = \left\{\sigma^i a \mid i = 1, 2, \ldots, |\sigma|\right\}.$$

From the orbit, we get a cycle permutation $\left(a, \sigma^2 a, \sigma^3 a, \ldots, \sigma^{|\sigma|-1} a\right)$. If the set of integers represented by the orbit of a under $\langle\sigma\rangle$ is not all of the set $\{1, \ldots, n\}$, choose an element b not in the orbit of a. The orbit of b will give rise to another cycle permutation $(b, \sigma^2 b, \sigma^3 b, \ldots)$. Keep going in this way. The orbits partition $\{1, 2, \ldots, n\}$ into a finite union of sets

$$\{1, 2, \ldots, n\} = \operatorname{Orb}(a) \cup \operatorname{Orb}(b) \cup \cdots \cup \operatorname{Orb}(z).$$

Corresponding to this partition, we get the disjoint cycle decomposition of σ:

$$\sigma = \left(a, \sigma^2 a, \sigma^3 a, \ldots\right)\left(b, \sigma^2 b, \sigma^3 b, \ldots\right) \cdots \left(z, \sigma^2 z, \sigma^3 z, \ldots\right).$$

(2) Since the cycles act on disjoint sets of numbers, the ordering of the cycles does not matter.

(3) Suppose that α and β are disjoint cycles. Then, since they commute by part (2), we see that $(\alpha\beta)^k = \alpha^k \beta^k = e \Longleftrightarrow \alpha^k = e$ and $\beta^k = e$. The result is a consequence of this fact. See the exercise that follows. ▲

Exercise 3.1.9 *Finish the proof of part (3) of Proposition 3.1.1.*

Example. What is the order of σ if

$$\sigma = \begin{pmatrix} 1 & 2 & 3 & 4 & 5 \\ 2 & 1 & 5 & 3 & 4 \end{pmatrix} = (12)(354)?$$ ▲

Answer. The order of σ is $6 = \operatorname{lcm}[2, 3]$.

Check.

i	σ^i
1	$(12)(354) = (354)(12)$
2	$(12)^2(354)^2 = (354)^2 = (345)$
3	$(12)^3(354)^3 = (12)$
4	$(12)^4(354)^4 = (354)$
5	$(12)^5(354)^5 = (12)(345)$
6	$(12)^6(354)^6 = (1)$

Question. What is σ^{-1}?

To figure this out, recall our other notation for σ.

$$\sigma = \begin{pmatrix} 1 & 2 & \cdots & n \\ \sigma(1) & \sigma(2) & \cdots & \sigma(n) \end{pmatrix}.$$

To find the inverse function, just switch row 1 and row 2 to get

$$\sigma^{-1} = \begin{pmatrix} \sigma(1) & \sigma(2) & \cdots & \sigma(n) \\ 1 & 2 & \cdots & n \end{pmatrix}.$$

Then we should reorder the columns to put the numbers on the top row in the usual order.

Example. Consider

$$\sigma = \begin{pmatrix} 1 & 2 & 3 & 4 & 5 \\ 2 & 1 & 5 & 3 & 4 \end{pmatrix} = (12)(354)$$

and find that

$$\sigma^{-1} = \begin{pmatrix} 2 & 1 & 5 & 3 & 4 \\ 1 & 2 & 3 & 4 & 5 \end{pmatrix}.$$

Reordering the columns gives

$$\sigma^{-1} = \begin{pmatrix} 1 & 2 & 3 & 4 & 5 \\ 2 & 1 & 4 & 5 & 3 \end{pmatrix} = (12)(345).$$

Another way to find the inverse is to use the fact that σ is a product of disjoint cycles. So then we just need to know the formula for the inverse of a cycle. One answer is to raise the cycle to the power $p =$ (length of cycle – 1). Another answer is to reverse the order of the numbers in the cycle:

$$(a_1 a_2 \cdots a_r)^{-1} = (a_r a_{r-1} \cdots a_2 a_1).$$

For our example, using the fact that disjoint cycles commute, we get the inverse of $\sigma = (12)(354)$:

$$\sigma^{-1} = (12)^{-1}(354)^{-1} = (12)(354)^2 = (12)(453) = (12)(345).$$

▲

Exercise 3.1.10 *Let* $\tau = \begin{pmatrix} 1 & 2 & 3 & 4 & 5 & 6 & 7 \\ 5 & 4 & 3 & 2 & 7 & 1 & 6 \end{pmatrix}$.

(a) Find the disjoint cycle decomposition of τ.
(b) Find τ^{-1}.
(c) Find the order of the permutation τ.

Next we want to define even permutation. This will allow us to define the alternating groups. It also arises in many contexts – for example, in the definition of the determinant, as we shall see in Section 7.3.

In order to define the notion of even permutation, we need to be able to write any permutation σ as a product of **transpositions** (ab), also known as **2-cycles.** If you get an even number of transpositions, then σ is an **even permutation.** Otherwise σ is odd. By Lemma 3.1.2 below, even though there are many different products of transpositions representing a given permutation σ, the numbers of transpositions occurring must be the same mod 2. Thus the evenness or oddness of σ is a well-defined concept.

In order to write an arbitrary permutation σ as a product of **transpositions** or 2-cycles, we just need to write any cycle as a product of transpositions since σ is a product of cycles. This decomposition of σ as a product of transpositions is not unique. First consider an example.

Example. We can write

$$(12345) = (15)(14)(13)(12)$$

or

$$(12345) = (12)(23)(34)(45).$$

To see that this works, you just have to see where both sides send $x \in \{1,2,3,4,5\}$. Make a table:

x	$(12345)x$	$(15)(14)(13)(12)x$	$(12)(23)(34)(45)x$
1	2	2	2
2	3	3	3
3	4	4	4
4	5	5	5
5	1	1	1

It follows that (12345) is an even permutation. In general, a cycle of odd length is an even permutation. ▲

Lemma 3.1.1 *Every transposition can be written as a product of an odd number of adjacent transpositions, that is, those of the form* $(a\ a+1)$.

Instead of proving the lemma in general, we give an example.

Example. $(14) = (12)(23)(34)(32)(21)$. ▲

Exercise 3.1.11 *Check the preceding example and then find the analogous decomposition of* (13) *and* (27) *in* S_7. *What is the general formula for* (ab) *with* $a < b$?

Lemma 3.1.2 *The even or oddness of a permutation is well defined. Equivalently, suppose* $\sigma = \alpha_1\alpha_2\cdots\alpha_r = \beta_1\beta_2\cdots\beta_s$, *where* α_i *and* β_j *are transpositions for all* i,j. *Then* $r \equiv s \pmod 2$.

Our discussion of this lemma follows that of Birkhoff and Maclane [9, p. 145]. Before attempting a proof of this lemma, we need a certain useful polynomial.

Definition 3.1.3 *We can associate the* ***very useful polynomial*** *in n indeterminates to our problem. Define*

$$V(x_1,\ldots,x_n) = \prod_{1\le i<j\le n} (x_i - x_j)\,.$$

Define the ***action*** *of permutation $\sigma \in S_n$ on the polynomial by*

$$(\sigma V)\,(x_1,x_2,\ldots,x_n) = V(x_{\sigma(1)},x_{\sigma(2)},\ldots,x_{\sigma(n)}).$$

Birkhoff and Maclane show (see [9, p. 448]) that our very useful polynomial also occurs when one solves cubic polynomial equations using repeated roots. The polynomial appears in the formula for the discriminant of higher degree polynomials over $\mathbb{Q}$. If you wish to avoid this polynomial, you can find a different proof of Lemma 3.1.2 in Gallian [33]. We call the $x_1,\ldots,x_n$ indeterminates – not variables – because we are not viewing polynomials as functions. In Section 5.5 we will discuss this distinction again.

Exercise 3.1.12 *Does $\sigma\tau V = \sigma(\tau V)$?*

From the definition of $V(x_1,\ldots,x_n)$, we see that we have a definition of the sign of a permutation, $\mathrm{sgn}(\sigma)$:

$$V(x_{\sigma(1)},x_{\sigma(2)},\ldots,x_{\sigma(n)}) = \mathrm{sgn}(\sigma)\,V(x_1,\ldots,x_n), \quad \text{where } \mathrm{sgn}(\sigma) = \pm 1.$$

Lemma 3.1.2 will be proved if we can show that

$$\mathrm{sgn}(\sigma) = \begin{cases} 1, & \sigma \text{ even} \\ -1, & \sigma \text{ odd.} \end{cases} \tag{3.1}$$

Example. In the case $n=3$, our polynomial is: $V(x_1,x_2,x_3) = (x_1-x_2)(x_1-x_3)(x_2-x_3)$. Then if $\sigma = (12)$, we have

$$((12)V)(x_1,x_2,x_3) = (x_2-x_1)(x_2-x_3)(x_1-x_3) = -V(x_1,x_2,x_3).$$

If $\sigma = (13)$, then

$$\begin{aligned}((13)V)(x_1,x_2,x_3) &= (x_3-x_2)(x_3-x_1)(x_2-x_1) = (-1)^3 V(x_1,x_2,x_3)\\ &= -V(x_1,x_2,x_3).\end{aligned}$$

▲

The following proposition implies equation (3.1) for $\mathrm{sgn}(\sigma)$. This will complete the proof of Lemma 3.1.2.

Proposition 3.1.2 ***(Properties of*** **$\mathrm{sgn}(\sigma)$*)*.** *Suppose that α,β,σ are in S_n. Then we have the following properties.*

(1) $\mathrm{sgn}(\alpha\beta) = \mathrm{sgn}(\alpha)\mathrm{sgn}(\beta)$.
(2) $\mathrm{sgn}(\text{tranposition}) = -1$.
(3) $\mathrm{sgn}(\sigma) = (-1)^{\#(\text{tranpositions in } \sigma)}$.

Proof.

(1) $\operatorname{sgn}(\alpha\beta) = \dfrac{V(x_{\alpha\beta(1)}, \ldots x_{\alpha\beta(n)})}{V(x_1, \ldots, x_n)} = \dfrac{V(x_{\alpha(\beta(1))}, \ldots x_{\alpha(\beta(n))})}{V(x_{\beta(1)}, \ldots x_{\beta(n)})} \dfrac{V(x_{\beta(1)}, \ldots x_{\beta(n)})}{V(x_1, \ldots, x_n)}$
$= \operatorname{sgn}(\alpha)\operatorname{sgn}(\beta)$.

(2) From Lemma 3.1.1 above and part (1), we just need to consider adjacent transpositions. Any transposition like $(a\ a+1)$ replaces the term $(x_a - x_{a+1})$ with its negative. The other affected terms of V are of the form $(x_a - x_j)$ and $(x_{a+1} - x_j)$ with $j > a + 1$. These do not change the sign as they are just interchanged. Thus $\operatorname{sgn}(a\ a+1) = -1$.

(3) This follows from parts (1) and (2). ▲

Now – at long last – we can consider the alternating groups.

Definition 3.1.4 *The* ***alternating group*** *is* $A_n = \{\sigma \in S_n \mid \sigma \text{ is even}\}$.

Proposition 3.1.3 *(Facts about the Alternating Group).*

(1) A_n is a subgroup of S_n.

(2) $|A_n| = \dfrac{|S_n|}{2} = \dfrac{n!}{2}$.

Proof.

(1) Use the finite subgroup test. So you just need to see that the product of two even permutations is even. Suppose $\sigma = \alpha_1\alpha_2\cdots\alpha_r$ and $\tau = \beta_1\beta_2\cdots\beta_s$, where the α_i and β_j are all transpositions. Here both r and s are even. Then $\sigma\tau$ is the product of $r + s$ transpositions, and $r + s$ is also even.

(2) There are as many even permutations as odd ones in S_n. For we have a 1–1, onto function

$$T: A_n \to S_n - A_n \quad \text{defined by } T(\sigma) = (12)\sigma.$$ ▲

Exercise 3.1.13 *Do the odd permutations in S_n form a subgroup of S_n? Why?*

Example: The Tetrahedral Group. Tet $= A_4$ is now our favorite new group. It is identifiable with the group of proper rotations of a tetrahedron. Here "proper"means it can be done by orientation-preserving 3-space rotation. See Figure 3.1 for a graph representing a tetrahedron that has been flattened. ▲

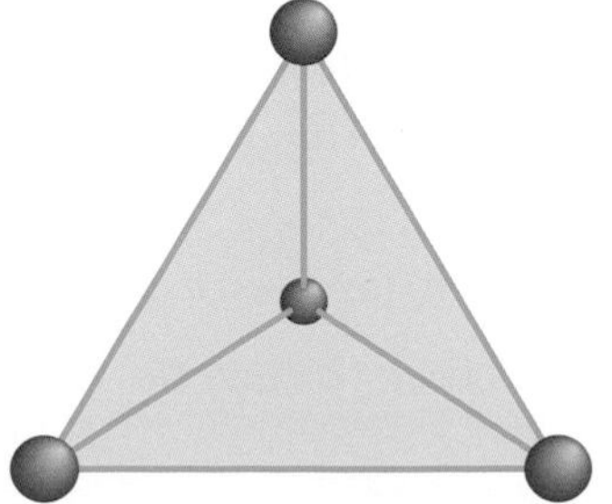

Figure 3.1 Tetrahedron

We can identify elements of the group Tet of proper rotations of the tetrahedron with permutations in S_4 easily enough. But why must they be even? Note that any one of four faces of the tetrahedron can be placed on the ground. Then you can rotate in three ways. Thus the order of the group Tet looks like 12, that is, the same as the order of A_4. So we have some evidence that Tet can be identified with A_4.

Exercise 3.1.14 *Show that the two groups* Tet *and* A_4 *are really the same by numbering the vertices of the tetrahedron and figuring out what permutation of the vertices corresponds to the various motions in* Tet.

Exercise 3.1.15 *Work out the Cayley table for* A_4. *Of course the program Group Explorer will do this for you but please show some of the computations here: that is, multiply out at least half of the products in the table using the disjoint cycle notation to make life easier. You could also use the presentation in Exercise 3.1.20 below.*

Many authors enlarge the tetrahedral group by allowing improper rotations. They thus get a group of order 24 which is S_4.

As we said earlier, the tetrahedral group is the symmetry group of methane CH_4 – a main component of natural gas production and a big contributor to global warming.

The following exercise shows that our very useful polynomial is actually a determinant – the Vandermonde determinant. Thus we could deduce Proposition 3.1.2 from properties of determinants to be found in Exercise 7.3.1.

Exercise 3.1.16 *Prove that our very useful polynomial from Definition 3.1.3 is the* ***Vandermonde determinant*** *defined by*

$$V(x_1, x_2, \ldots, x_n) = \det \begin{pmatrix} x_1^{n-1} & x_1^{n-2} & \cdots & x_1 & 1 \\ x_2^{n-1} & x_2^{n-2} & \cdots & x_2 & 1 \\ \vdots & \vdots & \ddots & \vdots & \vdots \\ x_n^{n-1} & x_n^{n-2} & \cdots & x_n & 1 \end{pmatrix}.$$

Then it is clear from properties of the determinant (see Section 7.3) that the transposition (ab), *for* $a, b \in \{1, 2, \ldots, n\}$, *switches rows a and b and thus multiplies the determinant by* -1. *Try the case* $n=3$ *first.*

Hint. *Use induction on n.*

Step 1. Subtract $x_n \cdot$ (the 2nd column) from the 1st column.

Step 2. Subtract $x_n \cdot$ (the 3rd column) from the 2nd column.

Keep going until

Step $(n-1)$. *Subtract* $x_n \cdot$ (the nth column) *from the* $(n-1)$th *column.*

Then expand by minors of the last row and use the induction hypothesis.

Exercise 3.1.17 *Consider the permutation* $\sigma = \begin{pmatrix} 1 & 2 & 3 & 4 & 5 \\ 5 & 4 & 1 & 2 & 3 \end{pmatrix}$. *Write* σ *as a product of disjoint cycles. Is* σ *even or odd? What is the order of* σ? *Find the inverse of* σ. *Compute* σ^3.

Exercise 3.1.18 *State whether each of the following statements is true or false and give a brief reason for your answer.*

(a) The permutation $\tau = (123)(321)$ has order 3.
(b) The product of two odd permutations is odd.

Exercise 3.1.19 *Show that, for $n \geq 3$, the group A_n is generated by 3-cycles (abc).*

Hint. *We know that S_n is generated by 2-cycles. So we just need to think about products of two 2-cycles. There are three cases to consider. $(ab)(ab)$, $(ab)(ac)$, $(ab)(cd)$, where a, b, c, d are pairwise distinct integers from $\{1, \ldots, n\}$. For the last case, look at: $(ab)(bc)(bc)(cd)$.*

Exercise 3.1.20 *Show that the tetrahedral group A_4 has generators $R = (234)$ and $F = (12)(34)$, with relations $R^3 = F^2 = (FR)^3 = I$.*

3.2 Isomorphisms and Cayley's Theorem

At last we define what we mean when we say that two groups are mathematically the same or isomorphic.

Definition 3.2.1 *Suppose that G and G' are groups. A function $T: G \to G'$ is called a* ***group isomorphism*** *iff T is 1–1 and onto and T preserves the group operations, that is, $T(xy) = T(x)T(y)$. Then we say that groups G and G' are* **isomorphic** *and write $G \cong G'$.*

Example 1. Any cyclic group of order n is isomorphic to the group of integers mod n under addition mod n. We have already realized this, while using the finite analog of the logarithm, first to investigate whether elements of cyclic groups can be squares and then in proving the fifth fact about powers in cyclic groups in Section 2.5. To prove that $C_n \cong \mathbb{Z}_n$ – now that we have a proper definition of isomorphic groups – suppose the cyclic group $C_n = \langle a \rangle = \{a^x \mid x \in \mathbb{Z}\}$. Define the function $T: \mathbb{Z}_n \to C_n$ by $T(x \bmod n) = a^x$. The map is well defined and 1–1 by the first fact about powers in cyclic groups in Section 2.5. The map is onto either by the pigeonhole principle or by noting that every element of the cyclic group $\langle a \rangle$ is a power of a. To see that the function preserves the group operation, use Proposition 2.4.1 to show that

$$T(x + y \pmod{n}) = a^{x+y} = a^x a^y = T(x)T(y).$$

Of course, here the first group has the operation of addition and the second group has multiplication as its operation. Nevertheless we can now identify the two groups as far as this book is concerned. We can now translate any question about a cyclic group into a question about the integers mod n under addition. We have our finite log table. ▲

Example 2. Two groups of the same order need not be isomorphic. For example, the symmetric group S_3 is not isomorphic to $\mathbb{Z}_6$ with the operation of addition mod 6. Any group isomorphic to a commutative group would have to be commutative. ▲

Proposition 3.2.1 *(Facts About Group Isomorphisms). Suppose that G and G′ are groups and the mapping $T: G \to G'$ is a group isomorphism from G onto G′. Then we have the following facts.*

(1) If e is the identity of G and e′ is the identity of G′, then $T(e) = e'$.
(2) For all $x \in G$, we have $T(x^{-1}) = T(x)^{-1}$.
(3) The order of $T(x) \in G'$ is the same as the order of $x \in G$.

Proof.

(1) We leave the proof of part (1) as an exercise.
(2) The equalities

$$e' = T(e) = T(xx^{-1}) = T(x)T(x^{-1})$$

and

$$e' = T(e) = T(x^{-1}x) = T(x^{-1})T(x)$$

imply the desired result.
(3) Suppose $n = |T(x)|$ and $d = |x|$ are both finite. Then, using fact 3 about powers in finite cyclic groups from Section 2.5,

$$T(e) = e' = T(x)^n = T(x^n)$$

implies $x^n = e$, as T is 1–1. Thus $d = |x|$ divides $n = |T(x)|$. To go the other way, we can use the fact that T^{-1} is also a group isomorphism by Exercise 3.2.5. So we see in the same way that that n divides d. Why can we do this? It follows that $n = d$. What happens if x has infinite order? ▲

Exercise 3.2.1 *Prove part (1) of Proposition 3.2.1.*

Exercise 3.2.2 *Answer the questions in the proof of part (3) of Proposition 3.2.1.*

Exercise 3.2.3 *Show that $\mathbb{Z}_6$ under addition is isomorphic to $\mathbb{Z}_7^*$ under multiplication.*

Exercise 3.2.4 *Show that finite isomorphic groups must have the same order.*

Exercise 3.2.5

(a) Show that if $T: G \to H$ is a group isomorphism then $T^{-1}: H \to G$ is also a group isomorphism.
(b) Suppose $T: G \to H$ is a group isomorphism and $S: H \to K$ is also a group isomorphism. Show that the composition $S \circ T: G \to K$ is a group isomorphism.

Exercise 3.2.6 *Show that if G and H are isomorphic groups, then G commutative implies H is commutative also.*

Exercise 3.2.7 *Consider the group D_3. Show that D_3 is isomorphic to a subgroup of S_6 as follows. Denote the elements of D_3 as $\{g_1, g_2, \ldots, g_6\}$. Then consider the map taking an* *element $a \in D_3$ to the permutation corresponding to row a of the Cayley table of D_3. This is the permutation $\sigma = \sigma(a)$ given by*

$$\begin{pmatrix} g_1 & g_2 & g_3 & g_4 & g_5 & g_6 \\ ag_1 = g_{\sigma(1)} & ag_2 = g_{\sigma(2)} & ag_3 = g_{\sigma(3)} & ag_4 = g_{\sigma(4)} & ag_5 = g_{\sigma(5)} & ag_6 = g_{\sigma(6)} \end{pmatrix}.$$

Write down the permutation $\sigma(a)$ explicitly for each element a of D_3.

The preceding exercise leads to the following theorem when generalized to any finite group.

Theorem 3.2.1 *(Cayley's Theorem).* *If G is a finite group of order n, then G is isomorphic to a subgroup of S_n, the group of permutations of n objects.*

Proof. For any $g \in G$, recall (from Definition 2.3.2) that we have the left multiplication map L_g of G defined by setting $L_g(x) = gx$, for all $x \in G$. See Definition 2.3.2. We showed in Section 2.3 that L_g is 1–1 and onto and thus permutes the elements of G.

We can enumerate $G = \{g_1, \ldots, g_n\}$ and write $L_a(g_i) = ag_i = g_{\sigma_a(i)}$, for a unique $\sigma_a \in S_n$. Then define $T: G \to S_n$ by $T(a) = \sigma_a$. We need to show that T is well defined, 1–1, and preserves the group operations. Certainly T is well defined. Suppose $\sigma_a = \sigma_b$. Then if $g_1 = e$, the identity of G, we have $a = g_{\sigma_a(1)} = g_{\sigma_b(1)} = b$. So T is 1–1. Next consider

$$g_{\sigma_{ab}(i)} = abg_i = a(bg_i) = a(g_{\sigma_b(i)}) = g_{\sigma_a(\sigma_b(i))}.$$

It follows that $T(ab) = \sigma_{ab} = \sigma_a \circ \sigma_b = T(a) \circ T(b)$.

The theorem is proved once you do the exercise below. ▲

Exercise 3.2.8 *Show that if T is the function defined in the proof of Theorem 3.2.1, then the image $T(G) = \{T(g) \mid g \in G\}$ is a subgroup of S_n.*

Hint. *You can use the finite subgroup test (i.e., Proposition 2.4.4) here.*

Cayley's theorem does not say that the group G of order n is isomorphic to S_n. This is impossible – unless $n = 1$. For S_n has order $n!$ while G has order only n. The symmetric group is huge compared with G, even for $n = 5$ when $n! = 120$. Under the isomorphism of Cayley's theorem, the group S_3 is isomorphic to a subgroup of S_6, which is a group of order 720. In general, Cayley's theorem realizes S_n as a subgroup of the much larger group $S_{n!}$. You could then use Cayley's theorem again to view both groups as subgroups of $S_{(n!)!}$. And you could keep going with that somewhat insane idea.

Exercise 3.2.9 *Consider the group $\mathbb{Z}_5$ of integers mod 5 under addition. Enumerate the elements as $\{[1], [2], [3], [4], [5]\}$. Find the permutations in S_5 coming from the proof of Cayley's theorem. For example, $\sigma_3([x]) = [3 + x \pmod 5]$. Then $\sigma_3 = \begin{pmatrix} 1 & 2 & 3 & 4 & 5 \\ 4 & 5 & 1 & 2 & 3 \end{pmatrix} =$ (14253). Note that the fact that we got a 5-cycle is not a shock as the element in S_5 must have the same order as that in $\mathbb{Z}_5$ to which it corresponds (by part (3) of Proposition 3.2.1).*

Exercise 3.2.10 *Show that the group S_3 is isomorphic to the following group of matrices under the operation of matrix multiplication as defined in formula (1.3) of Section 1.8 and the statement that follows it:*

$$G=\left\{\begin{pmatrix}1&0&0\\0&1&0\\0&0&1\end{pmatrix},\begin{pmatrix}1&0&0\\0&0&1\\0&1&0\end{pmatrix},\begin{pmatrix}0&1&0\\1&0&0\\0&0&1\end{pmatrix},\begin{pmatrix}0&0&1\\0&1&0\\1&0&0\end{pmatrix},\begin{pmatrix}0&1&0\\0&0&1\\1&0&0\end{pmatrix},\begin{pmatrix}0&0&1\\1&0&0\\0&1&0\end{pmatrix}\right\}.$$

Exercise 3.2.11 *Show that isomorphism between groups is an equivalence relation.*

Definition 3.2.2 *If G is a group and $T: G \to G$ is a group isomorphism, we say that T is a* **group automorphism**. *The set of all automorphisms of G is called* $\mathrm{Aut}(G)$ *and it forms a group under composition of functions.*

Definition 3.2.3 *Take a fixed element $g \in G$. Define the* **conjugation function** $C_g(x) = g^{-1}xg$ *for all $x \in G$. If $y = g^{-1}xg$ for some $g \in G$, we say that x and y are* **conjugate elements** *of G. If we have two groups G and H such that $H = x^{-1}Gx$, for some x in a group containing G and H, we say that G and H are* **conjugate groups**.

Exercise 3.2.12 *Show that conjugation is a group automorphism.*

Of course, if the group is Abelian, conjugation is not very interesting because it is the identity map taking x to x. We call the automorphism C_g an **inner automorphism**. The rest of the automorphisms of G are called **outer automorphisms**.

Note that this idea of conjugate is quite different from that of complex conjugate. That is a horse of a different color which goes under the heading of field conjugation – really an element of a Galois group, that of the field automorphisms of $\mathbb{C}$ fixing elements of $\mathbb{R}$. Such things will be discussed in Section 7.6 for finite fields.

It is often useful to glomp together conjugate elements of a group. This leads to the following definition.

Definition 3.2.4 *The* **conjugacy class** *of an element x of the group G is $\{x\} = \{g^{-1}xg \mid g \in G\}$.*

Exercise 3.2.13 *Show that there is an equivalence relation on a group G obtained by saying $x, y \in G$ are equivalent iff x and y are conjugate. Then show that the equivalence classes for this equivalence relation are the conjugacy classes.*

Exercise 3.2.14 *Find all the conjugacy classes in A_4.*

Hint. *Look at cycles and note that $\sigma(a_1 a_2 \cdots a_n)\sigma^{-1} = (\sigma(a_1)\sigma(a_2)\cdots\sigma(a_n))$, for any permutation $\sigma \in S_n$.*

Exercise 3.2.15 *Find all the conjugacy classes in $GL(2,\mathbb{C})$, the group of 2×2 complex matrices with nonzero determinant. Conjugate matrices were well studied in your linear algebra book. They correspond to matrices that give the same linear mapping of $\mathbb{C}^2$ but with respect to different bases of $\mathbb{C}^2$. Such matrices are called "similar" in linear algebra. The rational canonical form of a matrix (as well as the Jordan form) represent the classes of similar matrices. See Section 7.2, Dornhoff and Hohn [25], or Strang [115, Chapter 5].*

Exercise 3.2.16 *Find four subgroups of the symmetric group S_4 that are isomorphic to S_3.*

Exercise 3.2.17 *State whether each of the following statements is true or false and give a brief reason for your answer.*

(a) The groups S_4 and D_{12} are isomorphic.
(b) Define $f: \mathbb{Z} \to \mathbb{Z}$ by $f(x) = 2x$, $\forall x \in \mathbb{Z}$. Consider $\mathbb{Z}$ to be a group under addition. Then f is a group isomorphism.
(c) Consider the integers $\mathbb{Z}$ and the rationals $\mathbb{Q}$ as groups under addition. These groups are isomorphic.
(d) All infinite groups are isomorphic.

Exercise 3.2.18 *Show that the group $\mathrm{Aut}(\mathbb{Z}_n)$ of automorphisms of $\mathbb{Z}_n$ under addition is isomorphic to the group $\mathbb{Z}_n^*$. This is a case in which all but the trivial automorphism are outer automorphisms.*

Hint. *Consider the map sending $\sigma \in \mathrm{Aut}(\mathbb{Z}_n)$ to $\sigma(1)$.*

Exercise 3.2.19 *Show that the group $\mathbb{R}$ of real numbers under addition is isomorphic to the group $\mathbb{R}^+$ of positive real numbers under multiplication.*

Hint. *Make use of e^x and $\log x$.*

Exercise 3.2.20 *Which of the following groups are isomorphic?*

(a) the group of symmetries of a square;
(b) the group of symmetries of a rectangle;
(c) the multiplicative group $\mathbb{Z}_{12}^*$;
(d) $\mathbb{Z}_4$ under addition;
(e) the Klein 4-group from Exercise 2.2.1.

3.3 Cosets, Lagrange's Theorem, and Normal Subgroups

In this section we want to generalize the concept of $\mathbb{Z}_n$ – the integers mod n. The elements of $\mathbb{Z}_n$ are sets of the form $a + n\mathbb{Z}$. We call such sets cosets. Since the group $\mathbb{Z}$ is Abelian it does not matter whether a is on the right or left of $n\mathbb{Z}$. When a group is not Abelian, it does matter. In any case, the point is that we want to view the coset as an object in its own right by equating everything in a given coset, just as we did for $\mathbb{Z}_n$.

Definition 3.3.1 *If H is a subgroup of the group G, we say that a* **left coset** *of H is a set of the form $gH = \{gh \mid h \in H\}$. One can similarly define a* **right coset** *of H to be a set Hg. We denote by G/H the set of all left cosets and $H\backslash G$ the set of right cosets.*

The left coset gH in the definition is the image of H under the left multiplication map $L_g : G \to G$ defined by $L_g x = gx$, for all $x \in G$. See Definition 2.3.2. This is why we call gH a "left" coset and not a "right" coset.

Example. Take the group $G = \mathbb{Z}$ under addition and the subgroup $H = n\mathbb{Z} = \{nq | q \in \mathbb{Z}\}$. We view the elements of $\mathbb{Z}_n$ as cosets:

$$[a] = a + n\mathbb{Z} = \{a + nq \mid q \in \mathbb{Z}\}.$$

As we said, because this particular group is Abelian, left cosets are the same as right cosets. ▲

Proposition 3.3.1 *(Facts About Cosets). Suppose H is a subgroup of the finite group G. Then we have the following facts. Fact (2) holds even if G and H are not finite.*

(1) $|H| = |Hg| = |gH|$ for all $g \in G$.

(2) We have an equivalence relation on G by defining $x \sim y$ iff $x^{-1}y \in H$. Then the equivalence classes are the left cosets which partition G into a disjoint union of left cosets.

(3) $|G/H| = \dfrac{|G|}{|H|}$.

Proof.

(1) The left multiplication map from Definition 2.3.2 is $L_g(x) = gx$, for all $x \in G$. This map takes H 1–1 onto the left coset gH. Thus $|H| = |gH|$.

(2) If $x^{-1}y = h \in H$, then $y = xh \in xH$, and conversely. To see that this relation is an equivalence relation, we just need to prove that it is reflexive, symmetric, and transitive. The proofs are as follows.

   (a) Reflexive $x \sim x$: $x^{-1}x = e \in H$.

   (b) Symmetric $x \sim y \Longrightarrow y \sim x$: $x^{-1}y = h \in H$ implies $h^{-1} = y^{-1}x \in H$.

   (c) Transitive $x \sim y$ and $y \sim z \Longrightarrow x \sim z$: $x^{-1}y = h \in H$ and $y^{-1}z = k \in H$ implies $hk = x^{-1}yy^{-1}z = x^{-1}z \in H$.

   It follows that equivalence classes for this equivalence relation are the left cosets.

(3) This follows from (1) and (2) since G is a disjoint union of $|G/H|$ left cosets each of which has the same number $|H|$ of elements. Thus $|G| = |G/H|\,|H|$. ▲

Example 1. Take the group $G = \mathbb{Z}_6$ under addition along with the subgroup $H = \langle 2 \pmod 6 \rangle = \{2, 4, 0 \pmod 6\}$. The cosets of H are H and $1 + H = \{1, 3, 5 \pmod 6\}$. We see that each coset has three elements, the cosets are disjoint, and every element of G is in some coset, just as Proposition 3.3.1 predicted. ▲

Example 2. Let $G = S_3$ and $H = \langle (123) \rangle = \{(1), (123), (132)\}$. A computation shows that the coset $(12)H = \{(12), (23), (13)\}$. Thus G is a disjoint union of H and $(12)H$. This is just a special case of part (2) of the preceding proposition. ▲

The preceding facts about cosets lead quickly to Lagrange's theorem.

Theorem 3.3.1 (*Lagrange's Theorem*). *Suppose that H is a subgroup of a finite group G. Then $|H|$ divides $|G|$.*

Proof. This follows from part (3) of the preceding proposition. ▲

Instead of calling this Lagrange's theorem, we should perhaps have called it Lagrange's corollary – given the length of the proof. However, it is important – especially for Section 4.5 – and I guess that justifies calling it a theorem.

Corollary 3.3.1

(1) If G is a finite group, the order of any element of G must divide the order of G.
(2) Any group G of prime order is cyclic.
*(3) **(Fermat's Little Theorem).** If $a \in \mathbb{Z}_p$, where p is prime, then $a^p \equiv a \pmod{p}$.*

Proof.

(1) We know that the order of the cyclic group generated by a, namely $\langle a \rangle$, is the same as the order of the element a of G by fact 2 in Section 2.5. The result thus follows from Lagrange's theorem.
(2) A prime p has no positive divisors but p and 1. Thus from part (1) if $|G| = p$, then the order of any element $a \neq e$ of G must be p. So $\langle a \rangle = G$.
(3) If $a \equiv 0 \pmod{p}$ the result is clear. Otherwise $a \pmod{p}$ is in the group $\mathbb{Z}_p^*$ which has order $p-1$. This implies (by part (1)) that $a^{p-1} \equiv 1 \pmod{p}$. Multiply the congruence by a to obtain the result. ▲

Exercise 3.3.1 *Find all the subgroups of the multiplicative groups $\mathbb{Z}_7^*, \mathbb{Z}_8^*, \mathbb{Z}_9^*, \mathbb{Z}_{10}^*$.*

Exercise 3.3.2 *Find all left cosets of the subgroup $H = \{1, 11 \pmod{20}\}$ in the multiplicative group $\mathbb{Z}_{20}^*$.*

Exercise 3.3.3 *List all the subgroups of the tetrahedral group A_4. Then draw the poset diagram for the subgroups of A_4. Note that A_4 has no subgroup of order 6 and thus that the converse of Lagrange's theorem is false.*

Hint. *Group Explorer will do this exercise for you but you should explain why your list is complete. The program will even arrange the multiplication table according to cosets of a subgroup.*

Definition 3.3.2 *A subgroup H of the group G is called a **normal subgroup** iff $g^{-1}Hg = H$ for all $g \in G$. The equality $g^{-1}Hg = H$ means $H = \{g^{-1}hg \mid h \in H\}$.*

It is clear that every subgroup of an Abelian group is normal. In the next section we shall see why normal subgroups are nice – they are the only type of subgroups that allow us to make the set of cosets gH into a group G/H using the same method that worked for $\mathbb{Z}_n$.

The use of the word "normal" is not intended to imply that most subgroups are normal or that such normal subgroups are particularly ordinary. As usual in mathematics, the only meaning the word "normal" has is contained in its definition, Definition 3.3.2 here.

Exercise 3.3.4 *Show that a subgroup H of G is normal iff $gH = Hg$ for all $g \in G$, that is, iff every right coset is a left coset.*

Exercise 3.3.5 *Find all normal subgroups of S_3.*

Exercise 3.3.6 *Are the following statements true or false?*

(a) $\mathbb{Z}_7$ is a normal subgroup of $\mathbb{Z}_{14}$.
(b) $\langle F \rangle$ is a normal subgroup of D_3, using the notation of Section 2.1.
(c) $\langle R \rangle$ is a normal subgroup of D_3, using the notation of Section 2.1.

Exercise 3.3.7 *Show that if H is a subgroup of group G and $|G/H| = 2$, then H is a normal subgroup of G. This shows that the alternating group A_n is a normal subgroup of the symmetric group S_n.*

Example. Consider the multiplication table for the dihedral group D_3 arranged according to left cosets of the subgroup $\langle R \rangle$ – using the notation of Section 2.1. Recall that $RF = FR^2$.

Multiplication table for D_3

$\cdot$	I	R	R^2	F	FR	FR^2
I	I	R	R^2	F	FR	FR^2
R	R	R^2	I	FR^2	F	FR
R^2	R^2	I	R	FR	FR^2	F
F	F	FR	FR^2	I	R	R^2
FR	FR	FR^2	F	R^2	I	R
FR^2	FR^2	F	FR	R	R^2	I

There are graphs associated to the cosets G/H and a generating set S of G. These are called **Schreier graphs.** The vertices are the cosets gH, $g \in G$. Draw an arrow from coset gH to coset sgH for each $s \in S$. There are many examples of these graphs in Terras [116] or Terras [117]. Schreier graphs can have loops and multiple edges. ▲

Example. Let $G = K_4$, the Klein 4-group. Recall the multiplication table in Figure 2.14. Let $H = \{e, h\}$. Find G/H and draw the corresponding Schreier graph for the edge set $S = \{h, v\}$. To create the Schreier graph you take the Cayley graph $X(K_4, S)$ and glomp together vertices which are in the same H-coset. These graphs will be undirected since $x \in S$ implies $x^{-1} \in S$. See Figure 3.2, which shows the Cayley graph $X(K_4, S)$ on the left and the Schreier graph for G/H on the right. ▲

Exercise 3.3.8 *Show that an element g of the group G defines a function $L_g(xH) = gxH$, for all $xH \in G/H$. Then show that the function $L_g : G/H \to G/H$ is 1–1 and onto. Moreover*

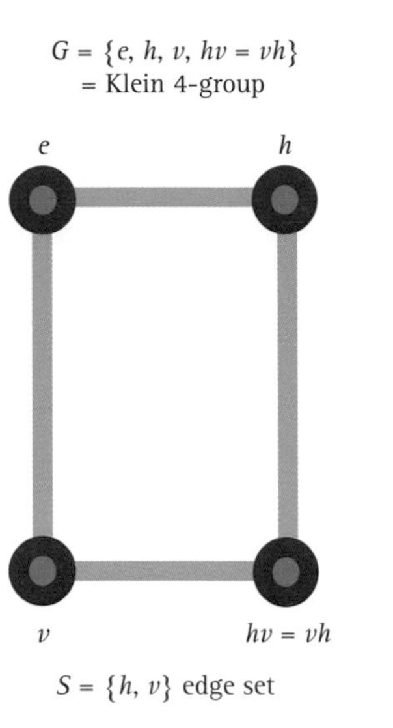

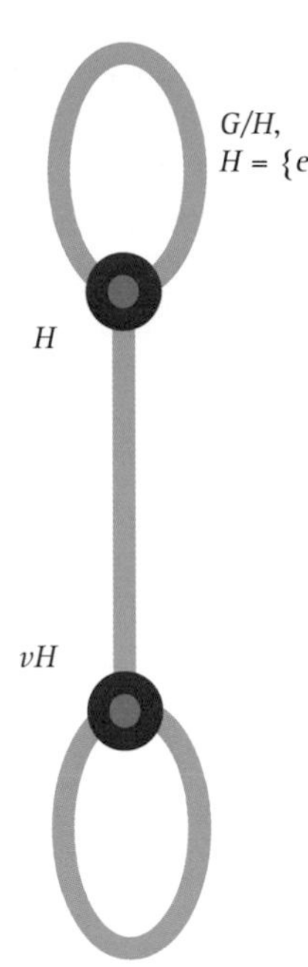

Figure 3.2 On the left is the Cayley graph for the Klein 4-group $K_4=\{e,h,v,hv\}$, with generating set $S=\{h,v\}$ using the notation of Figure 2.14. On the right is the Schreier graph for K_4/H, where $H=\{e,h\}$, with the same set $S=\{h,v\}$

*show that $L_{gg'} = L_g \circ L_{g'}$. We view L_g as defining a **group action** of the group G on G/H – a concept we will consider in Section 3.7.*

Exercise 3.3.9 *Draw the Schreier graph for the quotient G/H with*

$$G=\mathrm{Aff}(5)=\left\{\left|\begin{pmatrix} a & b \\ 0 & 1\end{pmatrix} a,b\in\mathbb{Z}_5,\ a\neq 0\right.\right\},$$

$$H=\left\{\left|\begin{pmatrix} 1 & b \\ 0 & 1\end{pmatrix} b\in\mathbb{Z}_5\right.\right\},\ \text{and}\ S=\left\{\begin{pmatrix} 2 & 0 \\ 0 & 1\end{pmatrix},\begin{pmatrix} -1 & 1 \\ 0 & 1\end{pmatrix}\right\}.$$

Exercise 3.3.10 *Just as in Example 3 of Section 2.3, consider the group G of non-singular 3×3 matrices with entries in the integers mod 2. Draw the Schreier graph for the quotient G/H with $G=GL(3,\mathbb{Z}_2)$,*

$$H=\text{the subgroup of matrices of the form}\begin{pmatrix} 1 & * & * \\ 0 & * & * \\ 0 & * & *\end{pmatrix},$$

*where * means the entry can be either 0 or 1, and generating set*

$$S=\left\{\begin{pmatrix} 0 & 1 & 1 \\ 0 & 1 & 0 \\ 1 & 0 & 0\end{pmatrix},\begin{pmatrix} 1 & 0 & 0 \\ 0 & 0 & 1 \\ 0 & 1 & 1\end{pmatrix}\right\}.$$

*Change the subgroup H to the subgroup consisting of transposes of the matrices in H. The **transpose** of a matrix $M=(m_{ij})$ is defined to be the matrix ${}^{T}M=(m_{ji})$. This produces a Schreier graph whose adjacency matrix has the same eigenvalues as the original. Recall that the **eigenvalues** $\lambda\in\mathbb{C}$ of matrix A are the solutions of $\det(A-\lambda I)=0$. But the two graphs are not isomorphic graphs (meaning not related by a 1–1,onto map between vertices preserving adjacency of vertices). See Terras [116] or Terras [117].*

The graphs in the preceding exercise helped to answer a question posed by Kac [51] – "Can you hear the shape of a drum?" This means can you distinguish two drums by their sounds (i.e., fundamental frequencies of vibration). Carolyn Gordon, David Webb, and Scott Wolpert used the same basic construction in 1992 to show that there are (non-convex) planar drums that cannot be heard. This means that the fundamental frequencies of vibration – eigenvalues of the Laplace operator – are the same for the two drums, although one drum cannot be turned into the other by a rigid motion.

It is also possible to think about **double cosets** of G modulo two subgroups H and K. These are sets $HgK=\{hgk|h\in H, k\in K\}$. The set of all such double cosets is denoted $H\backslash G/K$. The idea finds many applications in number theory and group representation theory. Once more the double cosets partition G into a disjoint union, but the double cosets may have different sizes. Double cosets are lurking behind Figure 0.4 in the preface.

Example. Let $G=S_3$, $H=\langle(12)\rangle, K=\langle(13)\rangle$. Find the double cosets $H\backslash G/K$. Then one double coset is $HK=\{(1),(12),(13),(12)(13)=(132)\}$, while the other double coset is $\{(23),(12)(23)=(123)\}$. ▲

Exercise 3.3.11 *Show that given a group G and two subgroups H and K, we can define an equivalence relation on G by saying $x, y\in G$ are equivalent and writing $y\sim x$ iff there are elements $h\in H, k\in K$ such that $y=hxk$. What are the equivalence classes?*

Exercise 3.3.12 *Draw the Schreier graph for $G=S_3$ and $H=\langle(12)\rangle$. Then do the same for $G=S_4$, $H=\langle(12)\rangle$.*

Exercise 3.3.13 *Find a normal subgroup of the affine group* Aff(7)*:*

$$\text{Aff}(7)=\left\{\begin{pmatrix} a & b \\ 0 & 1\end{pmatrix}\middle| a,b\in\mathbb{Z}_7,\ a\neq 0\right\}.$$

The operation is matrix multiplication.

3.4 Building New Groups from Old, I: Quotient or Factor Groups *G*/*H*

We want to imitate the construction of the integers modulo n. Suppose that H is a subgroup of the group G. Recall that a left coset gH is the set of all gh, for $h\in H$. Take $G=\mathbb{Z}$ under addition and the subgroup $H=n\mathbb{Z}$, consisting of all multiples of n. Then we have already made use of the cosets:

$$[0]=H,\quad [1]=1+H,\quad \ldots,\quad [n-1]=(n-1)+H.$$

These are the elements of $\mathbb{Z}_n$. We were able to add integers mod n by writing $[a]+[b]=[a+b]$. We had to show that this makes sense: that is,

$$[a]=[a'],[b]=[b'] \text{ implies } [a+b]=[a'+b'].$$

It is not possible to do the analog for *any* subgroup H of a group G. We will need to restrict our consideration to normal subgroups H of G – as in Definition 3.3.2. Abnormal subgroups need not apply for this construction job. Of course every subgroup of an Abelian group is normal. So let us consider a non-Abelian example.

Example. Consider the group S_n of permutations of n objects. Which subgroups H of S_n are normal? To answer this question, consider the following equation for an arbitrary cycle $(a_1 a_2 \cdots a_r)$ of length r:

$$\sigma \circ (a_1 a_2 \cdots a_r) \circ \sigma^{-1} = (\sigma(a_1)\, \sigma(a_2) \cdots \sigma(a_r)), \quad \forall \sigma \in S_n.$$

The equation is valid since the permutations only affect the elements of $\{1, 2, \ldots, n\}$ of the form $\sigma(a_j)$ for some $j = 1, 2, \ldots, r$. The element $\sigma(a_j)$ is sent to $\sigma(a_{j+1})$ by both sides for $j = 1, \ldots, r-1$. The element $\sigma(a_r)$ is sent to $\sigma(a_1)$.

This shows that if H is a normal subgroup of S_n and contains a cycle of length r, then H must contain all other cycles of length r. This means that the only normal subgroups of S_3 are the trivial subgroup $\{(1)\}$, the cyclic subgroup of order 3: $\langle (123) \rangle = \{(1), (123), (132)\}$, and S_3 itself. ▲

Exercise 3.4.1 *Find all the normal subgroups of* D_4.

Exercise 3.4.2 *Find all the normal subgroups of* A_4.

Exercise 3.4.3 *Let our group be the affine group*

$$G = \mathrm{Aff}(5) = \left\{ \begin{pmatrix} a & b \\ 0 & 1 \end{pmatrix} \middle| \, a, b \in \mathbb{Z}_5, \ a \neq 0 \right\}.$$

The operation is matrix multiplication. Find a normal subgroup of order 5 *and then another normal subgroup of order* 10. *Prove that the two subgroups are indeed normal.*

Now we need to multiply the cosets of a normal subgroup. We imitate what worked for the subgroup $n\mathbb{Z}$ of $\mathbb{Z}$ in Sections 1.6 and 2.3.

Definition 3.4.1 *If H is a normal subgroup of the group G, define the* **quotient (or factor) group** *G/H to be the set G/H of left cosets with multiplication $(aH)(bH) = abH$, $\forall a, b \in G$.*

Theorem 3.4.1 *If G is a group with normal subgroup H, then using the Definition 3.4.1, G/H is a group.*

Proof. Defining the multiplication of subsets S and T of G by:

$$ST = \{xy \mid x \in S, y \in T\}, \tag{3.2}$$

we can consider $(aH)(bH)$ as the product of two cosets. This is of course independent of the representatives a and b which were chosen. Then since H is normal, $(aH)(bH) = a(Hb)H = abHH = abH$, since $HH = H$. To see that G/H is a group, we need to find an identity element. That identity is H since $aHH = aH$ and $HaH = aHH = aH$. For any subset $S \subset G$, define $S^{-1} = \{x^{-1} | x \in S\}$. Then $(aH)^{-1} = H^{-1}a^{-1} = Ha^{-1} = a^{-1}H$ and $(a^{-1}H)(aH) = (Ha^{-1})(aH) = H$. The associative law is hopefully clear. ▲

Example. Recall the multiplication table of the example in the preceding section with $G = D_3$, $H = \langle R \rangle$, using the notation of Section 2.1.

Multiplication table for D_3

$\cdot$	I	R	R^2	F	FR	FR^2
I	I	R	R^2	F	FR	FR^2
R	R	R^2	I	FR^2	F	FR
R^2	R^2	I	R	FR	FR^2	F
F	F	FR	FR^2	I	R	R^2
FR	FR	FR^2	F	R^2	I	R
FR^2	FR^2	F	FR	R	R^2	I

Then we see that, since that multiplication table glomps the group elements according to cosets, the group $D_3/\langle R \rangle$ has the standard multiplication table for a group with two elements. The only coset is $F\langle R \rangle$. So we get the table shown. Of course you can see this easily directly since $F\langle R \rangle = \langle R \rangle F$, and $F^2 = I$.

Multiplication table for $D_3/\langle R \rangle$

$\cdot$	$I\langle R \rangle$	$F\langle R \rangle$
$I\langle R \rangle$	$I\langle R \rangle$	$F\langle R \rangle$
$F\langle R \rangle$	$F\langle R \rangle$	$I\langle R \rangle$

▲

Exercise 3.4.4 *Imitate the computations of the preceding example for* $\mathrm{Aff}(5)/H$, *where H denotes the normal subgroup of order 5 found in Exercise 3.4.3.*

Exercise 3.4.5 *Imitate the computations of the preceding example for $D_4/\langle R^2 \rangle$. Can you identify the group $D_4/\langle R^2 \rangle$ with one of the groups considered earlier?*

Exercise 3.4.6 *Suppose that H is a normal subgroup of G. Using the definition of set multiplication in equation (3.2), prove the following equalities which were used in our proof of Theorem 3.4.1.*

(a) $HH = H$.
(b) $(aH)(bH) = a(Hb)H$.
(c) $H^{-1} = \{h^{-1} | h \in H\} = H$.

Exercise 3.4.7 *Show that the center $Z(G)$ is a normal subgroup of the group G.*

Exercise 3.4.8 *Show that the intersection of two normal subgroups of G is a normal subgroup of G.*

Definition 3.4.2 *A* ***simple group*** *G is such that the only normal subgroups H of G are $H = \{e\}$ and $H = G$.*

It turns out that the alternating groups A_n are simple for $n \geq 5$. Many people have written thousands of pages on the **classification of finite simple groups**. Many finite simple groups are in infinite families of groups like A_n. Others are the so-called **sporadic groups**. The largest of those is called the monster group – with order 808 017 424 794 512 875 886 459 904 961 710 757 005 754 368 000 000 000. The choice of the word "simple" here is misleading, as cyclic groups with any non-trivial subgroups are certainly not simple groups, yet in many ways they are simpler than simple groups. For more information on the classification of finite simple groups see Wikipedia or Wilson [127].

The fact that A_5 is simple, plus a little Galois theory, implies that it is impossible to solve the general quintic equation over $\mathbb{Q}$ by repeated radicals. This can of course be done for quadratic equations – recalling the quadratic formula from junior high school or middle school. It can also be done for cubic and quartic equations. But once you have an equation involving a polynomial of degree 5 over the rationals, if you can find that its Galois group over $\mathbb{Q}$ is S_5 or A_5, then you know that formulas generalizing the quadratic formula involving higher order radicals than the square root will not suffice to find formulas for the roots of the polynomial. See Dummit and Foote [28, Chapter 14] or Gallian [33, Chapter 32] for more information on this subject. Dummit and Foote give the example of the quintic $x^5 - x + 1$ over $\mathbb{Q}$. We will not touch this subject since we will consider Galois theory only for finite fields in Section 7.6.

Exercise 3.4.9 *Show that a quotient G/H is cyclic if G is cyclic.*

Exercise 3.4.10 *Which Abelian groups are simple?*

Exercise 3.4.11 *If G is a group of order 15, show that G has an element of order 3.*

Hint. *There is no problem if G is cyclic – thanks to theorems in Section 2.5. What does Lagrange tell you about possible orders of elements of G?*

We will have more to say about groups of order ≤ 15 in Section 4.5.

Suppose you are given a group G and you hate G because it is not Abelian. You might want to make an Abelian group related to G. To do this one forms the **commutator subgroup** G', which is the subgroup of G generated by elements of the form $aba^{-1}b^{-1}$, for $a, b \in G$.

Exercise 3.4.12 *Show that the commutator subgroup G' is a normal subgroup of G and then show that the group G/G' is Abelian.*

Exercise 3.4.13 *Suppose that H is a subgroup of G but H is not a normal subgroup of G. Show that the multiplication of cosets given by $aHbH = abH$, for $a, b \in G$, does not turn G/H into a group.*

Exercise 3.4.14 *Suppose that H is a normal subgroup of the finite group G. If G/H has an element of order n, show that G has an element of order n.*

Hint. *First show that the order r of the coset gH in G/H divides the order of $g \in G$. Then look at g^r.*

Exercise 3.4.15 *Show that A_4 is not a simple group.*

3.5 Group Homomorphism

We have already defined group isomorphisms. Homomorphisms are similar except they need not be 1–1 and onto. The idea comes from C. Jordan in 1870. It is an extremely useful idea. Once you have seen the definition, you will find group homomorphisms all over this book.

Definition 3.5.1 *If G and G' are groups, a function $T: G \to G'$ is a* ***group homomorphism*** *iff $T(xy) = T(x)T(y)$ for all $x, y \in G$.*

Example 1. Let $G = \mathbb{Z}$ under addition and $G' = \mathbb{Z}_n$ under addition mod n for $n \geq 1$. Define $T(x) = x \pmod n$. Then $T: \mathbb{Z} \to \mathbb{Z}_n$ is a group homomorphism which is not an isomorphism. ▲

Example 2. Let $G = S_n$, for $n \geq 2$ and consider $\{\pm 1\}$ as a group under multiplication. The map sgn: $S_n \to \{\pm 1\}$ defined by equation (3.1) is a group homomorphism by Proposition 3.1.2. It cannot be 1–1 unless $n = 2$. ▲

Example 3. Let $\mathbb{R}$ be the group of real numbers under addition and let $\mathbb{C}$ be the group of complex numbers under addition: that is, $\mathbb{C} = \{x + iy \mid x, y \in \mathbb{R}\}$, where $i = \sqrt{-1}$. If $x + iy$ and $u + iv \in \mathbb{C}$, with $x, y, u, v \in \mathbb{R}$, we define the sum by

$$(x + iy) + (u + iv) = (x + u) + i(y + v).$$

We have a group homomorphism $T: \mathbb{R} \to \mathbb{C}$ defined by $T(x) = x$. It is 1–1 but not onto as $i \notin \mathbb{R}$. ▲

Exercise 3.5.1 *Show that the familiar exponential function $\phi: \mathbb{R} \to \mathbb{R}^+$ defined by setting $\phi(x) = e^x$, for every real number x, is a group homomorphism if we consider $\mathbb{R}$ as a group under addition and $\mathbb{R}^+$ as a group under multiplication. Is ϕ an isomorphism?*

Now we define an important subgroup associated to a group homomorphism.

Definition 3.5.2 *Suppose G, G' are groups and $T: G \to G'$ is a group homomorphism. The* ***kernel*** *of T is* $\ker T = \{g \in G | T(g) = e'\}$, *where e' is the identity of G'.*

What are the kernels of the homomorphisms in the three examples above?

Example 1. The kernel is $n\mathbb{Z}$. ▲

Example 2. The kernel is the alternating group A_n. ▲

Example 3. The kernel is $\{0\}$. ▲

The following lemma should be familiar from Proposition 3.2.1. The lack of injectivity (or 1–1 ness) does not hamper us in proving these basic facts.

Lemma 3.5.1 *Suppose G, G' are groups and $T: G \to G'$ is a group homomorphism. If e is the identity of G and e' is the identity of G', then $T(e) = e'$ and $T(x^{-1}) = T(x)^{-1}$.*

Proof. First note that $T(e)T(x) = T(ex) = T(x) = e'T(x)$ implies by the cancellation law that $T(e) = e'$.

Second note that $T(x^{-1})T(x) = T(x^{-1}x) = T(e) = e'$. Similarly $T(x)T(x^{-1}) = e'$. ▲

Lemma 3.5.2 *The group homomorphism $T: G \to G'$ is an isomorphism iff it is onto and the kernel is the trivial subgroup; that is,* ker $T = \{e\}$, *where e is the identity of G.*

Proof. Using Lemma 3.5.1, we need to show that T is 1–1 iff ker $T = \{e\}$. Suppose T is 1–1. Then clearly ker $T = \{e\}$. Conversely, suppose ker $T = \{e\}$. Then $Tx = Ty$ implies $T(xy^{-1}) = T(x)T(y)^{-1} = e'$, the identity of G'. But then $xy^{-1} \in \ker T = \{e\}$ and $xy^{-1} = e$, which implies $x = y$. Thus T is 1–1. ▲

Exercise 3.5.2 *Suppose that G, G' are finite groups and $T: G \to G'$ is a group homomorphism. Show that the order $|T(g)|$ divides the order $|g|$, for all $g \in G$.*

Example 1. Let $G = \mathbb{Z}$ under addition and $G' = \mathbb{Z}_n$ under addition mod n. Define $T(x) = x \pmod{n}$. The kernel of T is the additive subgroup $n\mathbb{Z}$. ▲

Example 2. Consider the group $\mathbb{R}^2$ of vectors $\begin{pmatrix} x \\ y \end{pmatrix}$, for x, y real. This is a group under componentwise addition: $\begin{pmatrix} x \\ y \end{pmatrix} + \begin{pmatrix} u \\ v \end{pmatrix} = \begin{pmatrix} x+u \\ y+v \end{pmatrix}$. Given a 2×2 matrix $M = \begin{pmatrix} a & b \\ c & d \end{pmatrix}$, we have a group homomorphism $L_M: \mathbb{R}^2 \to \mathbb{R}^2$ defined by

$$L_M \begin{pmatrix} x \\ y \end{pmatrix} = \begin{pmatrix} a & b \\ c & d \end{pmatrix} \begin{pmatrix} x \\ y \end{pmatrix} = \begin{pmatrix} ax+by \\ cx+dy \end{pmatrix}.$$

The kernel of L_M is called the nullspace of the matrix M (or the linear function L_M) in linear algebra. See Definition 7.2.3. ▲

Exercise 3.5.3 *Show that L_M in example 2 is indeed a group homomorphism. What is its kernel?*

Lemma 3.5.3 *If G, G' are groups and $T: G \to G'$ is a group homomorphism, then* ker T *is a normal subgroup of G.*

Proof. First let us show that ker T is a subgroup of G. We use the one-step subgroup test from Proposition 2.4.3. Suppose that $x, y \in \ker T$. This means that if e' is the identity of G', $T(x) = T(y) = e'$. But then $T(xy^{-1}) = T(x)T(y)^{-1} = e'$. So $xy^{-1} \in \ker T$.

So now we need only show that $a^{-1}(\ker T)a = \ker T$, for all $a \in G$. If $x \in \ker T$, look at $T(a^{-1}xa) = T(a)^{-1}T(x)T(a) = T(a)^{-1}e'T(a) = e'$. This means that $a^{-1}(\ker T)a \subset \ker T$. Multiply the set-theoretic equality by a on the left and a^{-1} on the right to get $\ker T \subset a(\ker T)a^{-1}$. Why is this legal? Then replace a by a^{-1} to see that $\ker T \subset a^{-1}(\ker T)a$. This finishes the proof that $a^{-1}(\ker T)a = \ker T$. ▲

Exercise 3.5.4 *Answer the "why?" in the middle of the second paragraph of the proof of the last lemma.*

Example 3. We define a homomorphism f taking the additive group $\mathbb{R}$ of real numbers to the multiplicative group $\mathbb{T}$ of complex numbers of absolute value 1, $\mathbb{T}=\{z\in\mathbb{C} \mid |z|=1\}$. The homomorphism is defined by $f:\mathbb{R}\to\mathbb{T}$, $f(x)=e^{2\pi ix}=\cos(2\pi x)+i\sin(2\pi x)$ **(Euler's identity)**. Here $i=\sqrt{-1}$. Note that $f(x+y)=f(x)f(y)$. **Question:** What is $\ker f$? **Answer:** $\ker f=\mathbb{Z}$. You see this from the diagram showing the location of the point $e^{2\pi ix}=\cos(2\pi x)+i\sin(2\pi x)$ in the plane. What are the polar coordinates of this point? The radius is 1 and the angle is $2\pi x$. ▲

Example 4. Suppose that $\mathcal{F}=\{F:\mathbb{R}\to\mathbb{R}\}$ and

$$\mathcal{D}=\{F:\mathbb{R}\to\mathbb{R}|f \text{ is everywhere differentiable}\}.$$

Both $\mathcal{F}$ and $\mathcal{D}$ are groups under pointwise addition – defined by $(f+g)(x)=f(x)+g(x)$, $\forall x\in\mathbb{R}$. Define $T:\mathcal{D}\to\mathcal{F}$ by $Tf(x)=f'(x)=\frac{df}{dx}$. The mapping T is a group homomorphism since the derivative of a sum is the sum of the derivatives (as shown in calculus or advanced calculus since calculus classes seem to be proofless). **Question:** What is the kernel of T? **Answer:** the functions on the real line with 0 derivative everywhere, that is, the constant functions (using the mean value theorem). ▲

Exercise 3.5.5 *Show that the determinant gives a group homomorphism* $\det: GL(n,\mathbb{R})\to\mathbb{R}^*$, *where* $\mathbb{R}^*=\mathbb{R}-\{0\}$ *under multiplication. Here, as in Example 3 of Section 2.3,* $GL(n,\mathbb{R})$ *is the general linear group of non-singular or invertible* $n\times n$ *real matrices under matrix multiplication defined as in formula (1.3) of Section 1.8 and the statement that follows it. Note that we will consider determinants in Section 7.3. You should also check that the general linear group is indeed a group. What is the identity?*

The following theorem goes back to C. Jordan in 1870. Emmy Noether proved more general versions of this and two other isomorphism theorems in 1927. See Exercises 3.5.9 and 3.5.13.

First we need a definition.

Definition 3.5.3 *Suppose that* G, G' *are groups and* $f: G\to G'$ *is a group homomorphism. The* ***image group*** *of* G *under* f *is*

$$f(G)=\{f(x) \mid x\in G\}.$$

It is an exercise to show that $f(G)$ is a subgroup of G', under the hypotheses of the definition. See Exercise 3.5.7.

Theorem 3.5.1 ***(First Isomorphism Theorem).*** *Suppose that* G, G' *are groups and* $f: G\to G'$ *is a group homomorphism. Then the quotient group* $G/\ker f$ *is isomorphic to the image group* $f(G)$ *from Definition 3.5.3 under the mapping*

$$F: G/\ker f\to f(G),$$

defined by $F(g\ker f)=f(g)$, $\forall g\in G$.

Proof. For simplicity, let $K = \ker f$.

F **is well defined.** This means $aK = bK$ implies $F(aK) = F(bK)$. If $aK = bK$, then there is an $x \in K$ such that $b = ax$. Thus, using the definition of $K = \ker f$, $F(bK) = F(axK) = f(ax) = f(a)f(x) = f(a)e' = f(a) = F(aK)$. Here e' is the identity in G'.

F **preserves group operations.** This means $F[(aK)(bK)] = F(aK)F(bK)$. Using our definition of multiplication of cosets in G/K, the definition of F, and the fact that f is a group homomorphism, we have $F[(aK)(bK)] = F(abK) = f(ab) = f(a)f(b) = F(aK)F(bK)$.

F **is one-to-one.** This means $F(aK) = F(bK)$ implies $aK = bK$. By definition $F(aK) = F(bK)$ implies $f(a) = f(b)$, which implies (using the fact that f is a group homomorphism) that $f(a^{-1}b) = e'$, the identity in G'. Thus, by the definition of the kernel, $a^{-1}b \in K = \ker f$. So $a^{-1}b = x \in K$. It follows that $b = ax$ and therefore that $bK = axK = aK$, since $xK = K$ for any x in K, as K is a subgroup. Thus F is 1-1. ▲

Exercise 3.5.6 *Show that if H is a normal subgroup of G, then the map $\pi : G \to G/H$ defined by $\pi(g) = gH$ is an onto group homomorphism. This map is often called the* ***natural projection*** *of G onto G/H.*

Exercise 3.5.7 *Show that, under the assumptions of Definition 3.5.3, the image group $f(G)$ does indeed form a subgroup of G'.*

Example 1. Recall that $\mathbb{Z}_m$ is the factor group $\mathbb{Z}/m\mathbb{Z}$ under addition mod m. We have a homomorphism (an example of the natural projection in Exercise 3.5.6) $f : \mathbb{Z} \to \mathbb{Z}_m$, given by $f(x) = x + m\mathbb{Z}$, for all $x \in \mathbb{Z}$. What is $\ker f$? It is the subgroup $m\mathbb{Z}$. In this case the first isomorphism theorem just says $\mathbb{Z}/m\mathbb{Z} \cong \mathbb{Z}_m$. This is not anything new. Visually, to get $\mathbb{Z}_m$ out of $\mathbb{Z}$, you roll up the infinite discrete line of integers into a finite circle $\mathbb{Z}/m\mathbb{Z}$ by identifying integers if their difference is a multiple of m. See Figure 3.3. ▲

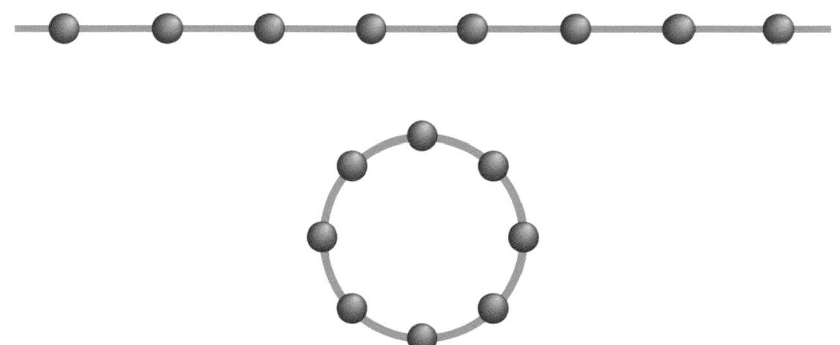

Figure 3.3 Roll up $\mathbb{Z}$ to get $\mathbb{Z}/n\mathbb{Z}$

Example 2. Consider the example again of a homomorphism taking the additive group $\mathbb{R}$ of real numbers to the multiplicative group $\mathbb{T}$ of complex numbers of absolute value 1, defined by

$$f(x) = e^{2\pi i x} = \cos(2\pi x) + i \sin(2\pi x).$$

Here $i = \sqrt{-1}$. We can identify $\mathbb{T}$ with the unit circle using the angle variable $2\pi x$. We saw that $\ker f = \mathbb{Z}$. It follows from the first homomorphism theorem that the additive quotient group $\mathbb{R}/\mathbb{Z}$ is isomorphic to the multiplicative group of complex numbers of absolute value 1: that is, $\mathbb{R}/\mathbb{Z} \cong \mathbb{T}$. This means that if you take the real line $\mathbb{R}$ mod the integers $\mathbb{Z}$, you roll $\mathbb{R}$ up into a circle. See Figure 3.4. ▲

Figure 3.4 Roll up the real line to get a circle $\mathbb{T} \cong \mathbb{R}/\mathbb{Z}$

Fourier series are representations of periodic functions or equivalently functions on the circle $\mathbb{R}/\mathbb{Z} \cong \mathbb{T}$. Jean-Baptiste Joseph Fourier invented the subject in the early 1800s while studying heat diffusion on a circular wire. Now when one uses computers for everything, the continuous circle should really be replaced by the finite circle $\mathbb{Z}/m\mathbb{Z}$. We will consider the finite Fourier transform in Section 4.2. See also my book [116, Part I]. The end result is that using group theory one can compute Fourier transforms fast enough to make them essential to modern signal processing. The most important algorithm in the subject is the fast Fourier transform or FFT (which was actually found by Gauss in 1805, while computing the orbit of the asteroid Juno). Another reference is Barry Cipra's article [16] discussing the top 10 algorithms of the twentieth century. One is the FFT. Fourier's work was really the beginning of applying group theory to practical problems. Mathematicians such as Laplace, Lagrange, and Legendre objected to his work as lacking in complete proofs. It took over a decade for Fourier's work on heat conduction to appear. Filling in gaps in the proofs would help to inspire the development of real analysis. The finite analog needed only the work of Gauss. But group theory itself was only an embryo at the time of Fourier.

Example 3. Suppose that $G = \langle a \rangle$ is a finite cyclic group of order n. Define the group homomorphism $f : \mathbb{Z} \to G$ by $f(k) = a^k, \forall k \in \mathbb{Z}$. It is easily checked that f is indeed a group homomorphism. What is the kernel of f?

$$\ker f = \{k \in \mathbb{Z} | a^k = a^0\} = \{nx | x \in \mathbb{Z}\} = n\mathbb{Z}.$$

Here we used some of the facts about powers in cyclic groups from Section 2.5. Thus, by the first isomorphism theorem, any cyclic group of order n is isomorphic to the group $\mathbb{Z}_n = \mathbb{Z}/n\mathbb{Z}$ (under addition mod n). ▲

Exercise 3.5.8 *State whether each of the following statements is true or false and give a brief explanation for your answer.*

(a) *Consider the integers $\mathbb{Z}$ as a group under addition and define the function $f : \mathbb{Z} \to \mathbb{Z}$ by $f(x) = x^3, \forall x \in \mathbb{Z}$. Then f is a group homomorphism.*

(b) Consider $\mathbb{Z}_3$ as a group under addition mod 3. *Define $f: \mathbb{Z}_3 \to \mathbb{Z}_3$ by $f(x) = x^3, \forall x \in \mathbb{Z}_3$. Then f is a group homomorphism.*

Exercise 3.5.9 ***(Third Isomorphism Theorem).*** *If H is a normal subgroup of K, and H and K are normal subgroups of G, then G/K is isomorphic to the group $(G/H)/(K/H)$.*

Hint. *Use the first isomorphism theorem starting out with a map $T: G/H \to G/K$, defined by $T(gH) = gK$, for $g \in G$.*

Exercise 3.5.10 *Consider the multiplicative group $\mathbb{R}^+$ of positive real numbers and the following functions f mapping $\mathbb{R}^+$ into $\mathbb{R}^+$. State whether each f is a group homomorphism. If so, find the image subgroup $f(\mathbb{R}^+)$ and kernel for each function f.*

(a) $f(x) = x^2$.
(b) $f(x) = 3x$.
(c) $f(x) = \sqrt{x}$.
(d) $f(x) = 1/x$.

Exercise 3.5.11 *Suppose that G is a finite group of prime order $|G| = p$ and $f: G \to G$ is a group homomorphism. If e is the identity of G, then either $f(x) = e$, for all $x \in G$, or f is a group isomorphism. Why? Is it true that in the second case $f(x) = x$, for all $x \in G$?*

Exercise 3.5.12 *Given an $m \times m$ real matrix A, define a function $E_A : \mathbb{R} \to \mathbb{R}^{m \times m}$ by*

$$E_A(t) = \sum_{k=0}^{\infty} \frac{t^n}{n!} A^n.$$

You may assume that this series converges for all real numbers t. Show that E_A is a group homomorphism from the additive group $\mathbb{R}$ to the multiplicative group $GL(m, \mathbb{R})$ of non-singular $m \times m$ real matrices. A non-singular matrix is a square matrix with nonzero determinant or, equivalently, a matrix with an inverse for multiplication – see Exercise 7.3.13. The image of E_A is called a ***one-parameter group****. The function is also named the matrix exponential* $\exp(tA)$. *It is useful in the solution of differential equations, and generalizations are important in the theory of Lie groups like $GL(n, \mathbb{R})$.*

Exercise 3.5.13 ***(Second Isomorphism Theorem).*** *Suppose that K and N are subgroups of the group G and that N is normal in G. Show that $K/N \cap K$ is isomorphic to KN/N.*

Hint. *Use the first isomorphism theorem applied to the map $T: K \to KN/N$ defined by $T(k) = kN$ for $k \in K$.*

Exercise 3.5.14 *Let G be a group and consider $\mathbb{Z}_n$ as a group under addition* $(\text{mod } n)$. *Show that any group homomorphism $T: \mathbb{Z}_n \to G$ is completely determined by the value $T(1 \text{ mod } n)$.*

Exercise 3.5.15 *For $n \geq 2$, consider $\mathbb{Z}_n$ as a group under addition* $(\text{mod } n)$ *and let $\mathbb{T}$ denote the multiplicative group of complex numbers of absolute value* 1. *Define the map $\tau: \mathbb{Z}_n \to \mathbb{T}$ by $\tau(x \ (\text{mod } n)) = e^{\frac{2\pi i x}{n}}$, for $x \in \mathbb{Z}$. Show that T is a 1–1 group homomorphism. Show that the image of τ is the group of nth roots of unity in $\mathbb{C}$, that is, all the complex roots of $x^n = 1$.*

3.6 Building New Groups from Old, II: Direct Product of Groups

In the following definition we are inspired by the group $\mathbb{R}^2$ consisting of vectors in the plane with addition defined by adding vectors in the usual way by adding componentwise.

Definition 3.6.1 *Given two groups G and H, we build a larger group consisting of ordered pairs called the* **direct product (or sum)** *of G and H denoted* $G \oplus H$ *(or* $G \times H$*), defined by* $G \oplus H = \{(g, h) | g \in G, h \in H\}$ *with the operation of componentwise multiplication:* $(g, h)(g', h') = (gg', hh')$.

It is pretty easily checked that the preceding definition makes $G \oplus H$ a group. The identity is (e, e'), where e is the identity of G, e' is the identity of G'. Inverses are given by $(g, h)^{-1} = (g^{-1}, h^{-1})$. The associative law follows from the associative laws for G and H.

Exercise 3.6.1 *Prove the claims in the preceding paragraph.*

Example 1. The usual **Euclidean plane,** $\mathbb{R}^2 = \mathbb{R} \oplus \mathbb{R}$, where we consider $\mathbb{R}$ to be a group under addition. Then $\mathbb{R}^2$ is a group under vector addition in the plane. Of course, it actually has more structure than that as it is a vector space. See Section 7.1. ▲

Example 2. The **Klein 4-group,** $\mathbb{Z}_2 \oplus \mathbb{Z}_2 \cong \mathbb{Z}_2^2$, which consists of ordered pairs of 0s and 1s. The elements are: $(0, 0), (0, 1), (1, 0), (1, 1)$. The operation is componentwise addition mod 2. Thus $(0, 1) + (1, 1) = (1, 0)$, for example. It is easily seen that each non-identity element has order 2. This group is isomorphic to $\mathbb{Z}_8^*$ as well as the group with multiplication table given in Figure 2.14. ▲

Similarly one can take direct products of any number of groups, just as happens in calculus, when you form $\mathbb{R}^n$ = n-dimensional Euclidean space, for $n \geq 3$. Of course, we can take direct products in which each component comes from a different group. There are lots of applications. Computers tend to like groups such as

$$\mathbb{Z}_2^n = \underbrace{\mathbb{Z}_2 \oplus \mathbb{Z}_2 \oplus \mathbb{Z}_2 \oplus \cdots \oplus \mathbb{Z}_2}_{n \text{ copies}}.$$

Binary error-correcting codes are subgroups of this group. We will consider them later. There are also applications in cryptography and genetics.

It is also possible to consider direct products of infinitely many groups. Then one would have vectors with infinitely many components. For example, sequences $\{x_n\}_{n\geq 1}$ of real numbers $x_n \in \mathbb{R}$ can be added via $\{x_n\} + \{y_n\} = \{x_n + y_n\}$ to get a group which might be called $\mathbb{R}^\infty$. We will not consider such groups here.

Cayley graphs attached to some of these groups are shown in Figures 3.5, 3.6, and 3.7. The last image is a four-dimensional hypercube or tesseract. These graphs are undirected because the generating sets S have the property that $s \in S$ implies $s^{-1} \in S$.

Next we consider a couple of questions.

Question 1. Let G and H be finite groups. Suppose $a \in G$ and $b \in H$. What is the order of the element $(a, b) \in G \oplus H$?

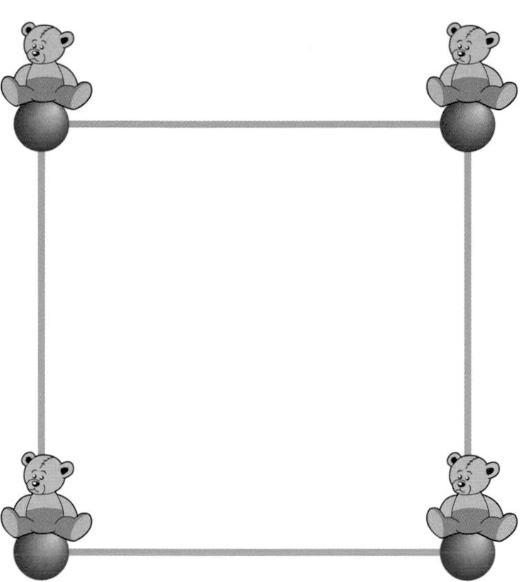

Figure 3.5 Cayley graph for $\mathbb{Z}_2 \oplus \mathbb{Z}_2$ with the generating set $\{(1,0),(0,1)\}$ and bears at the vertices

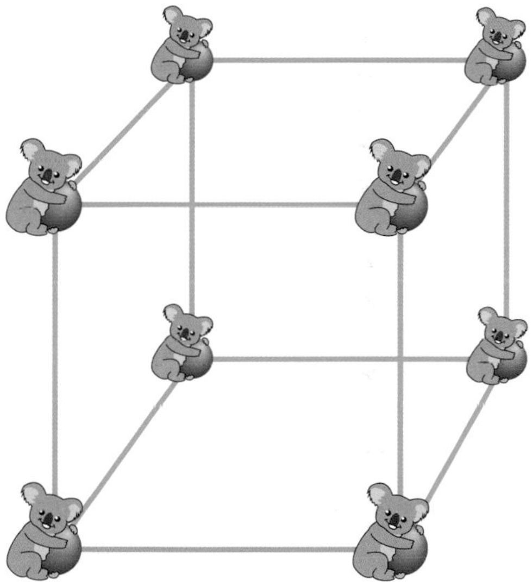

Figure 3.6 Cayley graph for $\mathbb{Z}_2 \oplus \mathbb{Z}_2 \oplus \mathbb{Z}_2$ with the generating set $\{(1,0,0),(0,1,0),(0,0,1)\}$ and koalas at the vertices

Answer. Recall Definition 3.1.2 of the least common multiple $r = \mathrm{lcm}[m,n]$ if $m=|a|$ and $n=|b|$. Then $|(a,b)| = \mathrm{lcm}(|a|,|b|)$.

Proof. First note that

$$(a,b)^k = (a^k, b^k) = (e,e') \Longleftrightarrow a^k = e = \text{identity of } G \text{ and } b^k = e' = \text{identity of } H. \quad (3.3)$$

Let k be the order of (a,b). The equivalence (3.3) implies $m=|a|$ divides k and $n=|b|$ divides k. So $r=\mathrm{lcm}[m,n]$ is less than or equal to k, the order of (a,b).

On the other hand, as $r=\mathrm{lcm}[m,n]$ is a common multiple of m and n, by (3.3), replacing k by r, we see that $(a,b)^r=(e,e')$ and thus k, the order of (a,b), divides r. So $k \le r \le k$, which means that $r=k$. ▲

Question 2. Suppose that we have two finite cyclic groups $G=\langle a\rangle$ and $H=\langle b\rangle$. When is $G \oplus H$ cyclic?

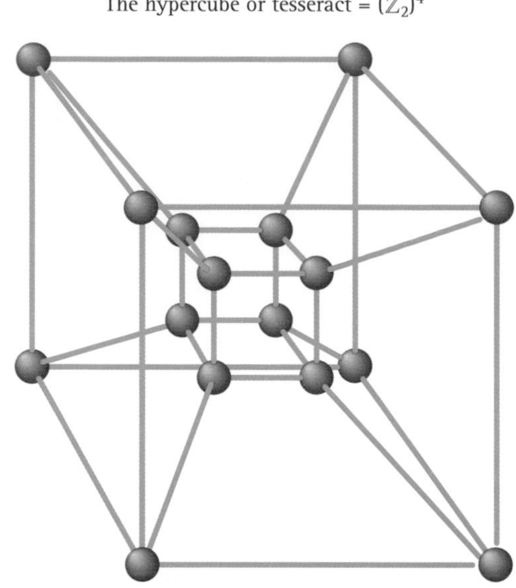

Figure 3.7 Cayley graph for $\mathbb{Z}_2 \oplus \mathbb{Z}_2 \oplus \mathbb{Z}_2 \oplus \mathbb{Z}_2$ with the generating set $\{(1,0,0,0),(0,1,0,0),(0,0,1,0),(0,0,0,1)\}$

Answer. Recall Definition 1.5.3 of the greatest common divisor. Then the answer to question 2 is that $G \oplus H$ is cyclic if and only if $\gcd(|H|,|G|)=1$.

Proof. Let $m=|G|$ and $n=|H|$. Suppose that $\gcd(m,n)=1$. Then (a,b) has order mn using the answer to question 1 and Exercise 3.1.7, which states that $mn=\gcd(m,n)\text{lcm}[m,n]$. This implies that $G \oplus H$ is cyclic. The converse is an exercise. ▲

Exercise 3.6.2 *Prove the converse part of the previous result.*

Example 3. Suppose that m and n are positive integers with $\gcd(m,n)=1$. We have a group homomorphism $f\colon \mathbb{Z} \to \mathbb{Z}_m \oplus \mathbb{Z}_n$, defined by $f(x)=(x+m\mathbb{Z}, x+n\mathbb{Z})$. Here the group operations are addition. It is easily seen that $f(x+y)=f(x)+f(y)$. We leave this to you to check as an exercise.

Note that $\gcd(m,n)=1$, $\ker f=\{x \mid x\equiv 0 \pmod m \text{ and } x\equiv 0 \pmod n\}=mn\mathbb{Z}$. The first isomorphism theorem (Theorem 3.5.1) then says that $\mathbb{Z}/mn\mathbb{Z}$ (or $\mathbb{Z}_{mn}$) is isomorphic to the image of f which is a subgroup of $\mathbb{Z}_m \oplus \mathbb{Z}_n$, but since both $\mathbb{Z}/mn\mathbb{Z}$ and $\mathbb{Z}_m \oplus \mathbb{Z}_n$ have mn elements, it follows that the image of f must be all of $\mathbb{Z}_m \oplus \mathbb{Z}_n$. Therefore we see that, as additive groups, $\mathbb{Z}_{mn} \cong \mathbb{Z}_m \oplus \mathbb{Z}_n$ if $\gcd(m,n)=1$. ▲

Exercise 3.6.3 *Assume* $\gcd(m,n)=1$. *Define* $f:\mathbb{Z}\to\mathbb{Z}_m \oplus \mathbb{Z}_n$ *by* $f(x)=(x+m\mathbb{Z}, x+n\mathbb{Z})$ *and show that* $f(x+y)=f(x)+f(y)$. *Then show that, as additive groups,* $\mathbb{Z}_m \oplus \mathbb{Z}_n$ *and* $\mathbb{Z}_{mn}$ *are isomorphic – using the first isomorphism theorem from the preceding section.*

In Section 6.2 we will see that the last exercise is equally true when we consider $\mathbb{Z}$ and $\mathbb{Z}_m \oplus \mathbb{Z}_n$ as rings (which means that the maps preserve the operation of multiplication as well as addition). This is called the **Chinese remainder theorem.** A version of it was found by the Chinese mathematician Sun Tsu in the first century AD. The main point of the theorem for the number theorist is the onto-ness of the map f which is proved very explicitly by constructing solutions to the simultaneous congruences

$$\left.\begin{array}{l} x \equiv a \pmod{m} \\ x \equiv b \pmod{n} \end{array}\right\}, \quad \text{when} \quad \gcd(m,n) = 1.$$

There is an old Chinese song which describes one construction of a simultaneous solution to three congruences. I explain it in my book [116, p. 14].

The Chinese remainder theorem has numerous computer applications: for example, in writing programs to multiply very large integers. One might also find it astounding that it seems to say that a group like $\mathbb{Z}_{mn}$, which we might think of as one-dimensional – at least when viewing its Cayley graph for the usual generating set – can be identified with a group $\mathbb{Z}_m \oplus \mathbb{Z}_n$, with a two-dimensional Cayley graph – at least when $\gcd(m,n) = 1$. Of course, groups are not vector spaces, so that dimension does not really mean anything. We define dimension of a vector space in Section 7.1. Moreover there are many ways to draw a given Cayley graph as well as many different Cayley graphs associated to a given group.

Consider the Cayley graph $X(\mathbb{Z}_{10} \oplus \mathbb{Z}_5, \{(\pm 1, 0), (0, \pm 1)\})$ in Figure 3.8. This can be viewed as a finite **torus.** The real torus (or doughnut) in Figure 3.9 is found by looking at the quotient of the plane $\mathbb{R} \oplus \mathbb{R}$ modulo $\mathbb{Z} \oplus \mathbb{Z}$. Note that we cannot claim that $\mathbb{Z}_{10} \oplus \mathbb{Z}_5$ is isomorphic to $\mathbb{Z}_{50}$, since $\mathbb{Z}_{10} \oplus \mathbb{Z}_5$ has no element of order 50.

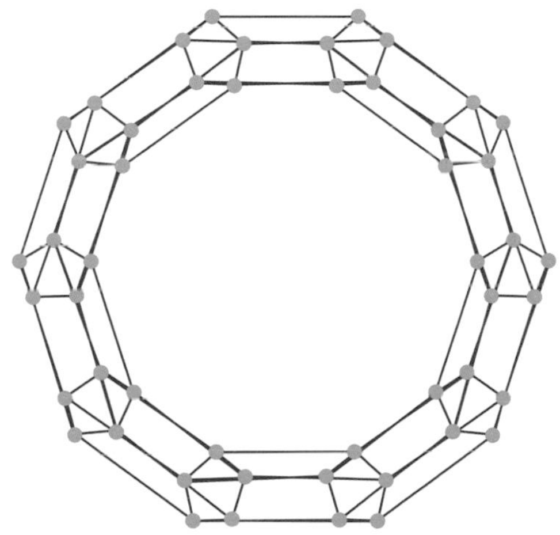

Figure 3.8 A finite torus, which is the Cayley graph $X(\mathbb{Z}_{10} \oplus \mathbb{Z}_5, \{(\pm 1, 0), (0, \pm 1)\})$

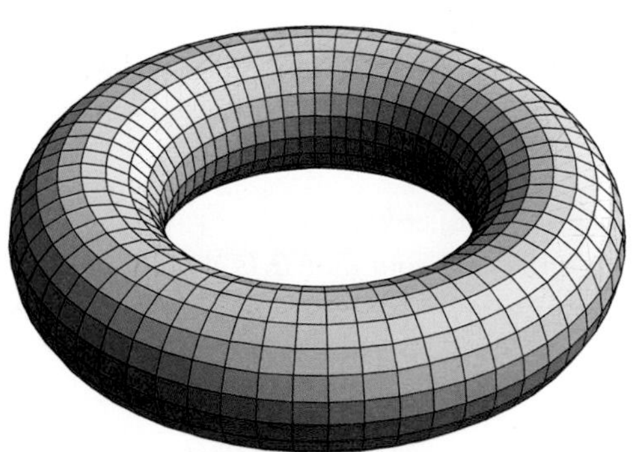

Figure 3.9 The continuous torus (obtained from the plane modulo its integer points; i.e., $\mathbb{R} \oplus \mathbb{R}$ modulo $\mathbb{Z} \oplus \mathbb{Z}$)

Exercise 3.6.4 *Suppose that G and H are groups. Define a map* $T\colon G \oplus H \to G$ *by* $T(x,y) = x$*, for all* $x \in G$ *and all* $y \in H$*. The map is called the projection onto the first coordinate.*

Show that T is a group homomorphism. What is ker T? *What does the first isomorphism theorem say?*

Exercise 3.6.5 *Give four examples of pairwise non-isomorphic groups of order* 8. *Explain.*

Exercise 3.6.6 *State whether each of the following statements is true or false and give a reason for your answer. In each of the last three statements the group operation on* $\mathbb{Z}_n$ *is addition* mod n.

(a) *If G and H are groups, then* $G \oplus H$ *is isomorphic to* $H \oplus G$.
(b) *The order of* $(3, 2)$ *in* $\mathbb{Z}_{12} \oplus \mathbb{Z}_4$ *is* 4.
(c) $\mathbb{Z}_5 \oplus \mathbb{Z}_5$ *has exactly six subgroups of order* 5.
(d) *The group of* 3×3 *matrices* $\begin{pmatrix} 1 & a & b \\ 0 & 1 & 0 \\ 0 & 0 & 1 \end{pmatrix}$, *with* $a, b \in \mathbb{Z}_n$ *is isomorphic to* $\mathbb{Z}_n \oplus \mathbb{Z}_n$, *for any* $n \geq 2$, *where the operation is matrix multiplication.*

Exercise 3.6.7 *Is the order of ab equal to the product of* $|a|$ *and* $|b|$ *for* a, b *in any finite group G? Explain.*

Exercise 3.6.8 *Show that the group* $\mathbb{C}$ *of complex numbers under addition is isomorphic to the group* $\mathbb{R} \oplus \mathbb{R}$.

There is one more group of order 8 (beyond those in Exercise 3.6.5). This group does not come out of products of groups or dihedral groups. It is the quaternion group. It is a subset of the quaternions invented in 1843 by William Rowan Hamilton to generalize the complex numbers $\mathbb{C} = \mathbb{R} \oplus i\mathbb{R}$ to four dimensions. Well, he actually wanted something three-dimensional over $\mathbb{R}$. But that proved to be impossible. He had to give up commutativity of multiplication also. He created the quaternions while walking with his wife along a canal in Dublin, Ireland. Then he carved the equations into a bridge – now a tourist destination, which I unfortunately missed when I was in Dublin. The space of **quaternions** is

$$\mathbb{H} = \mathbb{R} \oplus i\mathbb{R} \oplus j\mathbb{R} \oplus k\mathbb{R}, \quad \text{where } i^2 = j^2 = k^2 = ijk = -1. \tag{3.4}$$

See Stewart [114] for more of the story of Hamilton and his quaternions. At the moment we are just interested in 8 of these quaternions in order to get the group considered in the following exercise.

Exercise 3.6.9 *Consider the* ***quaternion group*** $Q = \{\pm 1, \pm i, \pm j, \pm k\}$, *with* $i^2 = j^2 = k^2 = ijk = -1$. *Create a multiplication table for this group of order* 8. *Show that Q is not isomorphic to* D_4. *Check that the Cayley graph of Q with generating set* $S = \{\pm i, \pm j\}$ *can be drawn as in Figure 3.10.*

Exercise 3.6.10 *Consider the quaternion group Q from the preceding exercise. Find three subgroups of Q of order* 4 *and state whether they are normal subgroups.*

Exercise 3.6.11 *Is* $\mathbb{Z}_4 \oplus \mathbb{Z}_8$ *isomorphic to* $\mathbb{Z}_{32}$? *Why?*

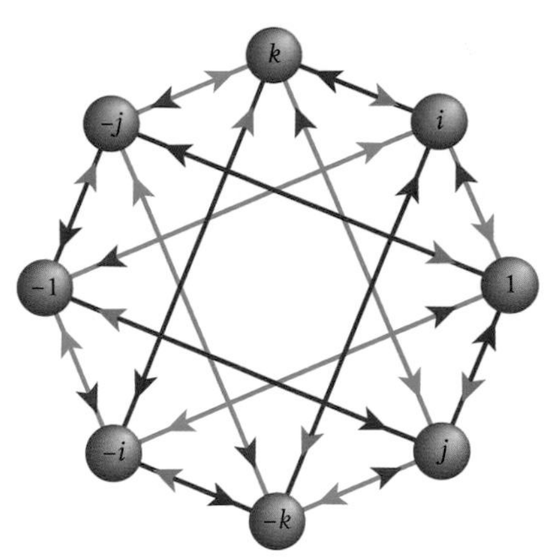

Figure 3.10 Cayley graph for the quaternion group with generating set $\{\pm i, \pm j\}$

The direct products considered above are the so-called **external direct products.** An internal direct product inside a group G will be a subgroup of G isomorphic to the external direct product of two other subgroups of G. Suppose H and K are subgroups of G. Then HK will be a subgroup of G as well as an **internal direct product** if $H \cap K = \{e\}$ and both H and K are normal subgroups of G. For then it can be shown that the map $T: HK \to H \oplus K$ defined by $T(hk) = (h, k)$ is well defined and a group isomorphism.

Exercise 3.6.12 *Prove that if H and K are normal subgroups of G such that $H \cap K = \{e\}$, then the map $T: HK \to H \oplus K$ defined by $T(hk) = (h, k)$ is well defined and a group isomorphism.*

Exercise 3.6.13 *Show that the groups G and H are both commutative if and only if $G \oplus H$ is commutative.*

Exercise 3.6.14 *Consider the groups $\mathbb{Z}_{60}$, $\mathbb{Z}_{30} \oplus \mathbb{Z}_2$. How many elements of orders 2, 3, 4, 5 does each group have?*

Exercise 3.6.15 *Suppose a and b are elements of the Abelian group G with finite orders $|a|$ and $|b|$ such that $\gcd(|a|, |b|) = 1$.*

(a) Show that $|ab| = |a|\,|b|$.
(b) Is it true that (under the hypotheses of this problem) the subgroup $H = \langle a, b \rangle$ is isomorphic to $\langle a \rangle \oplus \langle b \rangle$? Why?

Exercise 3.6.16 *If g is an element of some group G with finite order $|g| = r$, show that*

$$|g^t| = \frac{r}{\gcd(r, t)} = \frac{\operatorname{lcm}(r, t)}{t}.$$

Exercise 3.6.17 *Suppose a and b are elements of the Abelian group G with finite orders $|a|$ and $|b|$. Show that there is an element of G with order the least common multiple* $\operatorname{lcm}[|a|, |b|]$.

Hint. *Suppose we have the prime factorizations of the orders: $|a| = \prod p_i^{e_i}$ and $|b| = \prod p_i^{f_i}$, where the p_i are pairwise distinct prime numbers. Define*

$$c_i = \begin{cases} e_i & \text{if } e_i \leq f_i, \\ 0 & \text{otherwise.} \end{cases} \qquad d_i = \begin{cases} f_i & \text{if } f_i < e_i, \\ 0 & \text{otherwise.} \end{cases}$$

Then let $r = \prod p_i^{c_i}$ *and* $s = \prod p_i^{d_i}$. *It follows that* $a^r b^s$ *is the element we are seeking – using the preceding two exercises.*

Exercise 3.6.18 *Suppose that* $\gcd(m,n)=1$. *Define a map* $T: \mathbb{Z}_{mn}^* \to \mathbb{Z}_m^* \oplus \mathbb{Z}_n^*$ *by* $T(x \bmod mn) = (x \bmod m, x \bmod n)$. *Show that* T *is a well-defined group homomorphism between these multiplicative groups.*

Exercise 3.6.19 *What happens to the preceding exercise if we do not assume that* $\gcd(m,n)=1$?

3.7 Group Actions

We have already seen many examples of groups acting on sets. Here we want to prove something usually called Burnside's lemma though it was first found by A.-L. Cauchy in 1870 and then by F. G. Frobenius in 1900 and finally by W. Burnside in 1910. We will persist in naming the result after Burnside, since he did publicize it in his book on group theory. This lemma has applications in chemistry for counting certain kinds of chemical compounds. There are also applications to counting switching circuits, which are the sort of circuits that are the basis for computers. See Dornhoff and Hohn [25, Section 5.14]. The result leads immediately to work of Redfield in 1927 and Polyà in 1937 – which is often called Polyà enumeration theory. In Section 4.4 we consider an application to counting sudoku puzzles.

Definition 3.7.1 *A group* G *is said to* **act on a set** X **on the left** *if for every* $\sigma \in G$ *there is a function* $\Theta : G \times X \to X$, *written* $\Theta(\sigma, x) = \sigma \cdot x = \sigma x$, *for all* $\sigma \in G$, $x \in X$, *with the following two properties:*

1. $(\sigma\tau)x = \sigma(\tau x)$, *for all* $\sigma, \tau \in G$ *and* $x \in X$;
2. *if* e *is the identity of* G, *then* $ex = x$, *for all* $x \in X$.

The action given in the preceding definition is a **left group action.** One can similarly define a **right group action** of the group G on a set X written $x\sigma$ or x^σ, for $x \in X, \sigma \in G$. In the case of a right action, the two defining properties become $xe = x$, for the identity e in G and $(x\sigma)\tau = x(\sigma\tau)$, for all $x \in X$, $\sigma, \tau \in G$. You can essentially switch a left action to a right action by replacing $\sigma \in G$ with σ^{-1} in the formula for $\Theta(\sigma, x)$ in Definition 3.7.1.

Example 1. The dihedral group D_n acts on the regular n-gon. We examined the case $n=3$ in Figure 2.1. ▲

Recall that we noted that there is often confusion about actions being either left or right actions. It is a right–left confusion that many people (the present author included) have. The important property to check is that of property 1 in Definition 3.7.1. We saw already with Example 1 in Section 2.1 – the dihedral group D_3 – that confusion is quite natural. Recall the balls and buckets problem. Are you moving the balls or the buckets? This is really the question of right versus left actions.

Example 2. The group G acts on itself by left multiplication (see Definition 2.3.2), $L_g x = gx$ for all $g, x \in G$. The group also acts on itself on the right via $R_g x = xg$. ▲

Example 3. The symmetric group S_n acts on the set $X = \{1, 2, \ldots, n\}$ via $\sigma \cdot x = \sigma(x)$, for all $x \in X$. ▲

Exercise 3.7.1 *Show that the group G acts on the set of functions* $X = \{f: G \to \mathbb{C}\}$ *by setting* $(L_g f)(x) = f(g^{-1}x)$, $\forall x, g \in G$ *and* $f \in X$. *This is the left action. There is a similar right action. Define it.*

Exercise 3.7.2 *Define* $\mathcal{S}(X)$ *to be the set of* 1–1, *onto functions* $\sigma: X \to X$. *Show that* $\mathcal{S}(X)$ *is a group under the multiplication given by composition of functions. Then show that if group G acts on set X, via* $T_\sigma(x) = \sigma \cdot x$, *for* $\sigma \in G$ *and* $x \in X$, *we have a group homomorphism* $T: G \to \mathcal{S}(X)$ *given by* $T(\sigma) = T_\sigma$, *for all* $g \in G$. *Note that* $|X| = n$ *implies that we can identify* $\mathcal{S}(X)$ *with* S_n. *This generalizes Cayley's theorem.*

Exercise 3.7.3 *Check that the action of* S_n *on* $\{1, 2, \ldots, n\}$ *defined in Example 3 is indeed a left action.*

Example 4. The symmetric group S_n acts on **polynomials in n indeterminates** $x_1, \ldots, x_n$, with real coefficients. The action is given (as seen already in Definition 3.1.3) by

$$(\sigma P)(x_1, \ldots, x_n) = P\left(x_{\sigma(1)}, \ldots, x_{\sigma(n)}\right). \tag{3.5}$$

For example, suppose that $n = 4$,

$$P(x_1, x_2, x_3, x_4) = x_1 x_3 - x_2 x_4,$$
$$\sigma = (12)(34),$$
$$\tau = (123).$$

Then

$$\sigma P(x_1, x_2, x_3, x_4) = x_2 x_4 - x_1 x_3 = -P(x_1, x_2, x_3, x_4).$$

We find that

$$\tau P(x_1, x_2, x_3, x_4) = x_2 x_1 - x_3 x_4.$$

Now $\tau\sigma = (123)(12)(34) = (134)$. We compute $(\tau\sigma)P$ two ways. First

$$(\tau\sigma)\, P(x_1, x_2, x_3, x_4) = x_3 x_4 - x_2 x_1$$

and second

$$\tau\left(\sigma P(x_1, x_2, x_3, x_4)\right) = \tau\left(x_2 x_4 - x_1 x_3\right) = x_3 x_4 - x_2 x_1.$$

Mercifully they are indeed the same. Now prove that property 2 of a left action holds in general for the action in (3.5), as an exercise. ▲

Exercise 3.7.4 *Show that the action in equation (3.5) is indeed a left action.*

Example 5. Recall that the general linear group appeared in Exercise 3.5.5 as well as Example 3 of Section 2.3 (for 2×2 matrices with entries in the integers mod 2). The general linear group $GL(n, \mathbb{R})$ of non-singular real $n \times n$ matrices M acts on column vectors $x \in \mathbb{R}^n$ via $M \cdot x = Mx$. Here Mx means matrix multiply the $n \times n$ matrix M times the $n \times 1$ matrix x.

▲

Example 6. The symmetric group S_n acts on vectors $(v_1, \ldots, v_n) \in \mathbb{R}^n$ by

$$(\sigma)(v_1, \ldots, v_n) = \left(v_{\sigma^{-1}(1)}, \ldots, v_{\sigma^{-1}(n)}\right). \tag{3.6}$$

▲

Exercise 3.7.5 *Show that the action of S_n on $\mathbb{R}^n$ given in equation (3.6) is indeed a left group action.*

Hint. *One way to do this is to note that we can identify $\mathbb{R}^n$ with functions $f\colon \{1, \ldots, n\} \to \mathbb{R}$ by identifying the function f with the vector $(f(1), \ldots, f(n)) \in \mathbb{R}^n$. Then use the method of Exercise 3.7.1.*

We have already seen a special case of the following definition in the proof of Proposition 3.1.1.

Definition 3.7.2 *Assume that group G acts on set X. The* ***orbit*** *of $x \in X$ is $\mathrm{Orb}(x) = \{\sigma x \mid \sigma \in G\}$.*

Lemma 3.7.1 *Assume that group G acts on set X. We have an equivalence relation on the set X given by $x \sim y$ iff $y \in \mathrm{Orb}(x)$. Of course, the equivalence classes are the orbits.*

Exercise 3.7.6 *Prove the preceding lemma.*

Exercise 3.7.7 *Given a group G, we get a group action of $g \in G$ on G itself via conjugation: $T_g(x) = gxg^{-1}$, for any $x \in G$. Show that this does indeed define a group action of G on G. What are the orbits of this group action?*

The following definition is needed for our discussion of Burnside's lemma.

Definition 3.7.3 *Assume that group G acts on set X. The* ***stabilizer*** *of $x \in X$ is*

$$\mathrm{Stab}(x) = \{\sigma \in G \mid \sigma x = x\}.$$

The ***fixed points*** *of $\sigma \in G$ are*

$$\mathrm{Fix}(\sigma) = \{x \in X \mid \sigma x = x\}.$$

Exercise 3.7.8 *Show that $\mathrm{Stab}(x)$ is a subgroup of G.*

Exercise 3.7.9 *Consider Example 4 in this section – the symmetric group S_4 acting on polynomials in four indeterminates, with real coefficients. Find the orbit of the polynomial $P(x_1, x_2, x_3, x_4) = x_1x_3 - x_2x_4$ under S_4. What is the order of $\mathrm{Stab}(P)$ in S_4?*

Polynomials in n indeterminates over any field F that are fixed by every element of S_n are called **symmetric polynomials.** Favorites are the **elementary symmetric polynomials:**

$$s_1 = \sum_{i=1}^{n} x_i, \quad s_2 = \sum_{1 \le i < j \le n} x_i x_j, \quad s_3 = \sum_{1 \le i < j < k \le n} x_i x_j x_k, \quad \ldots, \quad s_n = x_1 x_2 \cdots x_n. \tag{3.7}$$

The **fundamental theorem on symmetric polynomials** says that any symmetric polynomial is a polynomial in the elementary symmetric polynomials. See Dummit and Foote [28] or Herstein [42] for more information.

Proposition 3.7.1 ***(Orbit/Stabilizer Theorem).*** *Assume that the finite group G acts on the finite set X. Then, for any* $x \in X$, $|\mathrm{Orb}(x)| = |G| \,/\, |\mathrm{Stab}(x)|$.

Proof. We can define a 1–1, onto function $F\colon G/\mathrm{Stab}(x) \to \mathrm{Orb}(x)$ via $F(\sigma \mathrm{Stab}(x)) = \sigma x$. We leave it as an exercise to prove this function is well defined, 1–1, and onto. ▲

Exercise 3.7.10 *Prove that the function F defined in the proof of Proposition 3.7.1 is well defined, 1–1, onto.*

Example. If H is a subgroup of the group G, then H acts on G by $h \cdot g = gh^{-1}$, for $h \in H$ and $g \in G$. The orbits are the left cosets gH. The orbit/stabilizer theorem states that $|G/H| = |G| \,/\, |H|$. We noticed this earlier in part (3) of Proposition 3.3.1. ▲

Exercise 3.7.11 *Find the order of the group of motions of a tetrahedron by computing* $|\mathrm{Orb}(f)|$ *and* $|\mathrm{Stab}(f)|$ *for any face f.*

The following lemma is usually given the name Burnside's lemma – but it does not appear to be due to Burnside, as we noted earlier in this section. This lemma says that the number of orbits of a finite group G acting on the finite set X is the average order of the sets $\mathrm{Fix}(\sigma)$, when averaged over $\sigma \in G$.

Lemma 3.7.2 ***(Burnside's Lemma).*** *Assume that the finite group G acts on the finite set X. Then the number of orbits of the group action is*

$$\#orbits = \frac{1}{|G|} \sum_{\sigma \in G} |\mathrm{Fix}(\sigma)|\,.$$

Proof. By Proposition 3.7.1

$$\#orbits = \sum_{x \in X} \frac{1}{|\mathrm{Orb}(x)|} = \sum_{x \in X} \frac{|\mathrm{Stab}(x)|}{|G|}.$$

It follows that

$$\#orbits = \frac{1}{|G|} \sum_{x \in X} |\mathrm{Stab}(x)| = \frac{1}{|G|} \sum_{x \in X} \sum_{\substack{\sigma \in G \\ \sigma x = x}} 1.$$

Now interchange the two sums to get

$$\#orbits = \frac{1}{|G|}\sum_{\sigma\in G}\sum_{\substack{x\in X\\ \sigma x=x}} 1 = \frac{1}{|G|}\sum_{\sigma\in G}|\mathrm{Fix}(\sigma)|\,.$$ ▲

Example. The group G acts on G by **conjugation**: $C_g(x)=gxg^{-1}$, for $x,g\in G$. We already studied conjugation in Section 3.2. The orbit of G is a **conjugacy class** $\{x\}=\{gxg^{-1}\mid g\in G\}$ as in Exercise 3.7.7. Then the stabilizer of $x\in G$ is $\mathrm{Stab}(x)=C_G(x)=\{g\in G\mid xg=gx\}=$ the **centralizer** of x in G. The center of G is $Z(G)=\{g\in G\mid gx=xg, \forall x\in G\}$. Suppose that $x_1,\ldots,x_t\notin Z(G)$ represent all the distinct conjugacy classes in G except those coming from the center. One has the **class equation**

$$|G|=|Z(G)|+\sum_{i=1}^{t}|\{x_i\}|=|Z(G)|+\sum_{i=1}^{t}\frac{|G|}{|C_G(x_i)|}. \tag{3.8}$$

This result is pretty obvious, since G is a disjoint union of its conjugacy classes as conjugacy is an equivalence relation. Thus the number of elements of G is the sum of the orders of the conjugacy classes. The conjugacy classes of elements of the center of G have just one element. The order of a conjugacy class is found using Proposition 3.7.1. ▲

We can use equation (3.8) to prove the following theorem of Cauchy.

Theorem 3.7.1 *(Cauchy).* *If a prime p divides the order $|G|$ of a finite group G, then G has an element of order p.*

Proof. There are two cases (and each case has two subcases). Both cases use induction on $|G|$. The theorem is certainly true if G has order 1 or 2 – or even any prime.
Case 1. The group G is Abelian. If G contains no proper subgroup then G is cyclic and has an element of order p by Theorem 2.5.1. Otherwise G has a proper subgroup H. If p divides $|H|$, we are done by the induction hypothesis. Otherwise, since p divides $|G|=|G/H|\,|H|$, it follows that p must divide the order of $|G/H|$. Thus G/H must contain an element of order p by the induction hypothesis. But this implies G must also contain an element of order p. For a proof, see the exercise below.
Case 2. The group G is not Abelian. Recall that $C_G(g)=\{x\in G\mid xg=gx\}$. Suppose that p does not divide $|G/C_G(x)|$ for some $x\in G$, $x\notin Z(G)$. Proposition 3.3.1 says $|G|=|G/C_G(x)|\cdot|C_G(x)|$. Thus p must divide $|C_G(x)|$, since p divides $|G|$. Then by induction, since $|C_G(x)|<|G|$ (Why?), we know that $C_G(x)$ has an element of order p and thus so does G.

Our last subcase is that p divides $|G/C_G(x)|=|G|/|C_G(x)|$ for all $x\in G$, $x\notin Z(G)$. Then by the class equation (3.8), p divides every term in the sum as well as the left-hand side, which implies that p divides $|Z(G)|$. Then by induction because G is not Abelian and $|Z(G)|<|G|$, we know that there must be an element of order p in the center of G. ▲

Exercise 3.7.12

(a) Prove the last statement in Case 1 of the proof of Theorem 3.7.1.
(b) Answer the "Why?" in Case 2 of the proof of Theorem 3.7.1.

Hint for (a). *gH has order $p\implies g^p=h\in H$ and $g^k\notin H$ if $1\le k<p$. If $|H|=n$, we know p does not divide n. What is the order of g^n?*

The preceding theorem is a special case of a theorem proved in 1872 by Ludwig Sylow, a high school teacher in Norway. To explain this theorem, some definitions are needed. For a prime p, a **p-group** means the group has order a power of p. One defines for a prime p, a **Sylow p-subgroup** of a finite group G to be a maximal p-subgroup of G. **Sylow's first theorem** says that, for any prime p, if for a finite group G, we write, $|G| = p^e n$, with $e \geq 1, \gcd(p, n) = 1$, then there is a subgroup H_i of G with $|H_i| = p^i$, for $1 \leq i \leq e$. In particular, there is a Sylow p-subgroup of G of order p^e. **Sylow's second theorem** says if H is a subgroup of the finite group G and $|H| = p^i$, for some prime p and some power i, then H is contained in some Sylow p-subgroup of G. **Sylow's third theorem** says if $|G| = p^e n$, with $e \geq 1, \gcd(p, n) = 1$, then all Sylow p-subgroups are conjugate and the number N_p of Sylow p-subgroups is a divisor of $|G|$ and $N_p \equiv 1 \pmod{p}$, for fixed p. We will not prove the Sylow theorems. You can find proofs in most algebra books: for example, Gallian [33] or Dummit and Foote [28].

Define the **normalizer** $N_G(H)$ of a subgroup H in a group G to be

$$N_G(H) = \{g \in G \mid gHg^{-1} \subset H\}.$$

The normalizer of a subgroup H is a subgroup of G. Then one can also show that if P denotes a p-Sylow subgroup of the finite group G, it follows that, if N_p is defined as in the preceding paragraph, $N_p = |G/N_G(P)|$.

Exercise 3.7.13 *Prove the last statements, assuming the Sylow theorems.*

Hint. *The conjugation mapping $c_g(x) = gxg^{-1}$, for $g \in G$, takes subgroups to conjugate subgroups. Then the stabilizer of subgroup H is $N_G(H)$. The result follows from the orbit/stabilizer theorem.*

Examples: Applications of Sylow Theorems

1. The group $S_3 = D_3 = \{I, R, R^2, F, FR, FR^2\}$ has three Sylow 2-subgroups $\langle F \rangle$, $\langle FR \rangle$, and $\langle FR^2 \rangle$. This group has one Sylow 3-subgroup $\langle R \rangle$. This checks with Sylow's third theorem, since $|S_3| = 2 \cdot 3$ and $N_2 \equiv 1 \pmod{2}$ means N_2 must be 3 as N_2 must also divide 6. Similarly $N_3 = 1$ checks with $N_3 \equiv 1 \pmod{3}$.
2. A_4 has an odd number of subgroups of order 4. This follows from $N_2 \equiv 1 \pmod{2}$.
3. There is only one group of order 15 and it is cyclic. Sylow says that both N_3 and N_5 must be 1. That means a group G of order 15 must have one normal subgroup A of order 3 and another normal subgroup B of order 5. These subgroups have to be cyclic. Moreover $A \cap B = \{e\}$. Then recall Exercise 3.6.12.
4. Suppose G is a group of order 10. Then Sylow's third theorem says it must have $N_2 = 1 + 2q$ Sylow 2-subgroups and $N_5 = 1 + 5r$ Sylow 5-subgroups, where N_2 and N_5 divide 10. This means that N_2 is either 1 or 5 while N_5 must be 1. This will help us to determine the groups of order 10 in Section 4.5. ▲

Exercise 3.7.14 *Finish the discussion of groups of order* 15.

One can use Burnside's lemma to count other sorts of things than possible groups of certain orders.

Example. Consider the number of necklaces that one can make using beads with two colors located at the vertices of a hexagon. This is equivalent to counting the chemical compounds that can be obtained by attaching H or CH_3 radicals to each carbon atom in a benzene ring (see Figure 0.1 in the preface). You might think at first that there are 2^6 ways to do this but these do not all give different compounds since rotation and flip do not change the chemistry or the necklace. Thus we need to consider the group D_6 acting on the necklaces. Distinct necklaces are distinct orbits (i.e., distinct equivalence classes). We need to make a table. To compute the numbers in the last column of the table, see equation (3.10) below. ▲

Group element σ	Number of such elements	$\lvert\text{Fix}(\sigma)\rvert$
I	1	2^6
F_1 flip across axis between opposite vertices	3	2^4
F_2 flip across axis between opposite sides	3	2^3
R, R^5, $R=$ rotate by $=\frac{2\pi}{6}=\frac{\pi}{3}$	2	2
R^2, R^4	2	2^2
R^3	1	2^3

From this, we find that the number of orbits is the sum of the numbers in the third column of the table divided by the order of the group, that is, $\frac{156}{12}=13$. The 13 necklaces are shown in Figure 3.11.

Let us say a bit more about these coloring problems. Suppose that G is a finite group permuting elements of a finite set X. And suppose C is a finite set of colors. Colorings of X are elements of the set $X^C=\{f\colon X\to C\}$. There are $|X|^{|C|}$ such colorings. An element $\sigma\in G$ acts on $f\in X^C$ via $(\sigma f)(x)=f(\sigma^{-1}x)$, for $x\in X$. Then the Burnside lemma says

$$\#\,(G\text{–inequivalent colorings})=\frac{1}{|G|}\sum_{\sigma\in G}|\text{Fix}(\sigma)|=\frac{1}{|G|}\sum_{\sigma\in G}|C|^{\ell(\sigma)}, \tag{3.9}$$

where $\ell(\sigma)$ is the number of cycles in the disjoint cycle decomposition of σ as a permutation of elements of X. Here it is assumed that we include all 1-cycles as well so that, for example, we write the identity in S_n as $I=(1)(2)\cdots(n)$. Thus $\ell(I)=n$. To prove equation (3.9), note that if $f\in X^C$ is fixed by $\sigma\in G$, then $f\circ\sigma^i=f$, for all powers i. If $u,v\in X$ are in the same cycle in σ, then $v=\sigma^i(u)$ for some i. Then $f(u)=f(v)$. So f must be constant on a cycle of σ. Conversely, if f is constant on any cycle of σ, then $f\in\text{Fix}(\sigma)$. Why? Moreover, it follows that

$$|\text{Fix}(\sigma)|=|C|^{\ell(\sigma)}. \tag{3.10}$$

Exercise 3.7.15 *Give an answer to the "Why?" in the previous paragraph. Then explain why equation (3.10) follows from this.*

Exercise 3.7.16 *In how many ways can the faces of a tetrahedron be colored with three different colors.*

Much more can be said about the use of group actions in counting colorings – a subject which is referred to as Pólya enumeration theory. See Dornhoff and Hohn [25] for more information on this theory and its extensions. This text also gives applications to the enumeration of logic circuits and switching functions (now called Boolean functions) $f\colon\{0,1\}^n\to\{0,1\}$. One wants to know, for example, how many essentially different devices are needed to implement these switching functions for $n=2^4$. The answer is that 222 devices suffice. See Dornhoff and Hohn [25, p. 263] for more information on this subject. The logic

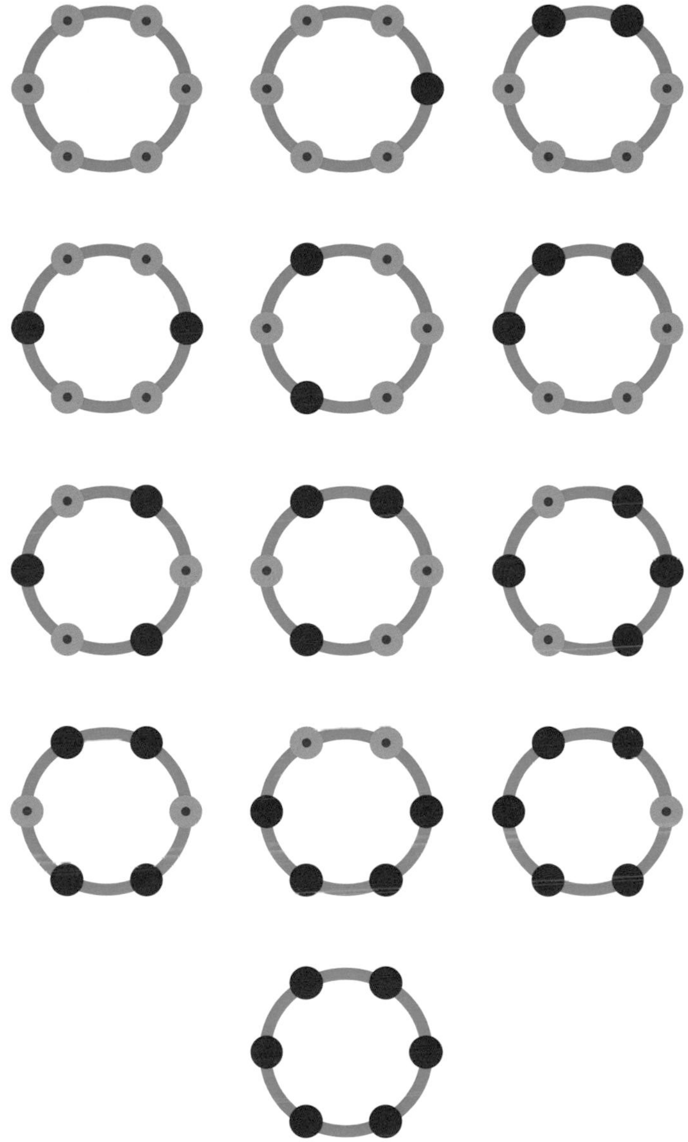

Figure 3.11 The 13 necklaces with six beads of two colors

circuits and Boolean or switching functions are idealizations of the circuits in the innards of our computers. Once these circuits were combinations of transistors, resistors, etc. connected by wires. Now one chip may contain millions – or billons – of these elements. From the 1970s to 2016, the number of transistors on an integrated circuit chip microprocessor CPU went from thousands to billions. See Wikipedia on transistor count or the IBM webpage. This frightening explosion of complexity requires algebra to deal with it. See also my book [116]. We will say a bit more about Boolean functions in Section 8.2.

Exercise 3.7.17 *Use mathematical induction to show that the coefficients of a monic polynomial of degree n are given by elementary symmetric polynomials in the roots defined by* *equation (3.7):*

$$\begin{aligned} f(y) &= (y - x_1)(y - x_2)\cdots(y - x_n) \\ &= y^n - s_1 y^{n-1} + s_2 y^{n-2} - s_3 y^{n-3} + \cdots + (-1)^n s_n. \end{aligned}$$

Exercise 3.7.18 *Find the number of tiny bracelets of four beads that can be made with two colors of beads.*

Exercise 3.7.19 *In how many ways can we paint a square floor made up of nine square tiles using purple and orange paint?*

Exercise 3.7.20 *In how many ways can you color a cube's faces with four colors?*

Exercise 3.7.21 *Consider the group of motions of the regular octahedron (a solid with eight faces consisting of equilateral triangles). What is the order of this group?*

Exercise 3.7.22 *What is the order of the group of motions of the regular dodecahedron (a solid with 12 faces consisting of regular pentagons). See Figure 3.12 for the dodecahedron graph. In this figure, one face is stretched out to contain the rest.*

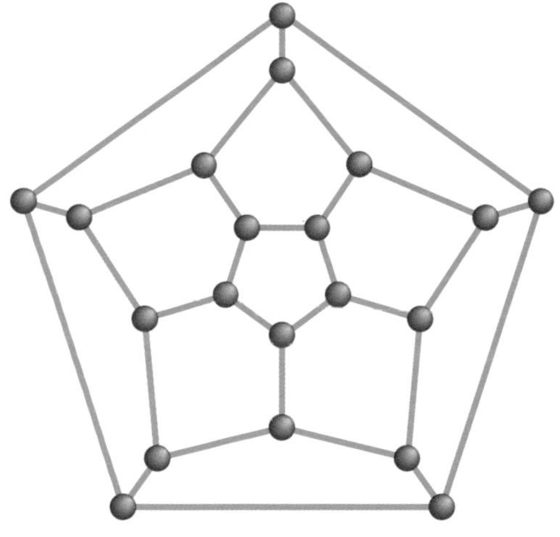

Figure 3.12 The dodecahedron graph drawn by Mathematica

Besides the regular Platonic solids of Figure 2.16, there are the 13 Archimedean solids whose faces are congruent regular polygons and whose vertices all have the same number of edges emanating from them. One example is shown in Figure 3.13.

Exercise 3.7.23 *Consider the group G of rotations of the Archimedean solid known as the cuboctahedron shown in Figure 3.13. This solid is made up of eight identical regular triangles and six identical squares with each vertex having four edges coming out of it. What is the order of the group G?*

Exercise 3.7.24 *Find all the conjugacy classes in the dihedral group D_3 and then do the same for D_4. Check the class equation (3.8) in each case.*

Exercise 3.7.25 *Show that if G is a group with center $Z = Z(G)$ then Z is a normal subgroup of G.*

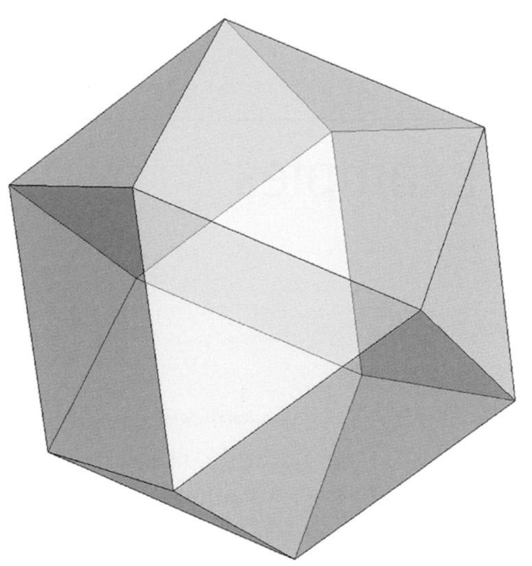

Figure 3.13 The cuboctahedron drawn by Mathematica

Exercise 3.7.26 *Show that if* G *is a group with center* $Z = Z(G)$ *such that* G/Z *is cyclic, then* G *is Abelian.*

4

Applications and More Examples of Groups

4.1 Public-Key Cryptography

One important application of group theory is to be found in an algorithm used to encrypt messages which is due to Rivest, Shamir, and Adleman in 1978. This method is called RSA cryptography. Perhaps a few definitions are in order. **Cryptography** is a part of the field of cryptology. **Cryptology** refers to the design of systems for encoding information that needs to be kept secret (cryptography) as well as the discovery of mechanisms for breaking into such systems (**cryptanalysis**). In this section we will give only a brief introduction to RSA cryptography. More information can be found in various books such as those of Kenneth H. Rosen [91], Neal Koblitz [55], or Ramanujachary Kumanduri and Cristina Romero [62]. See also the article of N. Koblitz [57] well as that of N. Koblitz and A. Menezes [58]. Or see Wikipedia as well as the Mathematica website for RSA encryption and decryption.

Quite often in modern times one wants to send a message (usually on the web) which can only be understood by the recipient. Public-key cryptography allows this to be done fairly easily and fairly securely (we hope). Think of your message as a number mod pq, where p and q are very large distinct primes that do not divide m. The encryption of the message m is just $m^t \pmod{pq}$ for some power t. To decrypt, one needs a power s so that $m^{ts} \equiv m \pmod{pq}$. From what we know about the group $\mathbb{Z}_{pq}^*$, we know that we need to solve $ts \equiv 1 \pmod{\phi(pq)}$, where $\phi(pq) = \left|\mathbb{Z}_{pq}^*\right|$. Here ϕ is Euler's phi-function considered in Section 2.3. To find s, it therefore seems that you need to know $\phi(pq) = (p-1)(q-1)$. See Exercises 2.3.10 and 3.6.18.

What happens is that anyone who wants to receive a secret message chooses p, q, t and publishes t and pq. The public key is (t, pq) and the secret key needed to decrypt the message is s. Anyone who wants to send a secret message m will compute $m^t \pmod{pq}$ and send this number. Why is it that a third party will be unable to compute m from this public knowledge? It is possible to test a large number to see that it is prime with a relatively small amount of computer time. The tests are probabilistic and thus there is a tiny chance of failure. See Kumanduri and Romero [62, Chapter 6] or Rosen [91]. The computational fact that enables RSA cryptography is that (at the moment) if p and q are huge primes, no one knows how to factor pq in a reasonable amount of computer time and then no one can compute $\phi(pq) = (p-1)(q-1)$. The security of the message thus depends on the difference in speed of primality testing versus factoring. If this situation should ever change, the RSA codes would not be secure.

The primes p and q must be VERY large for the RSA code to be secure. The RSA website (www.rsa.com) has a discussion. In 2011 the suggested size of the modulus $n = pq$ was 1024 bits (meaning 0s and 1s needed to express the number pq base 2) and thus since $2^{1025} - 1 \cong 3.5954 \times 10^{308}$ the number of decimal digits was something like 308. You would not like to see such a large number written down here. The RSA website said such a key was OK for corporate uses but not for extremely valuable keys. More recently one finds that the National Security Agency (NSA) is worried that someday soon quantum computers will exist and much larger primes would be necessary. See the Wikipedia article on key size.

It is somewhat frightening to one born in 1942 that a googol is only 10^{100}. I once thought that a 9-digit social security number was too large to factor easily. Thus, in the 1970s I was impressed when someone could factor their social security number on their HP-65 programmable calculator.

Other groups than $\mathbb{Z}_{pq}^*$ can be used for cryptography. In particular, one can use the group of points on an elliptic curve. We will say more about these things in Section 8.5.

Here we give a toy example of RSA cryptography with tiny primes and the encoding of letters modelled on examples in Rosen's book [91]. First we make a table of letters from A to Z and the corresponding pairs of numbers 00 to 25.

A	B	C	D	E	F	G	H	I	J	K	L	M
00	01	02	03	04	05	06	07	08	09	10	11	12
N	O	P	Q	R	S	T	U	V	W	X	Y	Z
13	14	15	16	17	18	19	20	21	22	23	24	25

We assume the public key is $(n = pq, t) = (37 \cdot 59, 23)$. We need $\gcd(\phi(37 \cdot 59), 23) = 1$ and that is indeed the case – since $\phi(37 \cdot 59) = 36 \cdot 58$ and $\gcd(36 \cdot 58, 23) = 1$. Note that $37 \cdot 59 = 2183$.

Before encrypting anything, we need to address the problem of modular exponentiation. We will need to find $1304^{23} \bmod 2183$, for example. To do this, you might want to compute the integer 1304^{23}. But that would be a really huge number since $\log_{10} 1304^{23} \approx 71.65$. In the old days (when I was young) this number would crash your calculator. Despite this humongous size, Mathematica says it can handle a number this large. It says `Mod[1304^23,2183]` $= 404$. On February 24, 2011, I saw that on a PC running Linux, Mathematica said it could deal with a number whose log base 10 is over 300 000 000.

Mathematica also has something called PowerMod. The command `PowerMod [1304,23,2183]` yields the same answer 404 and it should work better than just Mod since it is clever enough to compute inverses and square roots when they exist.

The number theory algorithm you would have to use if you were found on a desert island goes as follows. First you would write the exponent 23 in base 2 as $23 = 1 + 2 + 4 + 16$. Then you repeatedly square and reduce mod 2183. This gives the following computation. Since we keep reducing mod 2183 the numbers never have more than four digits.

i	$1304^i \pmod{2183}$
1	1304
2	$1304^2 \equiv 2024$
4	$2024^2 \equiv 234$
8	$234^2 \equiv 181$
16	$181^2 \equiv 16$

Then to obtain $1304^{23} = 1304^{1+2+4+16} = 1304 \cdot 1304^2 \cdot 1304^4 \cdot 1304^{16} \equiv 1304 \cdot 2024 \cdot 234 \cdot 16 \equiv 404 \pmod{2183}$. Mercifully we do not really need to do this if Mathematica (or some similar software) is available. More discussion of this algorithm can be found in Rosen [91].

Exercise 4.1.1 *Use the Mathematica command* `Mod[a,n]` *and repeated squaring to compute* $2104^{23} \pmod{2183}$. *Then check your answer using* `PowerMod[b,p,n]`. *Or do a similar computation using SAGE or whatever is your favorite program for computing modulo* n.

We want to encrypt a message: NEVER GIVE UP. Take the message and translate the letters to their numerical equivalent. Form blocks of two letters. Add Xs as necessary to make the last block of four numbers (two letters).

NE	VE	RG	IV	EU	PX
1304	2104	1706	0821	0420	1523

Blocks are used to avoid making the message easy to decipher by decoding using the frequency with which various letters should occur in a message.

To encipher a block B replace it by $E(B)=C=B^t \pmod{n}$. I am doing this using Mathematica. Of course there are many other programs that would work, but maybe not as easily.

So we need to compute stuff.

$$1304^{23} \equiv 404 \pmod{2183}, \quad 2104^{23} \equiv 1867 \pmod{2183}, \quad 1706^{23} \equiv 950 \pmod{2183},$$
$$821^{23} \equiv 1304 \pmod{2183}, \quad 0420^{23} = 964 \pmod{2183}, \quad 1523^{23} \equiv 1215 \pmod{2183}.$$

This means that the encrypted message is 0404 1867 0950 1304 0964 1215.

Exercise 4.1.2 *Check that the decryption of*

0404 1867 0950 1304 0964 1215

is indeed NEVER GIVE UP.

Exercise 4.1.3 *Suppose that the public key is still* $(n=pq,t)=(37\cdot 59, 23)$. *You need to decrypt the message:*

0404 1867 0551 0496 1337 0643 0026.

Hint. *First you must find* $s \equiv t^{-1} \equiv t^{e-1} \pmod{\phi(pq)}$. *Mathematica can do this via*

$$\mathtt{PowerMod[t,-1,EulerPhi[p*q]]}.$$

Once you replace the blocks B *by* $B^s \pmod{2183}$, *you will still need to translate the pairs of numbers to letters using the table above.*

Exercise 4.1.4 *Using the same public key as in the preceding exercise, what is the encrypted version of the message GROUPS RULE?*

Our encryption/decryption algorithm depends on the assumption that the block B in a message to encrypt or decrypt satisfies $\gcd(B,n)=1$, where $n=pq$, for distinct primes p and q. Thus, for example, any message including the block AA would be troublesome. It is possible to encrypt such a block B, but decryption would not seem to make sense. However Kumanduri and Romero [62, p. 137] note that as long as you have a message B that is not divisible by both p and q, you can decrypt. Sadly we did not really manage to insure this will not happen in our toy example since $pq < 2525$ and AA is a possible block. Nevertheless let us see how the argument goes. Suppose that $\gcd(B, p) = p$ and $B = kp$, where $\gcd(k, q) = 1$. Then you can still decrypt B^t as $B^{ts} \equiv B \pmod{pq}$, where $s \equiv t^{-1} \pmod{\phi(pq)}$. So we have two congruences:

$$B^{ts} \equiv B \pmod{q},$$
$$B^{ts} \equiv B \equiv 0 \pmod{p}.$$

Then it follows that $B^{ts} \equiv B \pmod{pq}$.

Exercise 4.1.5 *As we said in the preceding paragraph, our encryption/decryption algorithm depends on the assumption that the block b in a message to encrypt or decrypt satisfies* $\gcd(b, n) = 1$, *where* $n = pq$, *for distinct primes p and q. What is the probability that b does not satisfy this condition? Estimate this probability for the primes* $p = 37, q = 59$; *then for primes* $p, q > 10^{100}$.

Hint. *Note that the set of numbers* $0 \le b \le x$ *which are divisible by p is* $\left\{0, p, 2p, 3p, \ldots, \left\lfloor \frac{x}{p} \right\rfloor p\right\}$. *Then recall the inclusion–exclusion principle in Exercise 1.8.17.*

Exercise 4.1.6 *Suppose that the public key is still* $(n = pq, t) = (37 \cdot 59, 23)$. *You need to decrypt the message:*

0429 1384 1150 2037 1473.

Exercise 4.1.7 *Assume, as usual, that p and q are distinct large primes. Also assume that* $p > q$. *Show that finding* $\phi(pq)$ *is not easier than factoring* $m = pq$ *by showing that*

$$p + q = m - \phi(m) + 1 \quad \text{and} \quad p - q = \sqrt{(p+q)^2 - 4m}.$$

Then show that – once m and $\phi(m)$ *are known – you can easily find p, also q.*

There are also precautions that must be taken when choosing the primes p and q in order to avoid factoring tricks that might be applied to pq. Both $p - 1$ and $q - 1$ should have at least one large prime factor and $\gcd(p - 1, q - 1)$ should be small. Also p and q should not be too close together. There is an example showing why this last restriction is made in the exercise below based on a problem from Koblitz [55, p. 93].

Exercise 4.1.8 *Suppose that our large primes are p and q with* $p > q$. *If* $n = pq$, *then*

$$n = \left(\frac{p+q}{2}\right)^2 - \left(\frac{p-q}{2}\right)^2.$$

Let $a = (p + q)/2$ *and* $b = (p - q)/2$. *Both a and b are integers. Assume that b is small. Then* $b^2 = a^2 - n = pq/2$. *You can test integers* $a > \sqrt{n}$ *to see if* $a^2 - n$ *is a perfect square. That will give you a and b, which in turn give* $p = a + b$ *and* $q = a - b$.

(a) Use this method (which goes back to Fermat) to factor 777 923.
(b) Then use the same method to factor 869 107.

The protocols used are also important. If the same message is sent to more than one person using the same pq, the message g can be recovered without knowing $\phi(pq)$ – at least if two of the public keys t_i are relatively prime. This happens because we know g^{t_1} and

g^{t_2} (mod pq) and if t_1 and t_2 are relatively prime we know that there are integers u, v such that $t_1u + t_2v = 1$. But then $g = (g^{t_1})^u (g^{t_2})^v \pmod{pq}$.

Exercise 4.1.9 *Suppose that $pq = 2183$ and the public key for Spock is $t_1 = 23$ while that for Kirk is $t_2 = 17$. With $m = pq$, using our usual block encryption method from the preceding exercises, the encrypted message to Spock is $g^{23} \equiv a \pmod{m}$, while the message to Kirk is $g^{17} \equiv b \pmod{m}$, find g. Here the blocks for a are* 1635 2034 0061 2027 *and the blocks for b are* 0478 0961 1707 1562. *Figure out the message without factoring pq.*

Kumanduri and Romero [62, pp. 139ff] note that if you could find the decrypting exponent s somehow without knowing $\phi(pq)$ then you could factor $n = pq$. They use a probabilistic method that makes use of the fact that the congruence $x^2 \equiv 1 \pmod{pq}$ has two solutions more than the obvious two which are $\pm 1 \pmod{pq}$. If you have a solution x which is not $\pm 1 \pmod{pq}$, then you know that pq divides $(x-1)(x+1)$. This will allow you to factor $n = pq$ by computing $\gcd(n, x+1)$ or $\gcd(n, x-1)$. We will not go into the details of the use of the exponent s to produce a solution of $x^2 \equiv 1 \pmod{pq}$ but instead give an example or two.

Example. How to Factor *pq* if you Know the Decrypting Exponent as Well as the Encrypting Exponent. Suppose that $pq = 18\,833$, the encrypting exponent is $t = 47$, and the decrypting exponent is $s = 10\,895$. Then we factor $ts - 1 = 2^a b$, where b is odd. We find that $a = 6$ and $b = 8001$. Next we pick a random number $w = 63$. We should note that we needed to try six other random w before we found a "good" one. Now we create a sequence: $x_0 \equiv w^b \pmod{pq}$, $x_i \equiv x_{i-1}^2 \pmod{pq}$, for $i = 1, \ldots, a$. We know that $x_a \equiv 1 \pmod{pq}$. If we choose the first i such that $x_{i+1} \equiv 1 \pmod{pq}$, then assuming that x_i is not congruent to $\pm 1 \pmod{pq}$, we can find our p and q. In our case $x_1 \equiv 63^{8001} \equiv 11\,915 \pmod{18\,833}$. Then $x_2 \equiv x_1^2 \equiv 4071 \pmod{18\,833}$. And $x_3 \equiv x_2^2 \equiv 1 \pmod{18\,833}$. Thus we should look at $\gcd(4070, 18\,833) = 37$ or $\gcd(4072, 18\,833) = 509$. We have $18\,833 = 37 * 509$. Our algorithm fails when $x_0 \equiv 1 \pmod{pq}$ or $x_i \equiv -1 \pmod{pq}$. Kumanduri and Romero compute the probability that this happens for one choice of w is $\frac{1}{2}$. Thus the probability of failure for k random choices of w is $1/2^k$. In our example, we tried seven random values of w and $2^{-7} = 0.0078$. ▲

Exercise 4.1.10 *Find the four solutions $x \pmod{23 * 37}$ of the simultaneous congruences:*

$$x^2 \equiv 1 \pmod{23},$$
$$x^2 \equiv 1 \pmod{37}.$$

Note that these values of x indeed solve $x^2 \equiv 1 \pmod{23 * 37}$.

Exercise 4.1.11 *Imitate the example of factoring pq, assuming that you know both the encrypting and the decrypting exponents t,s, when $pq = 90\,743$, $t = 23$, and $s = 70\,247$.*

Exercise 4.1.12 *In our RSA cryptography, we choose our distinct large primes to be p and q, and then choose our encryption exponent to be t with $\gcd(t, \phi(pq)) = 1$. Thus the decryption exponent s satisfies $st \equiv 1 \pmod{\phi(pq)}$. Show that $t^{\phi(\phi(pq))} \equiv 1 \pmod{\phi(pq)}$ and thus that $s \equiv t^{-1} \equiv t^{\phi(\phi(pq))-1} \pmod{\phi(pq)}$.*

Exercise 4.1.13 *Prove that if p and q are distinct primes, then the multiplicative group $\mathbb{Z}_{pq}^*$ is isomorphic to the direct product $\mathbb{Z}_p^* \oplus \mathbb{Z}_q^*$.*

4.2 Chemistry and the Finite Fourier Transform

If you want to predict the behavior of a building in an earthquake, you need to know the fundamental frequencies of vibration of the building (or a finite element matrix approximation to the building) – as well as the likely frequencies of vibration of the earth on which the building sits caused by an earthquake. If you want to know the chemical constituents of a star, you need to know the spectrum. As Neil DeGrasse Tyson [121, p. 148] says: "In short, were it not for our ability to analyze spectra, we would know next to nothing about what goes on in the universe." Symmetry groups have much to say about spectra, as we shall see in this section. Fourier analysis helps to uncover spectra of symmetric objects. Here we will mostly discuss Fourier analysis on a finite cyclic group. This is sufficient for a molecule like benzene.

Let G be a finite Abelian group of order n, *with the group operation as addition*. Let $\widehat{G}$ denote the **dual group** of group homomorphisms $\chi : G \to \mathbb{T}$, where $\mathbb{T}$ is the circle group of complex numbers of norm 1: that is,

$$\mathbb{T} = \{z \in \mathbb{C} \mid |z| = 1\}.$$

Here $|x+iy|^2 = x^2 + y^2$, for $x, y \in \mathbb{R}$. The group operation on $\mathbb{T}$ is multiplication. We often call such χ a **character** of G. Then the dual group $\widehat{G}$ is the set of all such characters χ of G. What is the group operation on $\widehat{G}$? Suppose $\chi, \psi \in \widehat{G}$. Then if $g \in G$, define $(\chi\psi)(g) = \chi(g)\psi(g)$.

Exercise 4.2.1 *If* $i = \sqrt{-1}$ *define* $e^{ix} = \cos x + i \sin x$, *for* $x \in \mathbb{R}$. *Using properties of powers, show that* $T(x) = e^{ix}$, *for* $x \in \mathbb{R}$, *defines a group homomorphism from the additive group of real numbers to the multiplicative group* $\mathbb{T}$ *of complex numbers of norm* 1. *Find the kernel* K *of* T. *Use the first isomorphism theorem from Section 3.5 to identify* $\mathbb{T}$ *with* $\mathbb{R}/K$.

Exercise 4.2.2 *Show that the dual group* $\widehat{G}$ *corresponding to a finite Abelian group* G *is indeed a group.*

Before defining the Fourier transform, we consider a "new" kind of multiplication of functions. Suppose we are given two functions f, g mapping our group G into the complex numbers. We define the **convolution** of functions f and g to be $f * g$ where:

$$(f * g)(a) = \frac{1}{|G|} \sum_{b \in G} f(a-b)\, g(b), \quad \text{for } a \in G. \tag{4.1}$$

We read $f * g$ as f **splat** g. The operation is important for digital signal processing.

The following exercises give the algebraic properties of splat and show that it mirrors addition in the Abelian group G. Note that the function δ_0 has the property that $f * \delta_0 = (1/|G|)f$ and thus $|G|\delta_0$ is the identity for the operation splat. It is possible to define convolution of (integrable) functions $f : \mathbb{R} \to \mathbb{C}$, but you would need to replace the sum with an integral. However, the identity for splat does not exist as a function on the real line. This leads to the theory of distributions and the Dirac delta distribution; see Terras [118]. When functions on $\mathbb{R}$ are splatted together it tends to smooth out the corners. Convolution of our functions on a finite group like the additive group $\mathbb{Z}_n$ can mimic that behavior when they approximate functions on $\mathbb{R}$; see Exercise 4.2.12. Convolution of a nasty function on $\mathbb{R}$ that has lots of corners with a nice smooth function on $\mathbb{R}$ gives a nice function. In probability and statistics convolution of probability densities corresponds to the addition of independent random variables.

Exercise 4.2.3 *Show that convolution has the following properties for a finite Abelian group G under addition:*

(a) $f * g = g * f$;
(b) $f * (g + h) = f * g + f * h$;
(c) $f * (g * h) = (f * g) * h$.

Exercise 4.2.4 *Let a be an element of a finite Abelian group G under addition. Define the function* $\delta_a(x) = 1$ *if* $x = a$ *and* $\delta_a(x) = 0$ *otherwise. Let* $n = |G|$.

(a) Show that $\delta_a * \delta_b = \frac{1}{n}\delta_{a+b}$.
(b) Suppose $f: G \to \mathbb{C}$. *Show that* $(f * \delta_a)(x) = \frac{1}{n}f(x - a)$, *for all* $x \in G$.

Next we define the Fourier transform on a finite Abelian group G. This Fourier transform can be used to simplify convolution equations and analyze spectra of Cayley graphs for G – among other things. We define the (finite) **Fourier transform** of a function

$$f: G \to \mathbb{C}$$

at the character $\chi \in \widehat{G}$ by:

$$\widehat{f}(\chi) = |G|^{-1} \sum_{y \in G} f(y)\overline{\chi(y)}. \tag{4.2}$$

Since G is a finite Abelian group, there are no convergence problems. In fact, the theory of Fourier transforms on finite Abelian groups has many applications, since it is just what is needed for the fast Fourier transform or FFT – an idea which has decreased immensely the time needed to compute such transforms. For example, it is useful when one needs to multiply humongous integers. It was first found by Gauss in computing the orbit of the asteroid Juno in 1805. See Terras [116] for more information on finite and fast Fourier transforms. The continuous analog of the finite Fourier transform is a fundamental tool in applied mathematics. It will of course involve an integral and can be used to solve differential equations. See my book [118, Chapter 1] for that subject. The finite version is a much more easily computed animal.

From now on assume that the group G is cyclic of order n. So we will identify G with $\mathbb{Z}/n\mathbb{Z}$ under addition. Then we can show that the characters $\chi \in \widehat{G}$ can be identified with exponentials. Suppose $a, x \in \mathbb{Z}/n\mathbb{Z}$. We will identify elements of $\mathbb{Z}/n\mathbb{Z}$ and integers representing the cosets when we plug them into functions on $\mathbb{Z}/n\mathbb{Z}$. Define $\chi_a(b) = e^{2\pi i ab/n}$, for $a, b \pmod{n}$.

The space of functions $L^2(G) = \{f: \mathbb{Z}/n\mathbb{Z} \to \mathbb{C}\}$ is a vector space over $\mathbb{C}$ of dimension equal to n. (See Section 7.1 if you do not remember the definition of dimension of a vector space.) Moreover $L^2(G)$ has an **inner product** given by

$$\langle f, g \rangle = \sum_{x \in \mathbb{Z}/n\mathbb{Z}} f(x)\overline{g(x)}. \tag{4.3}$$

Exercise 4.2.5

(a) Show that as a vector space $L^2(\mathbb{Z}_n)$ *can be identified with* $\mathbb{C}^n$. *In fact, show that a convenient basis consists of the functions* δ_a, $a \in \mathbb{Z}_n$, *defined in Exercise 4.2.4.*
(b) Show that, under the operation of convolution, the space $L^2(\mathbb{Z}_n)$ *has an identity element e such that* $e * f = f$ *for all* $f \in L^2(\mathbb{Z}_n)$.
(c) Is $L^2(\mathbb{Z}_n)$ *a group under convolution?*

The space $L^2(\mathbb{Z}_n)$ is a vector space with a product defined by convolution. As such it is called the **group algebra** of the group $\mathbb{Z}_n$ under addition. Exercise 4.2.4 shows that the convolution of the functions δ_a defined in that exercise corresponds to addition of the elements a in $\mathbb{Z}_n$. The theory of the structure of such algebras was developed by J. H. M. Wedderburn. Emmy Noether noticed in 1929 that this structure theory could be used to do Fourier analysis on groups. However I personally find that Noether's methods make the subject less intuitive – especially for those of us introduced to classical Fourier analysis in that course which used the subject as a tool for the solution of partial differential equations.

Linear algebra (see Section 7.1) tells us that any basis of a vector space has the same number of elements and that an inner product space has an orthonormal basis. The additive characters of $\mathbb{Z}_n$ are orthogonal by the following lemma, which also says that to normalize them we must just multiply by $n^{-1/2}$.

Lemma 4.2.1 *(Orthogonality of Characters on $\mathbb{Z}/n\mathbb{Z}$). Suppose $a, x \in \mathbb{Z}_n$. With $\chi_a(b) = e^{2\pi iab/n}$, we have the following formula for inner products defined by (4.3)*

$$\langle \chi_a, \chi_b \rangle = \begin{cases} n & \text{if } a \equiv b \pmod{n}, \\ 0 & \text{otherwise.} \end{cases}$$

Proof. First what is $\langle \chi_a, \chi_b \rangle$? From the definitions, we obtain

$$\langle \chi_a, \chi_b \rangle = \sum_{x \in \mathbb{Z}/n\mathbb{Z}} \chi_a(x)\overline{\chi_b(x)} = \sum_{x \in \mathbb{Z}/n\mathbb{Z}} \chi_{a-b}(x).$$

If n divides $(a-b)$, then $\chi_{a-b}(x) = 1$ for all $x \in \mathbb{Z}/n\mathbb{Z}$ and the result is clear as we are summing n 1s. Otherwise, let $c = a-b$, which is not divisible by n. Call the sum on the right-hand side of the equality S. That is,

$$S = \sum_{x \in \mathbb{Z}/n\mathbb{Z}} \chi_c(x).$$

Note that

$$\chi_c(1)S = \chi_c(1) \sum_{x \in \mathbb{Z}/n\mathbb{Z}} \chi_c(x) = \sum_{x \in \mathbb{Z}/n\mathbb{Z}} \chi_c(x+1) = S.$$

The last equality holds since $x \mapsto x+1$ is a 1–1 map of $\mathbb{Z}/n\mathbb{Z}$ onto itself. But then S must be 0, as $\chi_c(1) \neq 1$ because c is not divisible by n. ▲

Exercise 4.2.6 *Show that if $G = \mathbb{Z}_n$, the dual group $\widehat{G}$ of additive characters of G is isomorphic to G.*

The following proposition says something about why the Fourier transform is useful. It changes the complicated multiplication of convolution to the simpler one of pointwise product. Moreover, it is an invertible transformation.

Proposition 4.2.1 ***(Some Properties of the Fourier Transform).*** *Let the group $G = \mathbb{Z}_n$ under addition. Use the preceding notation for the Fourier transform on G and convolution of functions on G.*

(1) Convolution

$$\widehat{f * g}(\chi) = \widehat{f}(\chi) \cdot \widehat{g}(\chi), \quad \text{for all } \chi \in \widehat{G}.$$

(2) Inversion

$$f(x) = \sum_{\chi \in \widehat{G}} \widehat{f}(\chi)\chi(x), \quad \text{for all } x \in I_K.$$

Proof.

(1) Note that

$$\widehat{f * g}(\chi) = n^{-2} \sum_{z \in G} \sum_{y \in G} f(z-y) g(y) \overline{\chi(z)}$$

$$= n^{-2} \sum_{y \in G} g(y) \sum_{w \in G} f(w) \overline{\chi(w+y)} = \widehat{f}(\chi) \cdot \widehat{g}(\chi),$$

where we set $w = z - y$ and reversed the order of summation.

(2) Observe that

$$\sum_{\chi \in \widehat{G}} \chi(x) \widehat{f}(\chi) = \tfrac{1}{n} \sum_{\chi \in \widehat{G}} \chi(x) \sum_{y \in G} f(y) \overline{\chi(y)}$$

$$= \sum_{y \in G} f(y) \tfrac{1}{n} \sum_{\chi \in \widehat{G}} \chi(x-y) = f(x),$$

since $\overline{\chi(y)} = \chi(y)^{-1} = \chi(-y)$, and Lemma 4.2.1 states that:

$$\frac{1}{n} \sum_{\chi \in \widehat{G}} \chi(x-y) = \begin{cases} 1, & x \equiv y \pmod{n}, \\ 0, & \text{otherwise.} \end{cases} \tag{4.4}$$

▲

Now we want to apply our transform to the study of benzene – C_6H_6. This is a molecule to be both admired and feared – admired for its stability, feared for its toxicity. See Figure 0.1 in the preface – a figure which is basically a Cayley graph of the group $\mathbb{Z}/6\mathbb{Z}$ under addition in which the generating set is $\{\pm 1 \pmod 6\}$. The **adjacency matrix** of this graph is the 6×6 matrix

$$A = \begin{pmatrix} 0 & 1 & 0 & 0 & 0 & 1 \\ 1 & 0 & 1 & 0 & 0 & 0 \\ 0 & 1 & 0 & 1 & 0 & 0 \\ 0 & 0 & 1 & 0 & 1 & 0 \\ 0 & 0 & 0 & 1 & 0 & 1 \\ 1 & 0 & 0 & 0 & 1 & 0 \end{pmatrix}.$$

This is an example of a circulant matrix – each row is a shift of the row above it: that is, the entries of a row are a cyclic permutation of the row above.

One can view the adjacency matrix as a matrix of the **adjacency operator** acting on complex-valued functions $f(x)$, for x in the Cayley graph $X(\mathbb{Z}_6, S)$, where $S=\{\pm 1 \pmod 6\}$. The action of the adjacency operator is given by

$$Af(x)=\sum_{s\in S} f(x+s)=n\,(\delta_S * f)\,(x), \tag{4.5}$$

where δ_S denotes the function that is 1 on S and 0 elsewhere on G. As in equation (4.1), $*$ denotes convolution of functions on G. The basis of $L^2(\mathbb{Z}_6)$ that gives rise to the adjacency matrix A is the set $\delta_{\{j\}}, j=1,2,\ldots,6$, where $\delta_j(x)=1$ if $x\equiv j \pmod 6$ and $\delta_j(x)=0$ otherwise.

Exercise 4.2.7 *Give the details needed to prove the claim made in the preceding paragraph.*

The spectral theorem from linear algebra says that $L^2(G)$ has an orthogonal basis of eigenfunctions of A. In fact we know these eigenfunctions in our special case. They are the characters of the additive group $\mathbb{Z}/6\mathbb{Z}$. To see this, just note that

$$A\chi_b(x)=\chi_b(x+1)+\chi_b(x-1)=(\chi_b(1)+\chi_b(-1))\,\chi_b(x).$$

Thus the eigenvalues of the adjacency matrix A are $\lambda_b=\chi_b(1)+\chi_b(-1)=2\cos(2\pi b/6)$, $b=0,1,2,3,4,5$. This is a case for which the finite Fourier transform is easier to use than Matlab, Mathematica, Scientific Workplace or whatever is your favorite program. So the **spectrum** – spec(A) for short – or the set of eigenvalues of the adjacency matrix coming from benzene is:

$$\begin{aligned}\operatorname{spec}(A)&=\{2\cos(0),2\cos(\pi/3),2\cos(2\pi/3),2\cos(\pi),\cos(4\pi/3),\cos(5\pi/3)\}\\&=\{-2,-1,-1,1,1,2\}.\end{aligned} \tag{4.6}$$

In order to think about the stability of benzene, one considers a system of vibrating springs arranged in a hexagon. The solutions can be expressed as linear combinations of states corresponding to the orthogonal eigenfunctions; see Terras [116, p. 214]. Exercise 4.3.8 involves a simpler system of springs.

In 1932 the chemist Hückel argued that the stability is governed by the **rest mass energy** defined by summing the largest half of the spectrum of A in (4.6):

$$E=\frac{2}{n}\sum_{\substack{\text{top } \frac{n}{2}\\ \lambda\in\operatorname{spec}(A)}}\lambda.$$

For benzene we get $E=\frac{1}{3}(2+1+1)\cong 1.3333$. If you compare this with the value of E for a four-vertex ring like cyclobutadiene, you see that it is indeed larger. We leave that as an exercise. A good reference for Hückel theory is Starzak [112].

Exercise 4.2.8 *Compute the spectrum of the adjacency matrix of the Cayley graph* $X(\mathbb{Z}_6,\{1,3,5 \pmod 6\})$.

Exercise 4.2.9 *Compute the rest mass energy for a molecule corresponding to the Cayley graph of* $\mathbb{Z}/4\mathbb{Z}$ *with generating set* $\{\pm 1 \pmod 4\}$.

It is possible to compute the rest mass energy for more complicated molecules such as buckminsterfullerene C_{60}. Here the Cayley graph is the soccerball (the old version) also known as the truncated icocahedron. Since the group involved is not Abelian, more complicated non-Abelian Fourier analysis is needed to see that the rest mass energy is 1.5. See Terras [116].

Exercise 4.2.10 *If $G=\mathbb{Z}/n\mathbb{Z}$, under addition, suppose the adjacency operator on the Cayley graph $X(G,S)$ is defined by (4.2) and the Fourier transform is defined by (4.5). Show that for any function $h(x)$, $x\in G$, the Fourier transform diagonalizes the adjacency operator, meaning that if we write $\widehat{h}=\mathcal{F}h$, then $\left(\mathcal{F}A\mathcal{F}^{-1}\widehat{h}\right)(\chi)=\mathcal{F}\delta_S(\chi)\cdot\widehat{h}(\chi)$, for all $\chi\in\widehat{G}$. If $\widehat{G}=\{\chi_1,\ldots,\chi_n\}$, use the basis of $L^2\left(\widehat{G}\right)$, given by the set of functions $\delta_{\{\chi_j\}}, j=1,2,\ldots,n$, where $\delta_{\{\chi_j\}}$ denotes the function that is 1 on the set $\{\chi_j\}$ and 0 otherwise. Then show that the matrix you get for $\mathcal{F}A\mathcal{F}^{-1}$ is diagonal, with jth diagonal entry $(\mathcal{F}\delta_S)(\chi_j)$.*

Exercise 4.2.11 ***(The Plancherel Theorem).*** *Show that if we define the norm of a function $f\in L^2(\mathbb{Z}_n)$ by $\|f\|_2^2=\sum_{x\in\mathbb{Z}_n}|f(x)|^2$ and we make the same definition of the norm of a function $F:\widehat{\mathbb{Z}_n}\to\mathbb{C}$, then*

$$\|f\|_2^2=\frac{1}{n}\left\|\widehat{f}\right\|_2^2.$$

Exercise 4.2.12 *Consider $\mathbb{Z}_{101}\to\mathbb{R}$ defined by*

$$f(x)=\begin{cases}1, & x\equiv\pm1,\pm2,\pm3,\pm4,\pm5 \pmod{101},\\ 0, & \text{otherwise}.\end{cases}.$$

Graph f as a function on the real line for $x\in[-50,50]\cap\mathbb{Z}$. Then graph $f*f$ and compare. Note that the function f is nonzero on $\{\pm1,\pm2,\pm3,\pm4,\pm5 \pmod{101}\}$. This set is called the support of f. What is the support of $f*f$?

The word “spectrum” is associated to spectroscopy – the study of the spectral properties of atoms and molecules. Spectroscopy will even allow the analysis of the elements in stars. For example, spectroscopy led to the discovery of helium in the spectrum of the sun before it was discovered on earth. See Neil DeGrasse Tyson [121, Chapter 15] for more of the fascinating story of spectroscopy and its use in cosmology. The subject really requires quantum mechanics, chemistry, and group representations (i.e., Fourier analysis on non-Abelian groups). I say a little more about the subject in [116].

There are many kinds of spectroscopy from shining a light on a prism to shining X-rays on a sample from calf thymus. The type known as X-ray diffraction spectroscopy was important for the discovery around 1953 of the structure of DNA as a double helix by Francis Crick and James Watson. That started with Photo 51, which can be found in a paper of Rosalind Franklin and Raymond Gosling. The photo shows a big X. See the Wikipedia article titled Photo 51. Sadly Rosalind Franklin died of ovarian cancer very soon after this work was done. One suspects the X-rays for that. The Wikipedia article on her life is a good start for trying to understand the history of this figure. The symmetry is quite visible in Photo 51 but the helix takes a leap of imagination – as well as a knowledge of the interpretation of such images in crystallography.

4.3 Groups and Conservation Laws in Physics

In this section algebra meets calculus to do a beautiful dance. To understand how invariance of a Lagrangian for a physical system under a group such as the group $O(3)$ of all 3×3 rotation matrices can lead to a conservation law such as the conservation of angular momentum one must know a bit about the calculus of variations. This is a subject that sadly seems to have disappeared from the undergraduate mathematics curriculum. However, many books for applied mathematics students contain a chapter devoted to variational calculus: for example, Courant and Hilbert [18], Cushing [21], and Greenberg [37]. Mathematica has a package called VariationalMethods which will solve calculus of variation problems. There are also books wholly devoted to the subject such as Gelfand and Fomin [34] and Smith [109]. This section will require a bit of calculus of the more advanced sort. My favorite advanced calculus reference is Lang [63].

It was Emmy Noether who showed that when groups of transformations leave the Lagrangian of a physical system invariant, there is a corresponding conservation law. This theorem of Emmy Noether was published in 1918. It led physicists L. M. Lederman and C. T. Hill to write (in [68]) that Noether's theorem is "certainly one of the most important mathematical theorems ever proved in guiding the development of modern physics."

Emmy Noether lived from 1882 to 1935 and was the creator of much of this algebra course. Sadly she lived in a time when women were not allowed to be professors in most universities and Hitler was forcing anyone who was Jewish to flee Europe or face death in a concentration camp. She died after only a few years in the US teaching at Bryn Mawr college and lecturing at the Institute for Advanced Study in Princeton, New Jersey. See Kramer [59, Chapter 28] for a short biography of Noether written by the mathematician Hermann Weyl. There is also an interesting discussion of the Institute for Advanced Study – which hosted many refugees from the Nazis such as Albert Einstein and Kurt Gödel – in Bardawil *et al.* [7].

An example of a calculus of variations problem is that of finding the curve connecting two points A and B in 3-space that minimizes distance. Such a curve is called a **geodesic**. This is the problem of finding a curve $p(t) = (x(t), y(t), z(t))$, for $t \in [a, b]$, such that $p(a) = A$, $p(b) = B$ and the following integral is minimized:

$$L[y] = \int_a^b \sqrt{x'(t)^2 + y'(t)^2 + z'(t)^2}\, dt.$$

You may feel that you already know the answer to this problem. If you replace the Euclidean distance with some other distance such as that on the surface of a sphere, then you may find the problem more interesting.

The simplest result in the calculus of variations goes as follows. Euler found it in 1744. Suppose that $F(x, y, z)$ is a nice function (e.g., possesses continuous second partials with respect to all variables). Consider the **functional**

$$J[y] = \int_a^b F(x, y, y')\, dx$$

on the domain

$$D = \{y \text{ continuously differentiable such that } y(a) = A \text{ and } y(b) = B\}. \tag{4.7}$$

Here "functional" just means a function whose variables are themselves functions. Then, if $J[y]$ is a maximum or minimum, we call y an **extremal** and it follows that y must satisfy the **Euler–Lagrange equation:**

$$\frac{\partial F}{\partial y} - \frac{d}{dt}\frac{\partial F}{\partial y'} = 0. \tag{4.8}$$

This is proved by a first derivative test argument.

Proof. A sketch of the proof which does a bit of cheating goes as follows. Let $f \in D$ as defined by (4.7). We want to create a "straight line" in this function space. To do so, fix a continuously differentiable function h on $[a, b]$ such that $h(a) = h(b) = 0$. Then for $t \in \mathbb{R}$, form the function $(y + th)(x)$. This new function of x satisfies the same boundary conditions as f and thus lies in our domain D. Thus if $J[y]$ is a minimum over the y in D, the new function $j(t) = J[y + th]$ has a minimum at $t = 0$. Then we can apply the ordinary first derivative test from calculus. We assume here that we can differentiate under the integral sign. And we use the chain rule for functions of several variables. This gives:

$$0 = j'(0) = \left.\frac{dJ[y + th]}{dt}\right|_{t=0} = \int_a^b \left.\frac{dF(x, y + th, y' + th')}{dt}\right|_{t=0} dx$$

$$= \int_a^b \left\{ \frac{\partial F(x, y, y')}{\partial y} h + \frac{\partial F(x, y, y')}{\partial y'} h' \right\} dx.$$

Then use integration by parts on the second term inside the {} to see that

$$0 = \int_a^b \left\{ \frac{\partial F(x, y, y')}{\partial y} - \frac{d}{dx}\frac{\partial F(x, y, y')}{\partial y'} \right\} h\, dx + \left.\frac{\partial F(x, y, y')}{\partial y'} h\right|_a^b.$$

Since $h(a) = h(b) = 0$, it follows that the boundary term vanishes. Thus

$$0 = \int_a^b \left\{ \frac{\partial F(x, y, y')}{\partial y} - \frac{d}{dx}\frac{\partial F(x, y, y')}{\partial y'} \right\} h\, dx$$

for all continuously differentiable functions h on $[a, b]$ such that $h(a) = h(b) = 0$. This implies that the term inside {} is identically zero, otherwise (using the continuity of the function in {}) one could create a function h to produce a contradiction. And that is indeed the Euler–Lagrange equation. If you want a totally air-tight argument, see the references. ▲

For the example of geodesics in 3-space, we need to consider the problem with more than one dependent variable. We ask for the curve $(y_1(x), y_2(x), y_3(x))$, $a \le x \le b$, maximizing or minimizing the functional

$$J[y_1, y_2, y_3] = \int_a^b F(x, y_1, y_2, y_3, y_1', y_2', y_3')\, dx$$

such that $(y_1(a), y_2(a), y_3(a)) = A$ and $(y_1(b), y_2(b), y_3(b)) = B$. Then one can use a similar argument to that just given to see that $y = (y_1, y_2, y_3)$ must satisfy a simultaneous **system of Euler–Lagrange equations:**

$$\left.\begin{aligned} \frac{\partial F}{\partial y_1} - \frac{d}{dt}\frac{\partial F}{\partial y_1'} &= 0, \\ \frac{\partial F}{\partial y_2} - \frac{d}{dt}\frac{\partial F}{\partial y_2'} &= 0, \\ \frac{\partial F}{\partial y_3} - \frac{d}{dt}\frac{\partial F}{\partial y_3'} &= 0. \end{aligned}\right\} \tag{4.9}$$

Exercise 4.3.1 *Obtain the system of equations (4.9) by a similar argument to the one that worked for one dependent variable.*

Now consider the problem of geodesics. What are the Euler–Lagrange equations? Note that $F = \sqrt{x'(t)^2 + y'(t)^2 + z'(t)^2}$ is independent of x, y, z. So the result is that $x'' = y'' = z'' = 0$. It follows that the tangent vector (x', y', z') is a constant vector. Thus the curve must be a straight line. Well, we knew that. Moreover, we still must prove that the straight line gives a minimum. We will make this an exercise after discussing the invariance of a functional under a transformation of the coordinates. This is directly related to the application of group theory found by Emmy Noether.

But before doing that, let us just note why this is useful for physicists. It is often possible to state the laws of physics as minimum principles for integrals on spaces of functions. For example, one can derive Newton's equations of motion from the principle of least action. What is action? Suppose we have a particle of mass m and position $(x(t), y(t), x(t))$ at time t acted on by a force F such as gravity. Then we make the following definitions.

Definition 4.3.1 *The **kinetic energy** of the particle is* $T = \frac{m}{2}\left((x')^2 + (y')^2 + (z')^2\right)$.

Definition 4.3.2 *The **potential energy** of the particle is U, where the force* $F = -\text{grad}\, U = -\left(\frac{\partial U}{\partial x}, \frac{\partial U}{\partial y}, \frac{\partial U}{\partial z}\right)$.

Definition 4.3.3 *The **Lagrangian** is* $L = T - U$.

Definition 4.3.4 *The **action*** $A = \int_{t_1}^{t_2} L\, dt$.

The **principle of least action** says that the particle will move so as to minimize action. The Euler–Lagrange equations say that this implies Newton's law – force = mass × acceleration.

In the situation of several dependent variables we get three Euler–Lagrange equations, one for each variable:

$$\frac{\partial}{\partial x}(T-U)-\frac{d}{dt}\frac{\partial}{\partial x'}(T-U)=0,$$

$$\frac{\partial}{\partial y}(T-U)-\frac{d}{dt}\frac{\partial}{\partial y'}(T-U)=0,$$

$$\frac{\partial}{\partial z}(T-U)-\frac{d}{dt}\frac{\partial}{\partial z'}(T-U)=0.$$

The end result is that $F=m(x'',y'',z'')=ma$, where a is acceleration.

Exercise 4.3.2 *Prove the last statement.*

Many of the favorite differential equations of mathematical physics can be derived from variational principles, such as Maxwell's equations, the Schrödinger equation. This makes the properties somewhat easier to understand I think – especially when one knows the following theorem of Emmy Noether.

Before stating her theorem, we need a definition. We say that a functional $J[y]=\int_a^b F(x,y,y')\,dx$ is **invariant under a transformation** $x^*=\Phi(x,y,y')$, $y^*=\Psi(x,y,y')$ if

$$\int_a^b F\left(x,y,\frac{dy}{dx}\right)dx=\int_{a^*}^{b^*} F\left(x^*,y^*,\frac{dy^*}{dx^*}\right)dx^*.$$

As an example, consider any functional which is independent of x, that is, $J[y]=\int_a^b F(y,y')dx$. Then if we consider the translation or shift transformation $x^*=x+t$, for some real t, we see that to shift the curve you get $y^*(x^*)=y(x^*-t)$. Then using the formula for substitution in a definite integral:

$$J[y^*]=J[y]=\int_{a^*}^{b^*} F\left(y^*,\frac{dy^*}{dx^*}\right)dx^*=\int_a^b F\left(y,\frac{dy}{dx}\right)dx.$$

We have already considered the three-dimensional version of the following exercise, but we did not prove that straight lines minimize distance. Thus it may be useful to do this exercise to obtain a simpler approach that works in other situations. For example, similar arguments can be used to see that great circles minimize distance on the sphere and to find geodesics in the non-Euclidean upper half plane. See Terras [118, pp. 111 and 151].

Exercise 4.3.3 *Consider a curve $p(t)=(x(t),y(t))$ in the plane, for $a\le t\le b$, connecting the points $A=(x(a),y(a))$ and $B=(x(b),y(b))$. The length of this curve is*

$$J[y]=\int_a^b \sqrt{x'(t)^2+y'(t)^2}\,dt.$$

Show that the minimum is achieved when the curve is a straight line. Thus straight lines are geodesics – no surprise.

Hint. *It is not hard to see that the arc length is invariant under translation of the point A to the origin and rotation of the point B to the y-axis. Then it is clear that* $\sqrt{(x'(t)^2) + (y'(t))^2} \geq |y'(t)|$. *The segment of the curve on the y-axis clearly minimizes the functional.*

Next we can finally state Emmy Noether's theorem.

Theorem 4.3.1 ***(Emmy Noether's Theorem).*** *Suppose that the functional*

$$J[y] = \int_a^b F(x, y, y')dx$$

is invariant under the one-parameter family of transformations defined for all values of the real parameter α *by*

$$x^\alpha = \Phi(x, y, y'; \alpha), \quad y^\alpha = \Psi(x, y, y'; \alpha)$$

such that $x^0 = x$ *and* $y^0 = y$, *for any choice of a and b. Then setting*

$$\phi(x, y, y') = \left.\frac{\partial\Phi\,(x, y, y'; \alpha)}{\partial\alpha}\right|_{\alpha=0} \quad \text{and} \quad \psi(x, y, y') = \left.\frac{\partial\Psi\,(x, y, y'; \alpha)}{\partial\alpha}\right|_{\alpha=0},$$

we have

$$\frac{\partial F}{\partial y'}\,\psi + \left(F - y'\frac{\partial F}{\partial y'}\right)\phi = \text{constant}$$

along each curve $y = y(x)$ *such that* $\frac{\partial F}{\partial y} - \frac{d}{dx}\frac{\partial F}{\partial y'} = 0$. *We are of course assuming that the functions* $\Phi(x, y, y'; \alpha)$ *and* $\Psi(x, y, y'; \alpha)$ *are continuously differentiable.*

Proof. To sketch the proof of Noether's theorem, proceed as follows. First we parameterize our curves: $(x^\alpha(t), y^\alpha(t))$, with the parameter t between 0 and 1. Then we can rewrite our functional using $'$ to mean d/dt,

$$G\,(x, y, x', y') = x'F\left(x, y, \frac{y'}{x'}\right)$$

to obtain

$$J[y] = J\,[x, y] = \int_0^1 G(x, y, x', y')dt.$$

Since $J\,[x^\alpha, y^\alpha] =$ constant, for all $\alpha \in R$, it follows (using integration by parts again) that

$$0 = \left.\frac{dJ\,[x^\alpha, y^\alpha]}{d\alpha}\right|_{\alpha=0} = \int_0^1 \left\{\left(\frac{\partial G}{\partial x} - \frac{d}{dt}\frac{\partial G}{\partial x'}\right)\phi + \left(\frac{\partial G}{\partial y} - \frac{d}{dt}\frac{\partial G}{\partial y'}\right)\psi\right\} dt$$

$$+ \left.\frac{\partial G}{\partial x'}\phi + \frac{\partial G}{\partial y'}\psi\right|_{t=0}^{1}.$$

Euler–Lagrange says

$$\left.\frac{\partial G}{\partial x'}\phi + \frac{\partial G}{\partial y'}\psi\right|_{t=0}^{1} = 0.$$

This implies the theorem using the definition of G and the fact that a and b are arbitrary. ▲

Exercise 4.3.4 *Fill in the details in the preceding argument.*

It is possible to extend Noether's theorem to the case of more than one dependent variable. Suppose we have a functional

$$J[y_1, \ldots, y_n] = \int_a^b F(x, y_1, \ldots, y_n, y_1', \ldots, y_n')\,dx,$$

and suppose that J is invariant under the one-parameter family of transformations:

$$\begin{aligned} x^\alpha &= \Phi(x, y_1, \ldots, y_n, y_1', \ldots, y_n'; \alpha) \\ y_i^\alpha &= \Psi_i(x, y_1, \ldots, y_n, y_1', \ldots, y_n'; \alpha), \quad i = 1, \ldots, n \end{aligned}$$

such that $x^0 = x$ and $y^0 = y$. Then Noether's theorem says that along each extremal $y_1, \ldots, y_n$ for $J[y_1, \ldots, y_n]$, we have

$$\sum_{i=1}^{n} \frac{\partial F}{\partial y_i'}\psi_i + \left(F - \sum_{i=1}^{n} y_i' \frac{\partial F}{\partial y_i'}\right)\phi = \text{constant}, \tag{4.10}$$

where

$$\begin{aligned} \phi(x, y_1, \ldots, y_n, y_1', \ldots, y_n'; \alpha) &= \left.\frac{\partial \Phi(x, y_1, \ldots, y_n, y_1', \ldots, y_n'; \alpha)}{\partial \alpha}\right|_{\alpha=0} \\ \text{and} & \\ \psi_i(x, y_1, \ldots, y_n, y_1', \ldots, y_n'; \alpha) &= \left.\frac{\partial \Psi_i(x, y_1, \ldots, y_n, y_1', \ldots, y_n'; \alpha)}{\partial \alpha}\right|_{\alpha=0}. \end{aligned} \tag{4.11}$$

Exercise 4.3.5 *Give a "proof" of Noether's theorem for functions of more than one dependent variable using the same argument we gave for the case of one dependent variable.*

It is also possible to extend all of our theorems in this section, including Noether's theorem, to the case of more than one independent variable. Then the Euler–Lagrange equations become partial differential equations.

Examples

1. Consider a system of particles as in Definitions 4.3.1–4.3.4. Suppose that the system is conservative: that is, U does not depend on time t explicitly. Then Noether's theorem implies that the total energy $T + U$ is constant throughout the motion.
2. Similarly, if the action integral is invariant under the group of translations of the x, y, z variables, then Noether's theorem says that the total momentum $m(x', y', z')$ is conserved.
3. If the action integral is invariant under rotations, then angular momentum is constant.

4. The functional associated with the electromagnetic field and Maxwell's equations is invariant under many transformations of 4-space and this leads to 15 conservation laws. ▲

Let us give a few details for Example 1. The functional to be minimized is the **action**

$$A = \int_{t_1}^{t_2} L\,dt, \quad \text{where } L = T - U.$$

Here, for a system of n particles at position (x_i, y_i, z_i) at time t, the **kinetic energy** is

$$T = \frac{1}{2}\sum_{i=1}^{n} m_i \left((x_i')^2 + (y_i')^2 + (z_i')^2 \right).$$

The **potential energy** is

$$U = U(t, x_1, \ldots, x_n, y_1, \ldots, y_n, z_1, \ldots, z_n),$$

so that the force acting on the ith particle is

$$-\left(\frac{\partial U}{\partial x_i}, \frac{\partial U}{\partial y_i}, \frac{\partial U}{\partial z_i} \right).$$

In Example 1, we are assuming that U does not depend on time t explicitly. That is, our independent variable is t and our shift transformation function Φ is $t^\alpha = \Phi(t, x, y, z, x', y', z'; \alpha) = t + \alpha$, for all $\alpha \in \mathbb{R}$. Our Ψ function is the identity, that is, $x_i^* = x_i, y_i^* = y_i, z_i^* = z_i$. It follows that the differentiated functions from (4.11) with respect to α are $\phi = 1$ and $\psi_i = 0$. Then Noether's theorem implies that

$$\sum_{i=1}^{n} \left(L_{x_i'} \cdot 0 + L_{y_i'} \cdot 0 + L_{x_i'} \cdot 0 \right) + \left(L - \sum_{i=1}^{n} \left(x_i' L_{x_i'} + y_i' L_{y_i'} + z_i' L_{z_i'} \right) \cdot 1 \right)$$

is constant along extremals of the action. It follows that the total energy $T + U$ is constant.

Exercise 4.3.6 *Check the last statement.*

Maybe you have been wondering: where are the groups in all of this? The answers for the examples go as follows.

(1) The group of translations of time is $\mathbb{R}$, the group of real numbers under addition.
(2) The group of translations of vectors in $\mathbb{R}^3$ is the additive group of vectors in 3-space. Of course this is not a one-parameter group. To say that a group G is a **one-parameter group** means that you can express any group element in the form $g = g(t),\ t \in \mathbb{R}$, where $g(t)$ is a nice function: for example, continuous, differentiable. But you can certainly build up $\mathbb{R}^3$ out of many one-parameter subgroups.
(3) The group is the rotation group $O(3) = \{3 \times 3$ real matrices U such that ${}^{\mathrm{T}}U = U^{-1}\}$, where ${}^{\mathrm{T}}U$ denotes the **transpose** of U. That is, if $U = (u_{ij})$, then ${}^{\mathrm{T}}U = (u_{ji})$. The group operation is matrix multiplication defined as in formula (1.3) of Section 1.8 and the sentence following it. Again the group is not a one-parameter group. You can express it using three parameters (the Euler angles).

(4) The group is the Lorentz group $O(3,1)$ consisting of 4×4 real matrices A such that ${}^{\mathrm{T}}AJA = J$, where J denotes the matrix

$$\begin{pmatrix} 1 & 0 & 0 & 0 \\ 0 & 1 & 0 & 0 \\ 0 & 0 & 1 & 0 \\ 0 & 0 & 0 & -1 \end{pmatrix}.$$

The group operation is matrix multiplication. This is discussed in the book of Gelfand and Fomin [34, p. 184 ff]. The Lorentz group is important for Einstein's theory of special relativity.

Exercise 4.3.7 *Show that the last two examples $O(3)$ and $O(3,1)$ are groups.*

Hint. *Recall that ${}^{\mathrm{T}}(AB) = {}^{\mathrm{T}}B\,{}^{\mathrm{T}}A$ and ${}^{\mathrm{T}}(A^{-1}) = \left({}^{\mathrm{T}}A\right)^{-1}$. It follows that we can write ${}^{\mathrm{T}}(A^{-1}) = \left({}^{\mathrm{T}}A\right)^{-1} = {}^{\mathrm{T}}A^{-1}$.*

The groups $O(3)$ and $O(1,3)$ are Lie groups – named for Sophus Lie. Such groups are much studied and are of great interest for physics and chemistry. The Lorentz group is fundamental for the theory of special relativity.

Exercise 4.3.8 *Consider the vibrating system consisting of two masses connected by springs to each other and two walls as in Figure 4.1. Assume the two bodies each have mass m and that the three springs each have stiffness constant k. Then the laws of physics tell us that the kinetic and potential energies of the system are*

$$T = \frac{1}{2}m\left(x''\right)^2 + \frac{1}{2}m\left(yx''\right)^2 \quad \text{and} \quad U = \frac{1}{2}kx^2 + \frac{1}{2}k\left(x-y\right)^2 + \frac{1}{2}ky^2.$$

Use the principle of least action to derive the system of differential equations that rule the motion of the system.

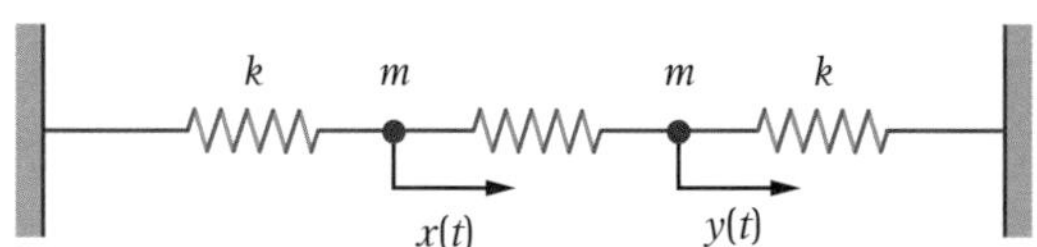

Figure 4.1 Vibrating system of two masses

I discuss a slightly more general vibrating system than that of the preceding problem in Terras [116, p. 215ff]. This system leads to an eigenvalue problem that helps to explain the importance of spectra for the understanding of vibrating systems. You can extend the theory to a system of vibrating masses arranged in a ring of six masses say. This gives a model for understanding benzene. See Starzak [112, Chapter 5] for more examples.

4.4 Puzzles

Next we want to consider an application of group theory to a puzzle – sudoku. There are many other puzzles one could choose, such as Rubik's cube (see Joyner [49]). However, it seems preferable to keep the puzzles as easy as possible. Even restricting our sudokus to smaller grids than are interesting to sudoku experts, we come face to face with some fairly big groups.

A good reference is Jason Rosenhouse and Laura Taalman [94]. Another is Crystal Lorch and John Lorch [71].

A classic sudoku puzzle is a 9×9 grid in which a number of clues are entered. The object is to fill in the rest of the grid so that each row has the numbers from 1 to 9 in some order, ditto for each column, ditto for each of the nine 3×3 blocks into which the grid is divided by two equally spaced vertical lines and two equally spaced horizontal lines. For the puzzle to be valid it must have one and only one solution. Often the puzzles are made to have some symmetry as well.

Here we will simplify our group theory by looking at shidoku or junior sudoku, which involves a 4×4 grid to be filled in with the numbers from 1 to 4. An example from Rosenhouse and Taalman [94, p. 160] is

	2	3	
3		1	
2	1		

The four blocks are outlined with double lines. This is a puzzle that has the maximal number of clues for a not totally trivial puzzle with a unique solution. The following example has the minimal number of clues in some sense, again from Rosenhouse and Taalman [94, p. 167]

1			
		3	
	2		
			4

We leave it to the reader to solve these puzzles.

Exercise 4.4.1 *Solve the preceding two shidoku puzzles.*

There are many questions one could ask. How many puzzles are there? What is the group acting on puzzles? How many orbits does this group have in the set of shidoku puzzles? Burnside's formula should help us once the group is identified. However brute force will also work. But we are impelled to take the group approach.

We shall say that a **shidoku grid** is a 4×4 grid, divided into 2×2 blocks, that has been completely filled in with the numbers $1, 2, 3, 4$ so that each row, column, and block contains all of the numbers $1, 2, 3, 4$. A **shidoku puzzle** is a grid with missing entries. What is the group acting on the shidoku grids? Well, clearly S_4 acts by permuting (or relabeling) the entries in the grid. Then there is the group K generated by the following operations:

(1) switch rows 1 and 2 or rows 3 and 4;
(2) switch columns 1 and 2 or columns 3 and 4;
(3) switch the top two blocks with the bottom two blocks;
(4) switch the right two blocks with the left two blocks;
(5) rotate the grid counterclockwise by $90°$;
(6) transpose the grid (as a 4×4 matrix) – meaning interchange column i with row i, $i = 1, 2, 3, 4$.

Exercise 4.4.2 *Show that it is not legal to switch rows 1 and 3.*

The shidoku group G is generated by these six operations on the grids plus S_4. To count the number of essentially different shidokus is to count the number of orbits of G on the shidoku grids. Burnside's lemma is set up to do this. The group G is the direct product of S_4 with K. The group G has order $3072 = 24 \times 128$. It is considered in great detail in the paper of Elizabeth Arnold, Rebecca Field, Stephen Lucas and Laura Taalman [1]. To use Burnside's lemma, one wants a smaller group.

Exercise 4.4.3

(a) Show that the group K generated by the six operations listed above has order 128.

(b) Since the action of S_4 commutes with the action of K, show that the shidoku group G is isomorphic to the direct product of S_4 with K.

Before attempting to use Burnside's lemma, let us first use brute force to find the number of inequivalent shidoku grids under the action of the shidoku group G. It is not really so hard to see that there are two orbits. Use the permutations of entries to see that we can take the upper left-hand block to be that in the grid below:

1	2		
3	4		

Then by permutation we can complete the first row and column as follows:

1	2	3	4
3	4		
2			
4			

Now there are essentially three things to try:

1	2	3	4
3	4	1	2
2	1		
4	3		

1	2	3	4
3	4	2	1
2	1		
4	3		

1	2	3	4
3	4	1	2
2	3		
4	1		

(4.12)

Each of these grids is equivalent to $96 = 24 \cdot 2 \cdot 2$ others under the action of the group G. Thus there are $288 = 3 \cdot 96$ total shidoku grids. However, there is still an equivalence.

Exercise 4.4.4 *Show that it is not possible to complete the following puzzle to a legal shidoku grid:*

1	2	3	4
3	4	2	1
2	3		
4	1		

Note that the second shidoku grid in (4.12) can be transformed to the third by a sequence of elements of G. Transpose the grid then permute the numbers 2 and 3. It follows that there are exactly two orbits of G on the shidoku grids. One has 96 elements and the other has 192 elements.

Rather than applying Burnside's lemma to the rather large group G, Arnold *et al.* [1] instead apply it to the subgroup $G_o = \langle s, t\rangle \times S_4$ of $H \times S_4$, where s denotes the operation of switching rows 3 and 4 of the shidoku grid, t denotes transpose of the grid, and $\langle s, t\rangle$ denotes the group generated by s and t. This group has order 192, which is the minimal order of a group which partitions the shidoku grids into the same number of orbits as G. Arnold *et al.* [1] then use the Burnside lemma to see that the number of orbits of G_o is indeed 2. They note that it is possible to visualize what is happening by creating a graph in which each vertex represents a shidoku grid. Then an edge corresponds to generators of the groups acting on the grid. For G_o they take generators $s, t, (12), (23), (34), (14)$. See Figure 6 of the paper [1].

We consider this Burnside's lemma calculation. First one needs to find the conjugacy classes for the group H generated by s, the operation of switching rows 3 and 4 of the shidoku grid, and t, the transpose of the grid. One finds that H is

$$H = \langle s, t\rangle = \{e, s, t, st, ts, sts, tst, stst\}, \text{ where } e \text{ is the identity.} \tag{4.13}$$

The defining relations for H are $s^2 = t^2 = e$ and $(st)^4 = e$. The group H has conjugacy classes $\{g\} = \{xgx^{-1} | x \in G\}$ represented by $e,\ s,\ t,\ st,\ (st)^2$.

Exercise 4.4.5

(a) *Prove equation (4.13) and show that* $(st)^4 = e$.

(b) *Since the group H is small, you should be able to identify it – especially after the next section. The problem is that you need different generators. Try* $x = st$ *and* t. *Then you will see the relations are* $t^2 = e$ *and* $x^4 = e$, *plus* $xtx = t$. *Thus you should be able to show that* H *is* D_4.

Exercise 4.4.6 *Show that the conjugacy classes in the group* $H = \langle s, t\rangle$ *defined in (4.13) are represented by* $e, s, t, st, (st)^2$, *where s denotes the operation of switching rows 3 and 4 of the shidoku grid, and t denotes transpose of the grid. Find the order of each conjugacy class. You may find it easier to compute everything using the preceding exercise to realize* H *as* D_4.

With the generators s, t, the group H is known as a Coxeter group. Many of our favorite groups turn out to be Coxeter groups: for example, all the dihedral groups, the symmetric groups, groups of reflections in Euclidean space or hyperbolic space. An example of the last group is $PSL(2, \mathbb{Z}) = SL(2, \mathbb{Z})/Z$, where $SL(2, \mathbb{Z})$ is the modular group of 2×2 integer matrices of determinant 1 and Z denotes the center of $SL(2, \mathbb{Z})$. A **Coxeter group** is defined to have generators $a_1, \ldots, a_n$ and relations $(a_i a_j)^{m_{ij}} = e$, the identity. Here $m_{ij} \in \mathbb{Z}^+ \cup \{\infty\}$, $m_{ii} = 1$ and $m_{ij} \geq 2$, for $i \neq j$. Thus all generators have order 2. If $m_{ij} = \infty$, there is no relation on $a_i a_j$. A reference for the subject is Björner and Brenti [10].

Exercise 4.4.7 *Given a Coxeter group, as described in the preceding paragraph, show that if for all* $i \neq j$ *we have* $m_{ij} = 2$, *then* $a_i a_j = a_j a_i$.

Exercise 4.4.8 *Show that the symmetric group S_n is a Coxeter group.*

Hint. *Use the transpositions $(a \quad a+1)$ to generate the symmetric group.*

To use Burnside's lemma, one must do a count. For each conjugacy class of H, one must count the invariant shidoku grids up to permutation of the entries. For example, consider the row swap s. The claim is that there is no shidoku grid for which a relabel will undo s. On the other hand if we consider t, we see that the following grid is invariant up to the permutation by (23).

$$\begin{array}{|cc|cc|} \hline 1 & 2 & 4 & 3 \\ 3 & 4 & 2 & 1 \\ \hline 4 & 3 & 1 & 2 \\ 2 & 1 & 3 & 4 \\ \hline \end{array} \quad \begin{array}{c} \xrightarrow{t} \\ \\ \xleftarrow{(23)} \end{array} \quad \begin{array}{|cc|cc|} \hline 1 & 3 & 4 & 2 \\ 2 & 4 & 3 & 1 \\ \hline 4 & 2 & 1 & 3 \\ 3 & 1 & 2 & 4 \\ \hline \end{array} .$$

One finds that the number of grids invariant up to permutation of the entries under the conjugacy class of t is $2 \cdot 4!$. This and the conjugacy class of the identity are the only conjugacy classes leaving grids invariant up to entry permutation. It follows that Burnside's lemma says that the number of orbits of the group $G_o = \langle s, t\rangle \times S_4$ acting on the grids is

$$\frac{1(12 \cdot 4!) + 2(0) + 2(2 \cdot 4!) + 2(0) + 1(0)}{8 \cdot 4!} = 2. \tag{4.14}$$

Exercise 4.4.9 *Show that conjugate elements of H correspond to the same number of G-invariant shidoku grids.*

Exercise 4.4.10 *Check the computation in equation (4.14).*

Since the number of G-inequivalent shidoku grids is only 2, you might worry that newspapers will run out of the classic sudoku 9×9 grids. However, Jarvis and Russell [48] found that the number of orbits of the larger group acting on 9×9 sudoku grids is 5 472 730 538 using Burnside's lemma and lots of computer time. So far, there is no simple way to find the number of orbits for classic sudoku. Of course, there are many puzzles corresponding to a single grid. McGuire *et al.* [76] show that 17 clues are necessary to produce a puzzle with a unique solution for classic sudoku. Gordon Royle's website has at least 50 000 classic sudoku puzzles with 17 clues. We ask for analogs for shidoku in the next problem (see Herzberg and Murty [43]).

Exercise 4.4.11 *What is the minimum number of clues necessary for a shidoku grid to have a unique solution?*

4.5 Small Groups

We seek to list the groups of small order up to isomorphism. Of course the program Group Explorer does this for us. But we want to convince ourselves that the list is complete. Unfortunately, we may not have the patience to give every detail of the proofs at this point. Moreover we will want to use the Sylow theorems which we stated in Section 3.7 – evilly without proof.

For orders p which are prime, we know by Corollary 3.3.1 of Lagrange's theorem that the group is cyclic C_p and that is all the possibilities for group orders 2, 3, 5, 7, 11, 13. Of course, there is only one group of order 1 as well.

It helps to know the **fundamental theorem of finitely generated Abelian groups**, which implies that a finite Abelian group is a direct product of cyclic groups. We will assume this theorem in this section. It has a very nice proof involving the analog of Gaussian elimination for matrices with integer entries. See Section 7.1 for a sketch of a proof.

We have seen in Exercise 2.4.14 that there are two groups of **order 4**: the cyclic group C_4 and the Klein 4-group $C_4 \oplus C_4$.

For **order 6**, there are two possibilities. If the group G of order 6 is not cyclic, it can only have elements of orders 1, 2, and 3 by Lagrange's theorem. Cauchy's theorem implies that there must be elements of orders 2 and 3. If G were Abelian, G would be cyclic by Exercise 3.6.17. Thus our group cannot be Abelian.

We know that G must have an element h of order 3. Then $H = \langle h \rangle$ is a normal subgroup of G, by Exercise 3.3.7. So G/H is cyclic of order 2. It follows that $G = \{e, h, h^2, g, gh, gh^2\}$, for some $g \in G - H$. One can show that $g^2 = e$ and then that G is isomorphic to S_3. To do this, use the fact that $g^2 \in \langle h \rangle = \{e, h, h^2\}$ and note that if $g^2 = h$ or h^{-1}, then g would have order 6 and G would be cyclic. Thus $g^2 = e$. Then note that $ghg \in \{h, h^2\}$ since the order of ghg is 3. But $ghg = h$ implies $gh = hg$ and the group is Abelian. Therefore $ghg = h^2$.

Exercise 4.5.1 *Complete the proof that any non-cyclic group of order* 6 *is isomorphic to* S_3.

Hint. *You know enough to do the multiplication table for G.*

If G is an Abelian group of **order 8**, there are as many possibilities as ways to write 8 as a product of positive integers. We have $8,\ 4 \cdot 2,\ 2 \cdot 2 \cdot 2$. Thus we have the Abelian groups $C_8, C_4 \oplus C_2, C_2 \oplus C_2 \oplus C_2$.

If G is a non-Abelian group of order 8, it turns out that there are two possibilities: the dihedral group D_4 and the quaternion group Q. To see this, you need to think about the possibilities for orders of elements. They cannot all have order 1 or 2 as then the group would be Abelian by Exercise 2.1.10. So there must be an element h of order 4. Then $H = \langle h \rangle$ is a normal subgroup of G, by Exercise 3.3.7. So G is the disjoint union of H and gH, for some $g \in G$. Moreover $g^2 \in H$.

Case 1. $ghg^{-1} = h^3$ and $g^2 = e$.

Case 2. $ghg^{-1} = h^3$ and $g^2 = h^2$.

It is a challenge to show that these are the only cases left to consider when G is a non-Abelian group of order 8. Then one must show that in the first case we get the dihedral group D_4 and in the second the quaternion group.

Exercise 4.5.2 *Explain the following statements arising from the preceding discussion of a non-Abelian group G of order* 8, *with element h of order* 4.

(a) We must have $ghg^{-1} = h^3$.
(b) It is not possible that there are two more cases with either $g^2 = h$ *or* $g^2 = h^3$.

Exercise 4.5.3 *Explain why the groups* $C_8, C_4 \oplus C_2, C_2 \oplus C_2 \oplus C_2$ *are not isomorphic.*

Exercise 4.5.4 *Prove that there are only two possibilities for a group of* **order** 9*: the cyclic group* C_9 *and the direct product* $C_3 \oplus C_3$. *In fact this generalizes to groups of order* p^2, *where* p *is a prime.*

Hint. *By Lagrange's theorem we are reduced to the case that all non-identity elements* a *of* G *have order* 3. *Then you can show that* $\langle a \rangle$ *is a normal subgroup. To see this, obtain a contradiction by assuming that* $\exists\, b \in G$ *such that* $\langle a \rangle \neq b \langle a \rangle b^{-1}$. *Then* b^{-1} *is in some coset of* $b \langle a \rangle b^{-1}$ *which will imply* $b \in \langle a \rangle$. *Moreover* $G/\langle a \rangle$ *has order* 3. *Use Exercise 3.6.12.*

For **order 10**, there are again two possibilities: C_{10} and D_5.

For **order 12** there are five possibilities: $C_{12}, C_2 \oplus C_6, D_6, A_4$, the semi-direct product $C_3 \rtimes C_4$ of C_3 and C_4. This last group $C_3 \rtimes C_4$ is a non-commutative group with generators a, b satisfying the relations $a^4 = b^3 = e$ and $bab = a$. On the other hand, the dihedral group D_6 has as generators a, b satisfying the relations $a^2 = b^6 = e$ and $bab = a^{-1}$.

Exercise 4.5.5 *Show that there are only two groups of order* 10. *Feel free to use the Sylow theorems (particularly the third).*

Exercise 4.5.6 *Prove that the following table has all the groups of orders* 14 *and* 15. *Again feel free to use the Sylow theorems (particularly the third).*

Finally we list the results of our thoughts about small groups in Table 4.1.

Table 4.1 *Representative non-isomorphic groups of orders* ≤ 15

order 1	C_1				
order 2	C_2				
order 3	C_3				
order 4	C_4	$C_2 \oplus C_2$			
order 5	C_5				
order 6	C_6	S_3			
order 7	C_7				
order 8	C_8	$C_2 \oplus C_4$	$C_2 \oplus C_2 \oplus C_2$	D_4	Q
order 9	C_9	$C_3 \oplus C_3$			
order 10	C_{10}	D_5			
order 11	C_{11}				
order 12	C_{12}	$C_2 \oplus C_6$	D_6	A_4	$C_3 \rtimes C_4$
order 13	C_{13}				
order 14	C_{14}	D_7			
order 15	C_{15}				

The only new group on our list is the **semi-direct product** $C_3 \rtimes C_4$. You can view this group as the points (x, y) with $x \in \mathbb{Z}_3$ under addition and $y \in \mathbb{Z}_4$ under addition, where the group operation $\star$ is the following for $x, u \in \mathbb{Z}_3$ and $y, v \in \mathbb{Z}_4$

$$(x, y) \star (u, v) = (x + (-1)^y u, y + v). \tag{4.15}$$

For a semi-direct product of two groups G and H to be defined, there must be a group action of H on G. In this case $h \in \mathbb{Z}_4$ acts on $g \in \mathbb{Z}_3$ by $h \cdot g = (-1)^h g$. For more information on semi-direct products, see Dummit and Foote [28].

Exercise 4.5.7 *Show that equation (4.15) makes the Cartesian product $\mathbb{Z}_3 \times \mathbb{Z}_4$ into a group.*

The group $C_3 \rtimes C_4$ is also called a **dicyclic group** (a special case of a metacyclic group). See M. Hall [38].

We include in Figures 4.2 and 4.3 the multiplication table and the Cayley graph $X(C_3 \rtimes C_4, \{a, b\})$ from Group Explorer. Note that in these figures the element a has order 4 and the element b has order 6.

e	a	a^2	a^3	b	ab	a^2b	b^2a	b^2	ab^2	a^2b^2	ba
a	a^2	a^3	e	ab	a^2b	b^2a	b	ab^2	a^2b^2	ba	b^2
a^2	a^3	e	a	a^2b	b^2a	b	ab	a^2b^2	ba	b^2	ab^2
a^3	e	a	a^2	b^2a	b	ab	a^2b	ba	b^2	ab^2	a^2b^2
b	ba	a^2b	ab^2	b^2	a	a^2b^2	a^3	a^2	ab	e	b^2a
ab	b^2	b^2a	a^2b^2	ab^2	a^2	ba	e	a^3	a^2b	a	b
a^2b	ab^2	b	ba	a^2b^2	a^3	b^2	a	e	b^2a	a^2	ab
b^2a	a^2b^2	ab	b^2	ba	e	ab^2	a^2	a	b	a^3	a^2b
b^2	b^2a	a^2b^2	ab	a^2	ba	e	ab^2	a^2b	a	b	a^3
ab^2	b	ba	a^2b	a^3	b^2	a	a^2b^2	b^2a	a^2	ab	e
a^2b^2	ab	b^2	b^2a	e	ab^2	a^2	ba	b	a^3	a^2b	a
ba	a^2b	ab^2	b	a	a^2b^2	a^3	b^2	ab	e	b^2a	a^2

Figure 4.2 Group Explorer's multiplication table for the semi-direct product $C_3 \rtimes C_4$

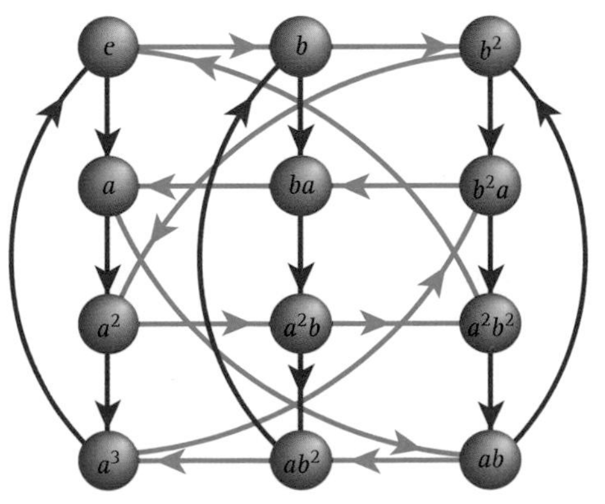

Figure 4.3 Group Explorer draws the Cayley graph $X(C_3 \rtimes C_4, \{a, b\})$

Exercise 4.5.8 *Count the elements of order* 2 *in the three non-Abelian groups of order* 12.

Exercise 4.5.9 *Show that the group* $\mathbb{Z}_3 \rtimes \mathbb{Z}_4$ *can be identified with a group of* 2×2 *complex matrices generated by the matrices* $\begin{pmatrix} 0 & i \\ i & 0 \end{pmatrix}$ *and* $\begin{pmatrix} w & 0 \\ 0 & w^2 \end{pmatrix}$*, where* $\mathrm{i} = \sqrt{-1} = \mathrm{e}^{\pi \mathrm{i}/2}$*, and* $w = \mathrm{e}^{2\pi \mathrm{i}/3}$*. Here* $\mathrm{i}^2 = -1$*, and* $\mathrm{i}^4 = 1$*, while* $w^3 = 1$*. The group operation is matrix multiplication.*

Exercise 4.5.10 *Consider the general linear group* $GL(3, \mathbb{Z}_2)$ *consisting of* 3×3 *matrices* g *whose entries come from* $\mathbb{Z}_2$ *such that* $\det(g) \neq 0$*. The group operation is matrix multiplication. Show that* $|GL(3, \mathbb{Z}_2)| = 168$*. Find the center of this group.*

Hint. *To find the order of* $GL(3, \mathbb{Z}_2)$*, note that the first column can be any vector in* $\mathbb{Z}_2^3$ *except* 0*. The second column can be any vector in* $\mathbb{Z}_2^3$ *except a scalar multiple of the first column. The third column must be outside the subspace of* $\mathbb{Z}_2^3$ *spanned by the first two columns.*

Exercise 4.5.11 *Show that* D_3 *is isomorphic to the* ***affine group*** $\mathrm{Aff}(3)$ *of matrices* $\begin{pmatrix} a & b \\ 0 & 1 \end{pmatrix}$*, with* $a, b \in \mathbb{Z}_3$ *and* $a \neq 0$*. The group operation is matrix multiplication.*

Exercise 4.5.12 *Show that the groups of order* 12 *on our list are not isomorphic.*

Exercise 4.5.13 *Consider the affine group* $\mathrm{Aff}(4)$ *of matrices* $\begin{pmatrix} a & b \\ 0 & 1 \end{pmatrix}$*, with* $b \in \mathbb{Z}_4$ *and* $a \in \mathbb{Z}_4^*$*, with group operation given by matrix multiplication. Which of the groups of order* 8 *in Table 4.1 is isomorphic to* $\mathrm{Aff}(4)$*?*

For more about the finite matrix groups in the exercises, see Terras [116]. One of the favorite graphs in this book is the Cayley graph attached to the affine group $\mathrm{Aff}(p)$, for prime p, with generating set

$$S_{a,\delta} = \left\{ \begin{pmatrix} y & x \\ 0 & 1 \end{pmatrix} \middle| \, x^2 = ay + \delta(y-1)^2 \right\}, \tag{4.16}$$

for δ a non-square in $\mathbb{Z}_p$ and $a \neq 0, 4\delta$. In Figure 4.4, we see the special case $X(\mathrm{Aff}(5), S_{1,2})$. The graph is obtained by putting a star on every face of a dodecahedron. We call these graphs "finite upper half plane graphs" and will consider them again in Section 8.3.

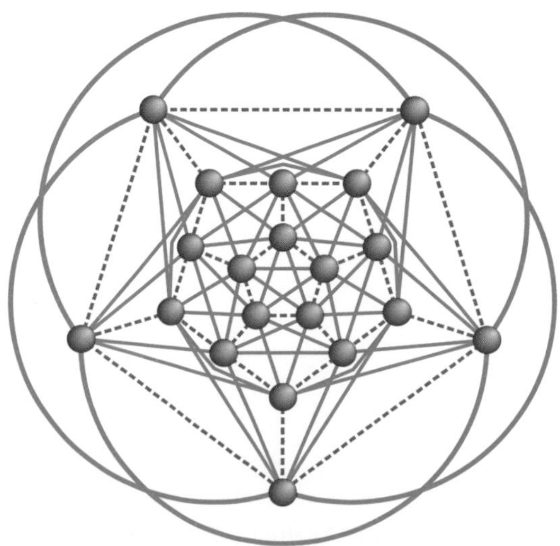

Figure 4.4 The Cayley graph $X(\mathrm{Aff}(5), S_{1,2})$, with generating set defined by equation (4.16), has edges given by solid green lines while the dashed magenta lines are the edges of a dodecahedron

Exercise 4.5.14 *Draw the Cayley graph* $X(\mathrm{Aff}(3), S_{1,2})$. *You should get an octahedron.*

Exercise 4.5.15 *Consider the* ***Heisenberg group*** Heis($\mathbb{R}$) *of matrices*

$$(x,y,z)=\begin{pmatrix}1 & x & z\\ 0 & 1 & y\\ 0 & 0 & 1\end{pmatrix}, \quad \text{with } x,y,z\in\mathbb{R}.$$

The group operation is matrix multiplication. Show that Heis($\mathbb{R}$) *is indeed a group and find its center. Then replace* $\mathbb{R}$ *by* $\mathbb{Z}_p$ *for prime* p. *Find the center of this finite group. If* $p=2$, *this finite Heisenberg group has order* 8. *To which group of order* 8 *is it isomorphic?*

The Heisenberg group with the field $\mathbb{R}$ replaced with a finite ring $\mathbb{Z}/q\mathbb{Z}$, $q=p^r$, p prime, has some interesting figures associated with the spectra of some of its Cayley graphs. One generating set that has been considered is the four-element set $S=\{(\pm1,0,0),(0,\pm1,0)\}$. Since the Heisenberg group is not commutative, one needs representation theory to study the spectra of the adjacency matrices of such graphs. This is described in Terras [116, Chapter 18]. Beautiful pictures come from separating the spectra corresponding to the representations of the Heisenberg group Heis($\mathbb{Z}/q\mathbb{Z}$) which are homomorphisms from Heis($\mathbb{Z}/q\mathbb{Z}$) into $GL(n,\mathbb{C})$. Figure 4.5 was obtained in this way by my student Marvin Minei. Taking larger and larger values of q leads to better and better approximations to a figure of D. R. Hofstadter, who was considering matrices analogous to those from graphs for the Heisenberg group over $\mathbb{Z}$. Hofstadter's butterfly is a fractal and appears to be the limiting figure of those for Heis($\mathbb{Z}/q\mathbb{Z}$) as $q\to\infty$. Hofstadter was interested in the subject thanks to an application in quantum physics. You can find more information about this on Wikipedia.

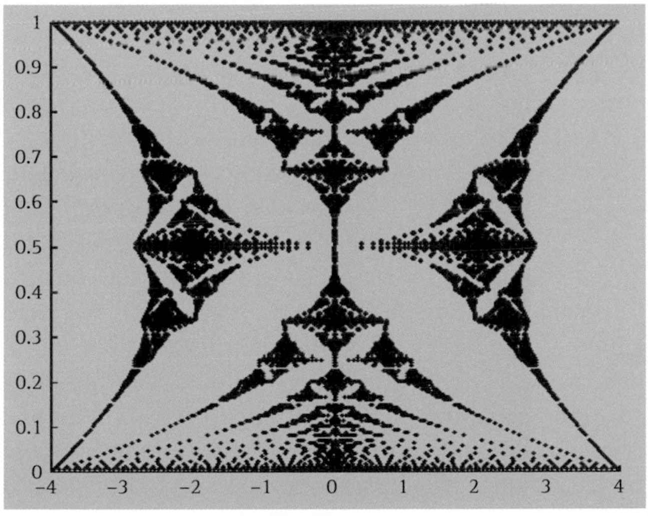

Figure 4.5 Butterfly from Cayley graph of Heis($\mathbb{Z}/169\mathbb{Z}$)

This section classified groups of order ≤ 15 and we did not give all the details. In particular, many things were exercises using Sylow theorems. Moreover, we did not give an exercise to show that the list for order 12 is correct. You can find these details in Dummit and Foote [28, pp. 184–185]. That is perhaps not so impressive. Why did we stop there? The answer is that it would have taken many pages to explain why there are nine non-isomorphic non-Abelian groups of order 16 (and five Abelian order 16 groups). We leave it to the interested reader to seek out the huge amount of information on the web concerning groups of small order. The small groups library for the computer program GAP has a list of representatives of isomorphism classes of groups of order ≤ 2000. See also Wikipedia. The groups of orders 2^n are particularly difficult to classify. It was only in the 1990s that it was found that there are 56 092 groups of order 256. Higher powers of 2 get into the millions and billions.

What is the good of such classifications, you may ask? Chemists are fond of the classification of the space groups of a crystal. Recall that there are 230 such groups (219 if we do not distinguish between mirror images). The types were enumerated by E. S. Federov, A. Schönfliess, and W. Barlow, independently in the 1890s. The symmetry groups of chemical quantities have an effect on the physical and spectroscopic properties of the molecules.

The buckyball or C_{60}, where now C stands for carbon and not cyclic, was discovered by R. Smalley and J. Kroto in 1985. It is a truncated icosahedron (a soccerball) whose faces consist of 20 hexagons and 12 pentagons. The icosahedral group A_5 leaves it invariant and accounts for some of its properties. Chung and Sternberg [15] apply the representation theory (i.e., non-Abelian Fourier analysis) for the icosahedral group to explain the spectral lines of the buckyball. They find that the stability constant for the buckyball is greater than that for benzene considered in Section 4.2.

The classification of groups is a major preoccupation of group theorists. From 1955 to the present at least 100 mathematicians have been working on classifying finite simple groups. Tens of thousands of pages of mathematics papers have been devoted to the project. Is the project finished? I have no idea. People seem to think there are no major gaps in the proofs. See Wikipedia and Wilson [127].

There are many types of finite simple groups: those occurring in infinite lists such as cyclic groups, alternating groups, Coxeter groups, groups of Lie type (e.g., groups like the **projective special linear group** $PSL(n,\mathbb{Z}_p) = SL(n,\mathbb{Z}_p)/Z$, where $SL(n,\mathbb{Z}_p)$ is the **special linear group** of $n \times n$ matrices of determinant 1 and entries in $\mathbb{Z}_p$ for prime p, and Z is its center – except when $n=2$ and $p=2$ or 3). Then there are 26 sporadic groups. The list of sporadic groups ends with the monster group of order $\cong 8 \cdot 10^{53}$. See Wikipedia or Wilson [127] for more information on the classification of finite simple groups.

Classifying infinite groups has also been a major project. The favorite sorts of continuous infinite groups are Lie groups. We saw examples of Lie groups after our consideration of Emmy Noether's theorem in Section 4.3. Another example is $SL(n,\mathbb{C})$, the **special linear group** of $n \times n$ complex matrices of determinant 1. The simple Lie groups over the complex numbers (one of which is $SL(n,\mathbb{C})$) have been classified. Here the meaning of the word "simple" is different from that of finite group theory. W. Killing did this classification in 1887. See Stewart [114]. Again there are exceptional groups on the list. Evidently Killing was not happy about that. In 1894 E. Cartan rederived Killing's theory in his PhD thesis and received most of the credit for the classification of Lie groups. Much of modern physics and number theory involves analysis on Lie groups over not just the complex numbers but also the reals and something called the field $\mathbb{Q}_p$ of p-adic numbers. Then vectors with entries from all of these groups over $\mathbb{R}$, $\mathbb{C}$, $\mathbb{Q}_p$ are put together to form one **adelic group**. Fourier analysis on such groups is used in Wiles' proof of Fermat's last theorem for example.

Groups are deeply embedded in various kinds of mathematics. For example, algebraic topology involves fundamental groups, homotopy groups, homology, and cohomology groups. We restrict ourselves here to the **fundamental group** of a finite graph X. The elements of this group are equivalence classes of closed directed paths on the graph starting and ending at some fixed vertex. Two paths in the graph X are **equivalent** iff one can be continuously deformed into the other. The product of two paths C, D in X means first go around C and then D. It turns out that the fundamental group of X is a free group on r generators, where r is the number of edges left out of a spanning tree T in X. A **spanning tree** T for a graph X means a connected graph with no closed paths such that T has the same vertices as X. For example, if X is the complete graph on four vertices K_4 (alias the tetrahedron graph), a spanning tree is pictured as the solid fuchsia edges in Figure 4.6 and thus there are three dashed purple edges left out of K_4. One closed path is also indicated by following the arrows around the outside triangle. You can create a topologically identical graph by collapsing the spanning tree to a point. This new collapsed graph is a bouquet of three loops as in Figure 4.7. It is fairly clear that each loop provides a generator of the fundamental group of the bouquet graph. So the fundamental group of K_4 is the free group on three generators.

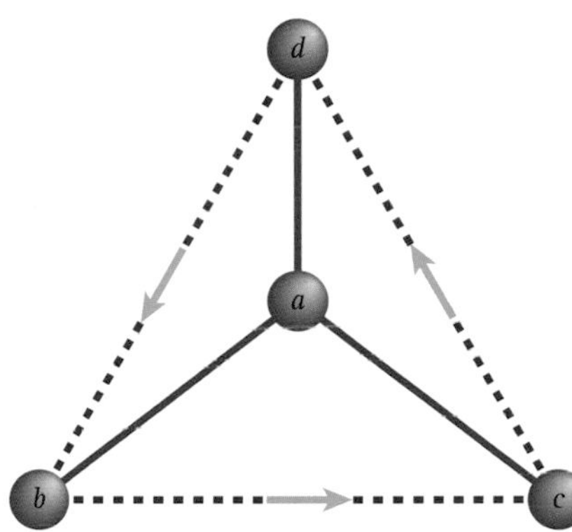

Figure 4.6 A spanning tree for the tetrahedron graph is indicated in solid fuchsia lines. Since the three dashed purple edges are left out, the fundamental group of the tetrahedron graph is the free group on three generators. The arrows show a closed path on the tetrahedron graph

Figure 4.7 The bouquet of three loops obtained by collapsing the tree in the tetrahedron graph of Figure 4.6 to point a

Free groups can be used to construct the group G with presentation $\langle S : R \rangle$ as a quotient of the free group generated by S modulo the normal subgroup generated by the relations R. There are some famous problems associated with such constructions. The **word problem** is that of deciding whether two words in the generators actually represent the same element of a group G defined by generators and relations. This problem was posed by Max Dehn in 1911. P. Novikov showed in 1955 that there is a finitely presented group whose word problem is undecidable.

Dehn was famous for solving the third of David Hilbert's 23 unsolved problems just after the problems were posed by Hilbert at the International Congress of Mathematicians in 1900. Dehn was the first to solve any of the problems, and some are still open. See Wikipedia for a list. There are now seven problems presented by the Clay Mathematics Institute in 2000 – of which only one has so far been solved.

There are more undecidable problems relating to group presentations than the word problem mentioned above. One is the problem of determining whether two groups defined by presentations are actually isomorphic. See Rabin [88].

We end our discussion of groups here with a passion flower from my garden with many different symmetries. The three-fold, five-fold, and ten-fold symmetries are obvious. I have tried to count the tiny hairlike petals and came up with a 90-fold symmetry.

Figure 4.8 A passion flower

Part II

Rings

5

Rings: A Beginning

5.1 Introduction

In the preface we gave a partial introduction to rings – sets with two operations, usually called $+$ and $\times$. Our main examples of rings that are not fields (i.e., closed under division by nonzero elements) are the ring $\mathbb{Z}$ of integers and the ring of polynomials $F[x]$ with coefficients in a field, like $F=\mathbb{Z}_p$, for prime p. These rings are called commutative rings since multiplication is commutative. This would be a good time for the reader to review what we said about the ring of integers $\mathbb{Z}$ in Chapter 1. Our main examples of fields are $\mathbb{Q}$, $\mathbb{C}$, and $\mathbb{Z}_p$ for prime p. Here we add a few pictures made using the arithmetic of finite rings. The distinction between a commutative ring like $\mathbb{Z}$ and a field like $\mathbb{Q}$ is just the fact that in $\mathbb{Q}$ we can divide by nonzero elements and remain in $\mathbb{Q}$, while division by $n \in \mathbb{Z}^+$ tends to ship us outside $\mathbb{Z}$. We will discover that many of the words we associated with groups like subgroup, group homomorphism, quotient group have ring analogs, and that will make the ring theory a bit easier to learn.

Figure 5.1 comes from making an $m \times m$ matrix of values of $x^2 + y^2 \pmod{m}$ for $x, y \in \mathbb{Z}/n\mathbb{Z}$. Then Mathematica does a `ListDensityPlot` of the matrix. There is a movie of such things on my website letting m vary from 3 to 100 or so.

Figure 5.1 The color at point $(x, y) \in \mathbb{Z}_{163}^2$ indicates the value of $x^2 + y^2 \pmod{163}$

A more complicated finite field picture is that of Figure 5.2. It is associated with 2×2 matrices with elements in the finite field $\mathbb{Z}_{11}$. We will explain it in Section 8.3. It should be compared with Figure 2.7.

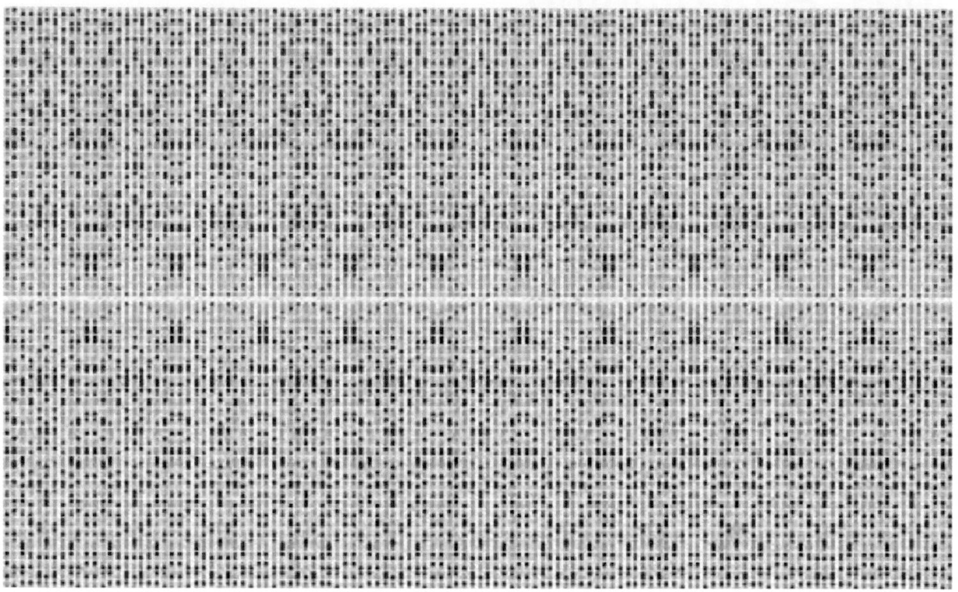

Figure 5.2 Points (x, y), for $x, y \in \mathbb{Z}_{11^2}, y \neq 0$, have the same color if $z = x + y\sqrt{\delta}$ are equivalent under the action of non-singular 2×2 matrices $g = \begin{pmatrix} a & b \\ c & d \end{pmatrix}$ with entries in $\mathbb{Z}_{11}$. The action of g on z is by fractional linear transformation $z \to (az + b)/(cz + d) = gz$. Here δ is a fixed non-square in the field $\mathbb{F}_{121}$ with 121 elements

Exercise 5.1.1 *Make a similar picture to Figure 5.1, replacing $x^2 + y^2 \pmod{m}$ with $x^3 + y^3 \pmod{m}$ for some odd integer m.*

Exercise 5.1.2 *The functions on $\mathbb{Z}_m$ given by $\chi_a(x) = e^{2\pi i(ax)/m}$ for x and $a \in \mathbb{Z}_m$ were considered in Section 4.2. From these functions one can build up trigonometric functions of two variables similar to those we graphed in Figure 2.5. Create a ListDensityPlot in Mathematica for*

$$\sin\left(4\pi\frac{x}{m}\right)\sin\left(\pi\frac{y}{m}\right) + \sqrt{\frac{2}{3}}\sin\left(\pi\frac{x}{m}\right)\sin\left(4\pi\frac{y}{m}\right), \quad \text{with } (x, y) \in \mathbb{Z}_m^2$$

and compare the result with Figure 2.5.

5.2 What is a Ring?

Our favorite ring for error-correcting codes will be $\mathbb{Z}_2$ or $\mathbb{Z}_p$, where p is a prime. Other favorites are the ring of integers $\mathbb{Z}$, the field of real numbers $\mathbb{R}$, the field of complex numbers $\mathbb{C}$, the field of rational numbers $\mathbb{Q}$.

Definition 5.2.1 *A **ring** R is an Abelian group under addition (denoted $+$) with a binary operation of multiplication (denoted $\cdot$) which is associative,*

$$a \cdot (b \cdot c) = (a \cdot b)c, \quad \text{for all } a, b, c \in R,$$

*and satisfies left and right **distributive laws:***

$$a \cdot (b + c) = a \cdot b + a \cdot c \quad \text{and} \quad (a + b) \cdot c = a \cdot c + b \cdot c, \quad \text{for all } a, b, c \in R.$$

We are assuming that multiplication is a binary operation as in Definition 1.8.11. Usually we will write $xy = x \cdot y$.

We will call the identity for addition 0. Multiplication in a ring need not be commutative. If it is, we say that the ring is a **commutative ring**.

Also, the ring need not have an identity for multiplication (except that some people do require this; e.g. Artin [2]). If the ring does have such an identity, we say it is a **ring with (two-sided) identity** for multiplication and we call this identity 1 so that $1 \cdot a = a \cdot 1 = a$, $\forall a \in R$. Some people (e.g., Gallian [33]) call the identity for multiplication a "unity." The word unity seems too close to the word unit which means that the element has a multiplicative inverse. See Definition 5.2.3.

If it exists, the identity for multiplication must be unique by the same argument that worked for groups. Most people might want to assume that 1 and 0 are distinct as well. Otherwise $\{0\}$ is a ring with identity for multiplication. That must be the silliest ring with identity for multiplication. However, it looks like some people do call this a ring with identity for multiplication. What can I say? The terminology is not set in stone yet. The subject is still alive. However, I will normally assume that $1 \neq 0$.

The examples $\mathbb{Z}_n$, for $n \geq 2$, $\mathbb{Z}$, $\mathbb{Q}$, $\mathbb{R}$, $\mathbb{C}$, are all commutative rings with our usual operations of addition and multiplication.

It is possible to drop the requirement that multiplication be associative. We will not consider non-associative rings here. See Exercise 5.2.14 for an example. Imagine the problems if you have to keep the parentheses in your products because $(ab)c \neq a(bc)$.

Example 1. A Non-commutative Ring. Consider the ring $\mathbb{R}^{2\times 2} = \left\{ \begin{pmatrix} a & b \\ c & d \end{pmatrix} \middle| \, a, b, c, d \in \mathbb{R} \right\}$, with addition defined by

$$\begin{pmatrix} a & b \\ c & d \end{pmatrix} + \begin{pmatrix} a' & b' \\ c' & d' \end{pmatrix} = \begin{pmatrix} a + a' & b + b' \\ c + c' & d + d' \end{pmatrix} \tag{5.1}$$

and multiplication defined by

$$\begin{pmatrix} a & b \\ c & d \end{pmatrix} \cdot \begin{pmatrix} a' & b' \\ c' & d' \end{pmatrix} = \begin{pmatrix} aa' + bc' & ab' + bd' \\ ca' + dc' & cb' + dd' \end{pmatrix}. \tag{5.2}$$

This ring is not commutative but it does have an identity for multiplication. What is the identity? ▲

Exercise 5.2.1 *Check the preceding statements about $\mathbb{R}^{2\times 2}$.*

Example 2: A Commutative Ring without an Identity for Multiplication. $2\mathbb{Z}$, the ring of even integers. ▲

Exercise 5.2.2 *Check that $2\mathbb{Z}$ is a ring without identity for multiplication.*

Proposition 5.2.1 ***(Properties of Rings).*** *Suppose that R is a ring. Then, for all $a, b, c \in R$, we have the following facts. Here $-a$ denotes the inverse of a under addition and $a - b = a + (-b)$.*

(1) $a \cdot 0 = 0 \cdot a = 0$, where 0 is the identity for addition in R.
(2) $a(-b) = (-a)b = -(ab)$.
(3) $(-a)(-b) = ab$.
(4) $a(b - c) = ab - ac$.
(5) If R has an identity for multiplication (which is unique and which we call 1) then $(-1)a = -a$, $(-1)(-1) = 1$.

Proof. First recall from Section 2.3 that, since R is a group under addition, the additive identity, 0, is unique as are additive inverses $-a$ of elements a.

(1) Using the fact that 0 is the identity for addition in R as well as the distributive laws, we have

$$0 + a \cdot 0 = a \cdot 0 = a \cdot (0 + 0) = a \cdot 0 + a \cdot 0.$$

Upon subtracting $a \cdot 0$ from both sides of the equation, we see that $0 = a \cdot 0$. You can make a similar argument to see that $0 \cdot a = 0$.

(2) We have

$$a(-b) + ab = a(-b + b) = a \cdot 0 = 0 \Longrightarrow a(-b) = -(ab).$$

Which ring axioms are being used at each point? We leave it as an exercise to finish the proof of (2).

(3) First note that $-(-a) = a$ since $a + (-a) = 0$. Then, by part (2) and the associative law for multiplication:

$$(-a)(-b) = -(a(-b)) = -(-ab) = ab.$$

(4) We have, using the distributive laws and part (2),

$$a(b - c) = ab + a(-c) = ab - ac.$$

(5) We leave this part as an exercise. ▲

Exercise 5.2.3 *Finish the proof of part (2) of the preceding proposition.*

Exercise 5.2.4 *Prove part (5) of the preceding proposition.*

Next we will define a subring in an analogous way to the way we defined a subgroup in Section 2.4. You should be able to write the definition yourself without looking at what follows. Just do not forget to say the subring is non-empty.

Definition 5.2.2 *Suppose that R is a ring. If S is a non-empty subset of R which is a ring under the same operations as R, we call S a* **subring** *of R.*

Proposition 5.2.2 *(Subring Test). A non-empty subset S of a ring R is a subring of R iff S is closed under subtraction and multiplication.*

Proof. $\Longleftarrow$ Assume S is closed under subtraction and multiplication. The one-step subgroup test from Proposition 2.4.3 implies that S is a subgroup of R under addition. Moreover S must be Abelian under addition since R is. Since S is closed under multiplication, we are done because the associative law for multiplication, plus the distributive laws, follow from those in R.

$\Longrightarrow$ Suppose S is a subring of R. Clearly S must be closed under subtraction and multiplication. ▲

Example 1. $\{0\}$ is a subring of any ring R.

To see this, apply the subring test. First note that $-0=0$ and thus $0-0=0+0=0$. Also $0 \cdot 0=0$, by multiplication rule (1). ▲

Example 2. $S=\{0,3,6,9 \pmod{12}\}=\{3x \mid x\in\mathbb{Z}_{12}\}$ is a subring of $\mathbb{Z}_{12}$.

To see this, use our subring test. Then $3x-3y=3(x-y)$ and $(3x)(3y)=3(3xy)$ are both in S. ▲

Example 3. $n\mathbb{Z}$ is a subring of $\mathbb{Z}$ for all $n\in\mathbb{Z}$.

To see this, use the subring test as in the preceding example. ▲

Example 4: The Gaussian Integers. $\mathbb{Z}[i]=\{a+bi \mid a,b\in\mathbb{Z}\}$ is a subring of $\mathbb{C}$. Here $i=\sqrt{-1}$.

Again we use the subring test. To see that $\mathbb{Z}[i]$ is closed under subtraction and multiplication, note that

$$
\begin{aligned}
(a+bi)-(c+di) &= (a-c)+(b-d)i\in\mathbb{Z}[i],\\
\text{and}\quad (a+bi)(c+di) &= (ac-bd)+(ad+bc)i\in\mathbb{Z}[i],
\end{aligned}
$$

since $a,b,c,d\in\mathbb{Z}$ implies $a-c,\ b-d,\ ac-bd,\ ad+bc\in\mathbb{Z}$. ▲

Example 5. The real numbers $\mathbb{R}$ form a subring of the complex numbers $\mathbb{C}$. ▲

Example 6. The ring $\mathbb{Z}$ is a subring of $\mathbb{Z}[\sqrt{-5}]=\{a+b\sqrt{-5} \mid a,b\in\mathbb{Z}\}$. ▲

Exercise 5.2.5

(a) Show that $2\mathbb{Z}\cup 5\mathbb{Z}$ is not a subring of $\mathbb{Z}$.
(b) Show that $2\mathbb{Z}+5\mathbb{Z}=\{2n+5m \mid n,m\in\mathbb{Z}\}=\mathbb{Z}$.
(c) Show that $2\mathbb{Z}\cap 5\mathbb{Z}=10\mathbb{Z}$.

Exercise 5.2.6 *Consider the set*

$$R = \left\{ \begin{pmatrix} a & b \\ 0 & c \end{pmatrix} \middle| \, a, b, c \in \mathbb{Z} \right\}.$$

Assume that addition is as in equation (5.1) above and multiplication is the usual matrix multiplication as in (5.2). Prove or disprove: R is a subring of the ring $\mathbb{Z}^{2\times 2}$ of all 2×2 matrices with integer entries under componentwise addition and the usual matrix multiplication.

We have already seen examples of the following definition – when we considered the unit groups $\mathbb{Z}_n^*$. As with this example, it is important that the inverse of the element of the ring be located in the *same* ring.

Definition 5.2.3 *Suppose R is a ring with identity for multiplication (which we call $1 \neq 0$). The* **units** *in R are the invertible elements for multiplication in R. The set of units is*

$$R^* = \{a \in R \mid \exists b \in R \text{ such that } ab = 1 = ba\}.$$

If $ab = 1 = ba$, write $b = a^{-1}$.

Proposition 5.2.3 *If R is a ring with identity for multiplication, the set of units R^* forms a group under multiplication.*

Proof. We need to check four things.

(1) R^* is closed under multiplication.
(2) The associative law holds for multiplication.
(3) R^* has an identity for multiplication.
(4) If $a \in R^*$, then $a^{-1} \in R^*$; that is, R^* is closed under inverse.

To prove (2), you just need to recall that the associative law holds in R.
To prove (3), just note that $1 \cdot 1 = 1$.
To prove (4), let $a \in R^*$. Then there is an element a^{-1} in R such that $aa^{-1} = a^{-1}a = 1$. But then $a = (a^{-1})^{-1}$ and thus $a^{-1} \in R^*$.
To prove (1), suppose $a, b \in R^*$. Then we have a^{-1} and b^{-1} in R and so – making use of the associativity of multiplication:

$$(ab)\, b^{-1}a^{-1} = abb^{-1}a^{-1} = 1.$$

Similarly $b^{-1}a^{-1}(ab) = 1$. It follows that $ab \in R^*$ with inverse $b^{-1}a^{-1}$. ▲

One moral of the preceding proof is that in a non-commutative ring, $(ab)^{-1} = b^{-1}a^{-1}$. We knew this already from Exercise 2.1.8.

Example 1. $\mathbb{Z}^* = \{1, -1\}$.

To see this, just note that if n and $\frac{1}{n}$ are both in $\mathbb{Z}$, then n must be 1 or -1. Otherwise, $|n| > 1$ and $0 < \frac{1}{|n|} < 1$. This contradicts Exercise 1.3.13. ▲

Example 2. $\mathbb{Z}_n^* = \{a \pmod{n} | \gcd(a,n) = 1\}$.
See Section 2.3 for the proof. ▲

Example 3. $\mathbb{Z}[x]$ is the ring of polynomials in one indeterminate x with integer coefficients. ▲

The elements of $\mathbb{Z}[x]$ have the form $f(x) = a_n x^n + a_{n-1}x^{n-1} + \cdots + a_1 x + a_0$, where $a_j \in \mathbb{Z}$. If $a_n \neq 0$, we say that the **degree** of f is $n = \deg f$. The zero polynomial is not usually said to have a degree (unless you want to say it has degree $-\infty$). We call x an "indeterminate" and not a "variable" because we must distinguish between polynomials and functions when we replace $\mathbb{Z}$ with $\mathbb{Z}_n$. We will say more about this later.

To add two of these polynomials, if the degree of f is n and the degree of g is $m \leq n$, put in some extra terms for g with coefficients that are 0, if necessary. Then you just add coefficients of like powers of x, that is:

$$f(x) = a_n x^n + a_{n-1}x^{n-1} + \cdots + a_1 x + a_0,$$
$$g(x) = b_n x^n + b_{n-1}x^{n-1} + \cdots + b_1 x + b_0$$

give

$$f(x) + g(x) = (a_n + b_n)x^n + (a_{n-1} + b_{n-1})x^{n-1} + \cdots + (a_1 + b_1)x + (a_0 + b_0). \quad (5.3)$$

Multiplication is more complicated but you have known how to do this since high school. We know that we want the operation to be associative and distributive. So suppose we are given polynomials

$$f(x) = a_n x^n + a_{n-1}x^{n-1} + \cdots + a_1 x + a_0,$$
$$g(x) = b_m x^m + b_{m-1}x^{m-1} + \cdots + b_1 x + b_0.$$

The product of $f(x)$ and $g(x)$ is

$$\begin{aligned} f(x)g(x) &= \left(a_n x^n + a_{n-1}x^{n-1} + \cdots + a_1 x + a_0\right)\left(b_m x^m + b_{m-1}x^{m-1} + \cdots + b_1 x + b_0\right) \\ &= a_n b_m x^{m+n} + a_n b_{m-1}x^{m+n-1} + \cdots + a_n b_1 x^{n+1} + a_n b_0 x^n \\ &\quad + \left(a_{n-1}x^{n-1} + \cdots + a_1 x + a_0\right)\left(b_m x^m + b_{m-1}x^{m-1} + \cdots + b_1 x + b_0\right) \\ &= a_n b_m x^{n+m} + (a_n b_{m-1} + a_{n-1}b_m)x^{n+m-1} + \cdots \\ &\quad + \left(\sum_{i+j=k} a_i b_j\right)x^k + \cdots + (a_1 b_0 + a_0 b_1)x + a_0 b_0. \end{aligned}$$

Thus the product of the polynomials $f(x)$ and $g(x)$ is

$$f(x)g(x) = \sum_{k=0}^{n}\left(\sum_{i+j=k} a_i b_j\right)x^k. \quad (5.4)$$

The sum and product are still in $\mathbb{Z}[x]$. Checking the other ring properties is a bit tedious. The zero polynomial has all its coefficients equal to 0. The additive inverse of $f(x)$ has as its coefficients the negatives of the corresponding coefficients of $f(x)$. The multiplicative identity is the degree 0 (constant) polynomial $f(x) = 1$. Checking the associative law for multiplication is the worst.

Assuming that no polynomial in the formula below is the zero polynomial, we have

$$\deg(fg) = \deg f + \deg g. \tag{5.5}$$

Definition 5.2.4 *Suppose that $f(x)$ is a polynomial with coefficients in some ring R. A **root** θ of f is an element of a possibly larger ring than R such that θ is a solution of the equation $f(\theta) = 0$.*

Thus, for example, $\sqrt{2} \in \mathbb{R}$ is a root of $x^2 - 2 = 0$ and $i \in \mathbb{C}$ is a root of $x^2 + 1 = 0$.

Exercise 5.2.7 *Complete the proof that the polynomial ring $\mathbb{Z}[x]$ is a commutative ring with identity for multiplication. Do your arguments work if you replace $\mathbb{Z}$ by any commutative ring R with identity for multiplication?*

Exercise 5.2.8 *What is the analog of formula (5.5) for* $\deg(f + g)$*?*

Hint. *Consider an inequality rather than an equality.*

Question. What is the group of units in the polynomial ring $\mathbb{Z}[x]$?

Given $f(x) \in \mathbb{Z}[x]$, suppose $fg = 1$, for some $g(x) \in \mathbb{Z}[x]$. This implies by formula (5.5) that $\deg f + \deg g = 0$. The only way that can happen is if $\deg f = \deg g = 0$. Thus the units of $\mathbb{Z}[x]$ are the nonzero constant polynomials that are units in $\mathbb{Z}$ itself, which implies

$$(\mathbb{Z}[x])^* = \mathbb{Z}^* = \{1, -1\}.$$

Of course, if $\deg f > 0$, you can still consider $1/f(x)$ but instead of a polynomial you get an infinite series – not a polynomial. For example, the geometric series is

$$\frac{1}{1-x} = \sum_{n=0}^{\infty} x^n.$$

Moreover, this is only a convergent series if $|x| < 1$. But algebra is not supposed to deal with convergence and limits. Instead an algebraist would view this as a "**formal power series**" with coefficients in some ring R: as an element of $R[[x]]$, whose elements look like $\sum_{n=0}^{\infty} a_n x^n$, with $a_n \in R$. This is in line with our insistence on viewing x in a polynomial $f(x)$ as an indeterminate rather than a function.

Exercise 5.2.9 *Find the group of units in $\mathbb{R}[x]$, the ring of polynomials with real coefficients, where the formulas for addition and multiplication are the same as those for $\mathbb{Z}[x]$, namely, formulas (5.3) and (5.4).*

Exercise 5.2.10 *Check that the set $C(\mathbb{R})$ consisting of all continuous real-valued functions on the real line forms a commutative ring if you define $(f + g)(x) = f(x) + g(x)$ for all $x \in \mathbb{R}$, and $(fg)(x) = f(x)g(x)$, $\forall x \in \mathbb{R}$. Here we assume that f, g are in $C(\mathbb{R})$. Does this ring have an identity for multiplication?*

Exercise 5.2.11 *Find the units in the ring $\mathbb{Z}^{2\times 2}$ of 2×2 matrices with integer entries and the usual matrix operations.*

Note that, for a ring R, we distinguish polynomials in $R[x]$ from functions $f\colon R \to R$. This is important when R is finite like $\mathbb{Z}_n$, for there are only finitely many functions but infinitely many polynomials.

Exercise 5.2.12 *How many functions are there mapping $\mathbb{Z}_n$ into $\mathbb{Z}_n$?*

Exercise 5.2.13 *Given two indeterminates, x and y, we can create the ring $R = (\mathbb{Z}[x])[y] = \mathbb{Z}[x, y]$. Define a* ***monomial*** *to be an element of R of the form $f(x, y) = cx^m y^n$, with $c \in \mathbb{Z}$. Define the* ***degree*** *of f to be $m + n$, assuming that $c \neq 0$. Then any polynomial in R is a sum of monomials. Define the* ***degree*** *of an arbitrary polynomial $p(x, y)$ in R to be the maximum degree of the monomials $cx^m y^n$ in p with $c \neq 0$. Show that $\deg(pq) = \deg p + \deg q$, for $p, q \in R$. What are the units in R?*

Exercise 5.2.14 ***(Example of a Non-associative Ring).*** *Consider the ring $M = \mathbb{R}^{n\times n}$ consisting of all $n \times n$ matrices with real entries. Addition is the usual componentwise addition of matrices. Define multiplication to be given by the* ***Lie bracket*** *$[A, B] = AB - BA$, for $A, B \in \mathbb{R}^{n\times n}$. Show that this multiplication is not associative but instead satisfies the* ***Jacobi identity:***

$$[A, [B, C]] + [B, [C, A]] + [C, [A, B]] = 0, \quad \text{for } A, B, C \in \mathbb{R}^{n\times n}.$$

The ring in the last exercise is the Lie algebra of the Lie group $GL(n, \mathbb{R})$ consisting of non-singular $n \times n$ real matrices with the operation of matrix multiplication. The dictionary relating Lie groups and Lie algebras is perhaps the most important tool in the study of Lie groups. See Terras [119].

Exercise 5.2.15 *Consider the ring $F[x_1, \ldots, x_n]$ of polynomials in n indeterminates $x_1, \ldots, x_n$ over a field F. Imitate Exercise 5.2.13 for this ring.*

Exercise 5.2.16 *Consider the ring of integer coefficient formal power series $\mathbb{Z}[[x]]$ in the indeterminate x, where we define the sum and product as follows for $\sum_{n=0}^{\infty} a_n x^n$ and $\sum_{n=0}^{\infty} b_n x^n$, with $a_n, b_n \in \mathbb{Z}$:*

$$\sum_{n=0}^{\infty} a_n x^n + \sum_{n=0}^{\infty} b_n x^n = \sum_{n=0}^{\infty} (a_n + b_n) x^n;$$

$$\sum_{m=0}^{\infty} a_m x^m \sum_{n=0}^{\infty} b_n x^n = \sum_{k=0}^{\infty} \left(\sum_{\substack{m+n=k \\ 0 \le m,n \le k}} a_m b_n \right) x^n.$$

Show that $\mathbb{Z}[[x]]$ is a commutative ring R with identity for multiplication such that $ab = 0$ implies either $a = 0$ or $b = 0$, for a, b in $\mathbb{R}$.

5.3 Integral Domains and Fields are Nicer Rings

In this section we want to consider rings that are more like the ring of integers $\mathbb{Z}$ or the ring of rational numbers $\mathbb{Q}$.

Definition 5.3.1 *If R is a commutative ring, we say $a \neq 0$ in R is a **zero divisor** if $ab = 0$ for some $b \in R$ such that $b \neq 0$.*

Example. In $R = \mathbb{Z}_6$, both 2 and 3 (mod 6) are zero divisors since $2 \cdot 3 \equiv 0 \pmod 6$. However, $R = \mathbb{Z}_5$ has no zero divisors. ▲

Definition 5.3.2 *If R is a commutative ring with identity for multiplication which has no zero divisors, we say that R is an **integral domain**. As usual we assume $1 \neq 0$.*

One might be forgiven for thinking that zero divisors are "bad" and thus integral domains are "good." Of course we have also been thinking $\mathbb{Z}_6$ is pretty nice and it is clearly not an integral domain. Perhaps we should just think that $\mathbb{Z}_5$ is way nicer than $\mathbb{Z}_6$.

Example 1. $\mathbb{Z}$ is an integral domain as are $\mathbb{R}$, $\mathbb{C}$, and $\mathbb{Q}$. We know this for $\mathbb{Z}$ as it was in our axioms of Section 1.3. We will not discuss the axioms for $\mathbb{R}$ and $\mathbb{C}$. Yes, it involves limits and that is calculus. See Cohen and Ehrlich [17] or Lang [63] or most advanced calculus texts for more information. As to the rationals, we assume you know how to deal with fractions and discuss $\mathbb{Q}$ more carefully in Section 6.4. If it seems bad to wait, feel free to jump ahead. ▲

Example 2. $\mathbb{Z}_n$ is not an integral domain if n is not a prime.

To see this, note that if n is not prime, then $n = ab$, where $0 < a, b < n$. But then neither a nor b can be congruent to 0 mod n and thus a and b are both zero divisors. ▲

Example 3. $\mathbb{Z}_p$ is an integral domain if p is a prime.

To see this, note that $ab \equiv 0 \pmod p \iff p$ divides ab. Then, by Euclid's lemma (see Lemma 1.5.1), this means p divides either a or b. So either a or b is congruent to $0 \pmod p$. ▲

Putting the two preceding examples together, we see that $\mathbb{Z}_n$ is an integral domain iff n is a prime.

The following lemma shows that cancellation is legal in an integral domain R even though inverses of nonzero elements may not exist in R.

Lemma 5.3.1 *(Cancellation Law in an Integral Domain). Suppose that R is an integral domain. If $a, b, c \in R$, $a \neq 0$, and $ab = ac$, then $b = c$.*

Proof. Since $ab = ac$, we see that $0 = ab - ac = a(b - c)$. Since $a \neq 0$ and R has no zero divisors, it follows that $b - c = 0$. Thus $b = c$. ▲

Exercise 5.3.1 *Consider the ring R of matrices of the form* $\begin{pmatrix} 0 & x \\ 0 & 0 \end{pmatrix}$, *for* $x \in \mathbb{Z}$, *with the usual componentwise addition and matrix multiplication. Show that R is indeed a ring. Is it an integral domain?*

Integral domains R are nice, but maybe not nice enough. Suppose we want to know that $a^{-1} \in R$ for any $a \in R - \{0\}$. Then we want a field. Of course, you can construct a field out of an integral domain by imitating the construction of $\mathbb{Q}$ out of $\mathbb{Z}$, but that is another story to be told in Section 6.4.

Definition 5.3.3 *A* ***field*** *F is a commutative ring with identity for multiplication such that any nonzero element* $a \in F$ *has a multiplicative inverse* $a^{-1} \in F$.

It follows from this definition that if F is a field, then the multiplicative group of units $F^* = F - \{0\}$, which is as big as the unit group could be. Yes, there is no way it is ever legal to divide by 0.

Proposition 5.3.1 *(Some Facts about Fields).*

(1) Any field F is an integral domain.
(2) Any finite integral domain D is a field.

Proof.

(1) If $a, b \in F$ such that $ab = 0$ and $a \neq 0$, then $b = a^{-1}ab = 0$. So F has no zero divisors.
(2) We just need to show that $D^* = D - \{0\}$ is closed under multiplication and multiplicative inverse. It is closed under multiplication because D has no zero divisors. Finiteness will force it to be closed under multiplicative inverse by the same argument that proved the finite subgroup test (Proposition 2.4.4) ▲

Exercise 5.3.2 *Show that the argument that proved the finite subgroup test (Proposition 2.4.4) finishes the proof of Proposition 5.3.1.*

Example 1. $\mathbb{Z}$ is not a field as the only units (i.e., invertible elements for multiplication) in $\mathbb{Z}$ are 1 and -1. ▲

Example 2. $\mathbb{Z}_p$ is a field iff p is prime. Why? Recall Example 3 above and the preceding proposition. ▲

Exercise 5.3.3 *State whether the following rings are integral domains. Then say whether they are fields. Give reasons for your answers.*

(a) The set $\mathbb{R}[x]$ *of all polynomials in one indeterminate with real coefficients and with addition and multiplication defined in (5.3) and (5.4) in Section 5.2;*
(b) $C(\mathbb{R}) = \{f : \mathbb{R} \to \mathbb{R} \mid f \text{ continuous}\}$ *with pointwise addition and multiplication of* $f, g \in C(\mathbb{R})$ *defined by* $(f + g)(x) = f(x) + g(x)$ *and* $(f \cdot g)(x) = f(x)g(x)$, *for all* $x \in \mathbb{R}$.

Example 3. The set of rational numbers $\mathbb{Q} = \left\{\frac{a}{b} \mid a, b \in \mathbb{Z}, b \neq 0\right\}$ is a field. See Section 6.4. ▲

Example 4. The set of real numbers $\mathbb{R}$ is also a field – as is the set of complex numbers $\mathbb{C}$. Again we leave this to the advanced calculus texts for $\mathbb{R}$. However, if you know that $\mathbb{R}$ is a field we make it an exercise to see that $\mathbb{C}$ is a field. If you think of $\mathbb{R}$ as the set of infinite decimals, you should be able to get pretty close to proving $\mathbb{R}$ is a field. You can add, subtract, multiply and divide them presumably. I prefer to think of real numbers as limits of Cauchy sequences of rationals myself. ▲

So we could view the finite field $\mathbb{Z}_p$ for p prime, as an analog of the field $\mathbb{R}$ of real numbers. But the picture of $\mathbb{R}$ is a continuous line without holes, while our picture of $\mathbb{Z}_p$ is a finite circle of points. If p is large, a penguin sitting on a finite circle might think it was a continuous line – just as the earth looks flat to the creatures living on it. It is certainly possible to create a finite analog of a sphere as well by looking at finite analogs of the rotation group and viewing the sphere as a quotient of $O(3,\mathbb{R})/O(2,\mathbb{R})$, where we view $O(2,\mathbb{R})$ as the rotations fixing the north pole on the sphere. Replace $\mathbb{R}$ by $\mathbb{Z}_p$ to get a finite analog.

Exercise 5.3.4 *Assuming $\mathbb{R}$ is a field, prove that $\mathbb{C}$ is a field.*

Question. Are there other finite fields besides $\mathbb{Z}_p$ for prime p?

Answer. Yes, for example, you can imitate the construction that gives the complex numbers $\mathbb{C}$. First note that $-1 \neq a^2$ for all $a \in \mathbb{Z}_3$. Thus we can consider i to be some creature outside $\mathbb{Z}_3$ such that $i^2 = -1$. Of course, we are not thinking that $i \in \mathbb{C}$. So if you want to be careful replace i by some letter with no particular meaning. I will usually use θ. But – reader beware – we cannot replace 3 by 5 in this construction since $2^2 \equiv -1 \pmod 5$.

Anyway, we can define

$$\mathbb{F}_9 = \mathbb{Z}_3[i] = \{a + bi \mid a, b \in \mathbb{Z}_3\}, \quad \text{where } i^2 = -1.$$

The order of $\mathbb{F}_9$ is 9 since there are three choices of a and three choices of b in $a + ib$. You add and multiply in $\mathbb{F}_9$ just as you would in the complex numbers, except that every computation is modulo 3. Thus for $a, b, c, d \in \mathbb{Z}_3$, we define

$$\begin{aligned}(a+bi) + (c+di) &= (a+c) + (b+d)i \quad \text{and} \\ (a+bi)\cdot(c+di) &= (ac-bd) + (ad+bc)i.\end{aligned}$$

Why does this make $\mathbb{Z}_3[i]$ a field? You get a ring by the same arguments that work to see the complex numbers form a ring. To see that $\mathbb{F}_9$ is a field, we need to see that if $a + ib$ is a nonzero element of $\mathbb{F}_9$, then it has a multiplicative inverse in $\mathbb{F}_9$. Again use the same argument that works for $\mathbb{C}$. That is,

$$\frac{1}{a+ib} = \frac{1}{a+ib}\,\frac{a-ib}{a-ib} = \frac{a-ib}{a^2+b^2} = \frac{a}{a^2+b^2} + i\frac{-b}{a^2+b^2}.$$

Why is $a^2 + b^2 \neq 0$? This is a little harder to prove than it would be if $a, b \in \mathbb{R}$. We know that $a + ib \neq 0 \Longrightarrow$ either a or b is not 0 in $\mathbb{Z}_3$. Suppose a is not 0 in $\mathbb{Z}_3$. Thus $a \equiv 1$ or $-1 \pmod 3$. So $a^2 \equiv 1 \pmod 3$. Then $b^2 \equiv 0$ or $1 \pmod 3$. It follows that $a^2 + b^2 \equiv 1$ or $2 \pmod 3$. Thus $a^2 + b^2$ is not $0 \pmod 3$. Since $\mathbb{Z}_3$ is a field, we know that $1/(a^2 + b^2) \in \mathbb{Z}_3$. The same argument works if b is not 0 in $\mathbb{Z}_3$.

Note that, for any element $z \in \mathbb{F}_9$, $3z = z + z + z = 0$. This happens because $z = x + iy$, with $x, y \in \mathbb{Z}_3$ and $3z = 3x + i3y = 0$. We say that $\mathbb{F}_9$ has characteristic 3.

Definition 5.3.4 *The* ***characteristic*** *of a ring R is the smallest $n \in \mathbb{Z}^+$ (assuming such an n exists) such that*

$$nx = \underbrace{x + x + \cdots + x}_{n \text{ times}} = 0, \quad \text{for all } x \in R.$$

If no such n exists, we say that R has ***characteristic*** 0.

Some authors (e.g., Birkhoff and Maclane [9]) do not say characteristic 0 but instead say characteristic ∞.

Lemma 5.3.2 *Suppose that R is a ring with identity* 1 *for multiplication. Then we have the following facts.*

(1) If the additive order of 1 *is not finite, then the characteristic of R is* 0.
(2) If the additive order of 1 *is n, then the characteristic of R is n.*

Proof.

(1) We leave this as an exercise.
(2) First note that the characteristic must be divisible by n. Why? On the other hand, if $\underbrace{1 + \cdots + 1}_{n \text{ times}} = 0$, then $\forall x \in R$

$$\left(\underbrace{1 + 1 + \cdots + 1}_{n \text{ times}}\right) x = \underbrace{x + x + \cdots + x}_{n \text{ times}} = 0,$$

This means that the characteristic is $\leq n$. It follows that the characteristic must equal n.

▲

Exercise 5.3.5 *Prove part (1) of the preceding lemma. Then answer the "Why?" in the proof of part (2) of that lemma.*

Example 1. $\mathbb{Z}$, $\mathbb{Q}$, $\mathbb{R}$, $\mathbb{C}$ all have characteristic 0. To see this, by the preceding lemma, you just need to note that no finite sum of 1s can equal 0 in these rings. ▲

Example 2. $\mathbb{Z}_p$ has characteristic p by the preceding lemma since that is the additive order of 1 in $\mathbb{Z}_p$. ▲

Example 3. $\mathbb{F}_9 = \mathbb{Z}_3 + i\mathbb{Z}_3$ has characteristic 3 since 3 is the additive order of 1 – again using the preceding lemma. $\mathbb{F}_9$ has nine elements. Thus the order of a finite field need not equal its characteristic. ▲

Example 4. $\mathbb{Z}_p[x]$, the ring of polynomials in one indeterminate with coefficients in $\mathbb{Z}_p$, has characteristic p. $\mathbb{Z}_p[x]$ has infinitely many elements. ▲

Exercise 5.3.6 *Prove the preceding statements about Examples 1–4.*

Recalling the definition of isomorphic groups, you should feel no surprise upon reading the following definition.

Definition 5.3.5 *Two rings R, S are* **isomorphic** *if there is a 1–1 and onto function* **(ring isomorphism)** *$f: R \to S$ preserving both ring operations. If rings R and S are isomorphic, we will write $R \cong S$.*

For most purposes we can identify isomorphic rings, just as we identified isomorphic groups earlier. In 1964 D. Singmaster asked how many rings of order 4 (up to ring isomorphism) are there? The solution was given by D. M. Bloom (11 rings of order 4, of which three have a multiplicative identity [108]). See Wikipedia, and the website on small rings from an abstract algebra class of Gregory Dresden [27]. Of course one needs to define direct sum of rings and quotient rings before trying to answer such questions. You may be able to guess what these things are from your knowledge of direct sum of groups and quotient groups, and group isomorphism. So, for example, Benjamin Fine [31] has shown that there are 11 non-isomorphic rings of order p^2 for any prime p.

There are two non-isomorphic rings of order p, if p is a prime. The one we know well is $\mathbb{Z}_p$. The other one has the same additive group but every product is defined to be 0, We call that ring $\mathbb{Z}_p(0)$. It seems like a pretty silly ring to me. I would ban it from this book if I had the power. You might defend the ring $\mathbb{Z}_p(0)$ by saying that it is isomorphic to the ring of matrices:

$$R = \left\{ \begin{pmatrix} 0 & x \\ 0 & 0 \end{pmatrix} \middle| x \in \mathbb{Z}_p \right\}, \tag{5.6}$$

where the ring operations are matrix addition and multiplication. We considered a similar ring in Exercise 5.3.1.

Exercise 5.3.7 *Prove that $\mathbb{Z}_p(0)$ and R defined by (5.6) are isomorphic rings.*

Instead of considering the rings of order p^2, for prime p, we note that there are four non-isomorphic rings of order pq, where p and q are distinct primes (see Fine [31]), but we will not give the proof here. These rings are $\mathbb{Z}_{pq}$, $\mathbb{Z}_{pq}(0)$, $\mathbb{Z}_p(0) \oplus \mathbb{Z}_q$, $\mathbb{Z}_p \oplus \mathbb{Z}_q(0)$. Here **direct sum of rings** $R \oplus S$, for rings R, S, is defined as for groups – with the sum and product defined componentwise:

$$(a, b) + (c, d) = (a + c, b + d) \text{ and } (a, b)(c, d) = (ac, bd), \text{ if } a, c \in R, b, d \in S.$$

We will give a detailed discussion of direct sums of rings in Section 5.4.

Exercise 5.3.8 *Find the characteristics of the four non-isomorphic rings $\mathbb{Z}_{pq}$, $\mathbb{Z}_{pq}(0)$, $\mathbb{Z}_p(0) \oplus \mathbb{Z}_q$, $\mathbb{Z}_p \oplus \mathbb{Z}_q(0)$ of order pq, where p and q are distinct primes.*

Wikipedia lists the numbers of non-isomorphic rings of various small orders, which it obtains from the On-Line Encyclopedia of Integer Sequences (sequence listed under A027623). However, as in the case of order p, some of the rings would no doubt seem silly to me. Somehow I do not have the same feeling about any of the small groups. Of course every Abelian group gives rise to a ring all of whose products are defined to be 0. Yes, I find these rings to be somewhat silly.

Lemma 5.3.3 *Suppose that R is an integral domain. Then the characteristic of R is either a prime or* 0.

Proof. If the additive order of 1 is not finite, the characteristic of R is 0, by the preceding lemma.

Suppose the additive order of 1 is $n \in \mathbb{Z}^+$. We must show that n is prime. We do a proof by contradiction.

Otherwise $n = ab$ for some integers a, b with $0 < a, b < n$. This means $0 = a \cdot b = (a \cdot b) \cdot 1 = (a \cdot 1) \cdot (b \cdot 1)$. Since R has no zero divisors, it follows that either $a \cdot 1 = 0$ or $b \cdot 1 = 0$. But this contradicts the minimality of $n = ab$. Therefore n is prime. ▲

Question. Which of the following five ring examples are fields – using the standard ring operations?

$\mathbb{Z}$; $\quad \mathbb{Z}[i] = \{a + bi \mid a, b \in \mathbb{Z}\}$, where $i \in \mathbb{C}$, $i^2 = -1$;

$\mathbb{Z}[x]$ = polynomials with integer coefficients in one indeterminate x;

$\mathbb{Z}[\sqrt{-5}]$; $\quad \mathbb{Z}_p$, for prime p.

Answer. Only the last example $\mathbb{Z}_p$, for prime p, is a field. In all other cases $\frac{1}{2}$ is not in the ring, even though 2 is.

We define subfield just as we defined subgroup or subring.

> **Definition 5.3.6** *A subset F of a field E is a* ***subfield*** *if it is a field under the operations of E. We also say that E is a* ***extension field*** *of F.*

Proposition 5.3.2 ***(Subfield Test).*** *Suppose that E is a field. Then a non-empty subset F of E is a subfield of E iff* $\forall a, b \in F$ *with* $b \neq 0$, *we have* $a - b$ *and* $ab^{-1} \in F$.

Proof. Just use the one-step subgroup test (from Proposition 2.4.3) on F to see that it is an additive subgroup of E and then use the same test again to see that $F - \{0\}$ is a multiplicative subgroup of $E - \{0\}$. The commutative and distributive laws are automatic. ▲

The following lemma gives an equation in characteristic $p \neq 0$ which some calculus students seem to believe is true in the real numbers. But that would mean most of the terms in the binomial theorem (see Exercise 1.8.12) somehow vanish miraculously.

Lemma 5.3.4 *Suppose that R is an integral domain of nonzero characteristic p, which is necessarily a prime. Then* $\forall x, y \in R$, *we have* $(x + y)^p = x^p + y^p$.

Proof. By the binomial theorem (whose proof works in any integral domain as shown in Exercise 1.8.12), we have

$$(x+y)^p = \sum_{k=0}^{p} \binom{p}{k} x^k y^{p-k}.$$

Here we interpret the terms in the sum as products of positive integers with elements of R as in Definition 5.3.4. To finish this proof, we must show that the prime p divides $\binom{p}{k}$ if $k=1,2,\ldots,p-1$. This follows from the fact that the binomial coefficient is an integer which is represented by the fraction:

$$\binom{p}{k} = \frac{p(p-1)\cdots(p-k+1)}{k(k-1)\cdots 2\cdot 1}.$$

Since p clearly divides the numerator, we just need to show that p does not divide the denominator. But that is true (by Euclid's Lemma 1.5.1 or unique factorization into primes) since p divides no factor in the denominator. This means that the binomial coefficients $\binom{p}{k}$ that are not congruent to 0 mod p are only those of the $k=0$ and $k=p$ terms in the sum representing $(x+y)^p$. ▲

Exercise 5.3.9 *Which of the following rings are integral domains and which are fields? Give a brief explanation of your answer.*

(1) $\mathbb{Z}[i]=\{a+bi \mid a,b\in\mathbb{Z}\}$, *where* $i\in\mathbb{C}$, $i^2=-1$. *The ring operations are the standard ones in* $\mathbb{C}$.
(2) $\mathbb{Z}/12\mathbb{Z}$ *with the usual ring operations* mod 12.
(3) $\mathbb{Z}_2^{2\times 2}$, 2×2 *matrices with coefficients in* $\mathbb{Z}_2$. *The ring operations are as in equations (5.1) and (5.2) of Section 5.2.*
(4) $\mathbb{Z}_{11}$ *with the usual ring operations* mod 12.
(5) $\mathbb{Z}\oplus\mathbb{Z}$ *with componentwise addition and multiplication:* $(a,b)+(c,d)=(a+c,b+d)$ *and* $(a,b)\cdot(c,d)=(ac,bd)$, *for* $a,b,c,d\in\mathbb{Z}$.
(6) $\mathbb{Q}$, *the rational numbers with the usual ring operations.*
(7) $R[x]$, *if* R *is a ring,* x *an indeterminate.*

Exercise 5.3.10 *Are the following rings integral domains? Are there elements of the rings that are neither units nor zero divisors?*

(a) $C[0,1]=\{$continuous real-valued functions $f\colon[0,1]\to\mathbb{R}\}$ *with addition and multiplication defined (as usual in calculus) pointwise, that is,*

$$(f+g)(x)=f(x)+g(x) \quad \text{and} \quad (fg)(x)=f(x)g(x), \quad \forall x\in[0,1].$$

(b) $C(\mathbb{Z}_n)=\{f\colon\mathbb{Z}_n\to\mathbb{R}\}$ *with addition defined as in part (a) but multiplication defined by convolution from equation (4.1) of Section 4.2.*

Exercise 5.3.11

(a) *List all the zero divisors – if any – in the seven rings from Exercise 5.3.9.*
(b) *List all the units in the rings R of part (a); that is, find* R^*.
(c) *What is the relation between the zero divisors and the units of R, if any?*

Exercise 5.3.12 *Show that if D is an integral domain of characteristic* 0 *and $D' = \langle 1 \rangle$ is the cyclic subgroup of the additive group of D generated by* 1, *then D' and $\mathbb{Z}$ are isomorphic rings. This means that you must find a 1–1 function T mapping $\mathbb{Z}$ onto D' which preserves addition and multiplication.*

In Section 6.1 we will have a similar exercise to the last one for fields of prime characteristic p. See Exercise 6.1.9.

A division ring is a non-commutative ring with identity such that the nonzero elements form a group under multiplication. An example is the quaternions defined in equation (3.4) of Section 3.6. Wedderburn proved that a finite division ring is a field. For a proof see Herstein [42].

Definition 5.3.7 *An* ***ordered field*** *F is a field with a subset $P \subset F$ of positive elements having the properties O1, O2, and O3, as in Section 1.3:*

O1 $F = P \cup \{0\} \cup (-P)$, *where* $-P = \{-x \mid x \in P\}$. *Moreover this is a disjoint union, that is:*

$$0 \notin P, \quad 0 \notin -P, \quad P \cap (-P) = \emptyset.$$

O2 $n, m \in P \Longrightarrow n + m \in P$.

O3 $n, m \in P \Longrightarrow n \cdot m \in P$.

Then we define, for $a, b \in F$, $a < b$ to mean that $b - a \in P$. This ordering will have the same properties as that of $\mathbb{Z}$ discussed in Section 1.3. The fields of rational numbers $\mathbb{Q}$ and real numbers $\mathbb{R}$ are ordered fields. The field of complex numbers $\mathbb{C}$ is not an ordered field.

Exercise 5.3.13 *Show that a subfield of an ordered field is an ordered field.*

Exercise 5.3.14 *Show that $\mathbb{C}$ is not an ordered field.*

5.4 Building New Rings from Old: Quotients and Direct Sums of Rings

We need to build quotient rings in the same way that we constructed $\mathbb{Z}/n\mathbb{Z}$. We will also imitate the construction of quotient groups in Section 3.4. To create a quotient group using a subgroup H of a group G, it was necessary that H be a normal subgroup. It turns out we will need a similar notion for the subring S of ring R. That is, we will need S to be an ideal in R as in the next definition.

Definition 5.4.1 *The non-empty subset A of a ring R is an* ***ideal*** *iff A is an additive subgroup of R such that $ra \in A$ and $ar \in A$, $\forall r \in R$ and $\forall a \in A$.*

Some would say "two-sided ideal" rather than ideal. Note that every ring R has two ideals: $\{0\}$ and R. Any other ideal is called a **proper** ideal.

Example. $n\mathbb{Z}$ is an ideal in $\mathbb{Z}$. We call $n\mathbb{Z}$ the **principal ideal generated by** n and write $n\mathbb{Z} = \langle n \rangle$. ▲

Definition 5.4.2 *Given a ring R and an element $a \in R$, the (2-sided)* ***ideal generated by*** *a, denoted $\langle a \rangle$, consists of elements ra and ar for all $r \in R$. Such an ideal $\langle a \rangle$ is called a* ***principal ideal****. Similarly, the ideal $\langle S \rangle$* ***generated by a subset*** *S of R is the smallest ideal containing S.*

Exercise 5.4.1 ***(The Ideal Generated by a Set).***

(a) Suppose that R is a commutative ring with identity. Let $S \subset R$. Show that

$$\langle S \rangle = \left\{ \sum_{i=i}^{n} r_i s_i \;\middle|\; \forall r_i \in R, \forall s_i \in S, \forall n \in \mathbb{Z}^+ \right\}$$

is indeed an ideal. We call it the ***ideal generated by*** *S.*

(b) Suppose $R = \mathbb{Z}$. Show that if $a, b \in \mathbb{Z}$, then $\langle \{a, b\} \rangle = \langle \gcd(a, b) \rangle$.

Ideals were introduced by Richard Dedekind in 1879. The main use for them in number theory is to get a substitute for prime numbers – the prime ideals we are about to define. This allows one to have unique factorization of ideals in rings of integers of algebraic number fields into products of prime ideals, though the unique factorization fails for actual algebraic integers in a ring like $\mathbb{Z}\left[e^{2\pi i/n}\right]$ for most values of n. Another way to say that is not all ideals in $\mathbb{Z}\left[e^{2\pi i/n}\right]$ are principal ideals. The concept of ideal was further developed by David Hilbert and Emmy Noether. We will find applications in error-correcting codes and random number generators. Ideal theory was not accepted by everyone. In particular, Leopold Kronecker objected to ideal theory just as he objected to the work of Karl Weierstrass in analysis and Georg Cantor in set theory. It appears at the moment, however, that history has found that Kronecker was the loser in all these wars.

To construct the quotient ring R/A, assuming that A is an ideal in the ring R, we create the set of additive **cosets** $[x] = x + A = \{x + a \mid a \in A\}$ for each $x \in R$. Once again, you can view these cosets as equivalence classes for the equivalence relation defined on elements $x, y \in R$ by $x \sim y$ iff $x - y \in A$.

Exercise 5.4.2 *Prove this last statement.*

Then we add and multiply cosets as we did for $\mathbb{Z}_n$:

$$[x] + [y] = [x + y], \qquad [x] \cdot [y] = [xy]. \tag{5.7}$$

This defines the **quotient ring (or factor ring)** R/A.

Theorem 5.4.1 *Suppose that A is a subring of the ring R. Then, with the definitions just given in (5.7), R/A is a ring iff A is an ideal in R.*

Proof. $\Longleftarrow$ If A is an ideal in R, we need to see that the operations defined in equation (5.7) make R/A a ring. Once we have checked that the operations are well defined, everything else will be easy. To check the operations make sense, suppose that $[x] = [x']$ and $[y] = [y']$. Then we must show that $[x + y] = [x' + y']$ and $[xy] = [x'y']$. In fact, we have already checked the additive part in Section 3.4, since A is automatically a normal subgroup of the additive group of R. Why?

So we will just check the multiplicative part. We need to prove that $xy - x'y' \in A$. To do this, we recall proofs of the formula for the derivative of a product. That means we should add and subtract $x'y$ (or xy'). This gives

$$xy - x'y' = xy - x'y + x'y - x'y' = (x - x')\,y + x'(y - y').$$

Since both $x - x'$ and $y - y'$ are in the ideal A, it follows that $(x - x')\,y$ and $x'(y - y')$ are both in A. But then the sum must be in A and we are done.

So now we know addition and multiplication in R/A are well defined. From the fact that R is a ring, it is easy to see that R/A must be a ring too. The identity for addition in R/A is $[0]$. The additive inverse of $[a]$ is $[-a]$. The associative laws in R/A follow from the laws in R, as do the distributive laws.

$\Longrightarrow$ Conversely, if R/A is a ring, the multiplication of cosets defined in (5.7) must be well defined. Since $[0] = A$, for any $a \in R$, we have $[0]\,[a] = [a]\,[0] = [0]$. This means $AR \subset A$ and $RA \subset A$. Of course A must also be closed under addition and subtraction as $[0] \pm [0] = [0]$ in R/A. Thus A must be an ideal in R. ▲

Example. Consider the ring $\mathbb{R}[x]$ of polynomials in the indeterminate x. An example of an ideal in this ring is the principal ideal generated by $x^2 + 1$:

$$A = \langle x^2 + 1\rangle = \left\{ f(x)(x^2 + 1) \mid f(x) \in \mathbb{R}[x] \right\}. \qquad ▲$$

Question. What is $\mathbb{R}[x]/\langle x^2 + 1\rangle$?

Answer. We can identify this quotient with the ring $\mathbb{C}$ of complex numbers. To give some evidence for this statement, let $\theta = [x] = x + A = x + \langle x^2 + 1\rangle$ in $\mathbb{R}[x]/A$. Then $\theta^2 + 1 = [x]^2 + [1] = [x^2 + 1] = [0]$. This means $\theta^2 = -1$ in our ring $\mathbb{R}[x]/A$. So θ behaves like i in $\mathbb{C}$.

In order to prove our statement identifying $\mathbb{C}$ and $\mathbb{R}[x]/\langle x^2 + 1\rangle$, we need to study polynomial rings a little more. See Section 5.5. In particular, we need the analog of the division algorithm for polynomial rings like $\mathbb{R}[x]$. Once we have that, we can identify cosets $[f(x)]$ in $\mathbb{R}[x]/A = \mathbb{R}[x]/\langle x^2 + 1\rangle = \mathbb{R}[x]/(x^2 + 1)\,\mathbb{R}[x]$ with cosets of the remainders $[r(x)]$ upon dividing $f(x)$ by $x^2 + 1$; that is, $f(x) = (x^2 + 1)\,q(x) + r(x)$, where $\deg r < 2$ or $r(x) = 0$. So the remainders look like $a + bx$, with $a, b \in \mathbb{R}$. This means elements of $\mathbb{R}[x]/A$ have the form $[a + bx] = [a] + [b][x] = a + b\theta$, which we can identify with a complex number $a + bi$, for $a, b \in \mathbb{R}$.

Our Goal. Replace $\mathbb{R}$ in the preceding construction with a finite field $\mathbb{Z}_p$. Then replace $x^2 + 1$ with any irreducible polynomial mod p, where irreducible means the analog of prime – a polynomial with no non-trivial factorization (see Definition 5.5.1). Then apply the result to error-correcting codes in Section 8.2. For example, take $p = 2$. Since $x^2 + 1 = (x + 1)^2$ in $\mathbb{Z}_2[x]$, we know that $x^2 + 1$ is not an irreducible polynomial in $\mathbb{Z}_2[x]$. An irreducible polynomial is our analog of a prime in the ring $\mathbb{Z}_2[x]$. An example of an irreducible polynomial in $\mathbb{Z}_2[x]$ is $x^2 + x + 1$. Why? It has no roots mod 2 and thus cannot have degree 1 factors as we will show in Section 5.5 on polynomial rings. This will imply that $\mathbb{Z}_2[x]/\langle x^2 + x + 1\rangle$ is a field with four elements {[0], [1], [x], [x+1]} where $[x]^2 + [x] + 1 = 0$.

The following theorem says that every ideal in the ring of integers is principal. We will have a similar theorem later about polynomial rings like $\mathbb{R}[x]$ or $\mathbb{Z}_p[x]$, p prime.

Theorem 5.4.2 *Any ideal A in the ring $\mathbb{Z}$ of integers is principal; that is, $A = \langle n \rangle = n\mathbb{Z}$ for some $n \in \mathbb{Z}$. In fact, if $A \neq \{0\}$, we can choose n to be the smallest positive element of A.*

Proof. The case $A = \{0\} = \langle 0 \rangle$ is clear. Otherwise $A \neq \{0\}$ and we let n be the least positive element of A. Then $\langle n \rangle \subset A$. Suppose that $a \in A$. The division algorithm says $a = nq + r$, with $0 \leq r < n$. Since A is an ideal and $r = a - nq$, we know that $r \in A$. But n is the least positive element of A. Therefore $r = 0$. This implies $A \subset \langle n \rangle$. So $A = \langle n \rangle$. ▲

Exercise 5.4.3 *In the preceding proof, why must the nonzero ideal A of $\mathbb{Z}$ have a positive element?*

Exercise 5.4.4 *As in the preceding example constructing the complex numbers, show that the quotient ring $\mathbb{Q}[x]/\langle x^2 - 2 \rangle$, where x is an indeterminate, can be identified with the ring $\mathbb{Q}\left[\sqrt{2}\right]$ consisting of numbers $a + b\sqrt{2}$, with $a, b \in \mathbb{Q}$, under the usual sum and product of real numbers.*

In pursuit of our goal, we ask two questions that lead us to two definitions.

Question 1. Suppose A is an ideal in R. When is the quotient ring R/A an integral domain?

Answer. When A is a prime ideal, which is defined as follows.

Definition 5.4.3 *Suppose that A is an ideal in the ring R, a commutative ring with identity for multiplication. We say that the proper ideal A is a* ***prime ideal*** *in R iff, for $a, b \in R$, $ab \in A \implies$ either a or $b \in A$.*

Lemma 5.4.1 *Suppose that A is an ideal in the ring R, a commutative ring with identity for multiplication. Then R/A is an integral domain iff A is a prime ideal.*

Proof. First note that R/A automatically has all the usual properties of an integral domain except for the lack of zero divisors. It inherits these properties from R. For example, the identities for addition and multiplication are $[0]$ and $[1]$, respectively. We get a zero divisor in R/A iff there are $a, b \in R$ such that $[a]\,[b] = [0]$ but $[a] \neq [0]$ or $[b] \neq [0]$. This means $ab \in A$ but $a \notin A$ or $b \notin A$. That is equivalent to saying that A is not a prime ideal. ▲

If R is a commutative ring with identity for multiplication, then R is an integral domain iff $\langle 0 \rangle$ is a prime ideal.

Example. Which nonzero ideals $\langle n \rangle$ in $\mathbb{Z}$ are prime ideals? The answer is the ideals $\langle p \rangle$ with p a prime. ▲

Proof. First note that $ab \in \langle n \rangle = n\mathbb{Z}$ is equivalent to saying n divides ab.

If n is not a prime, then $n = ab$ with $1 < a, b < n$. It follows that $ab \in \langle n \rangle$, but n cannot divide either a or b. Thus $\langle n \rangle$ cannot be a prime ideal in $\mathbb{Z}$.

If p is a prime and $ab \in \langle p \rangle$, then p divides ab. Euclid's Lemma 1.5.1, tells us that then p must divide either a or b. Thus either a or b must be in $\langle p \rangle$ and $\langle p \rangle$ is a prime ideal. ▲

But we really want the answer to the following question.

Question 2. Suppose A is an ideal in R, a commutative ring with identity for multiplication. When is the quotient ring R/A a field?

Answer. When A is a maximal ideal, which is defined as follows.

Definition 5.4.4 *Suppose A is an ideal in R, a commutative ring with identity for multiplication. We say that the proper ideal A is a* ***maximal ideal*** *in R iff, for an ideal B of R, the containment* $A \subset B \subset R \implies$ *either* $B = A$ *or* $B = R$.

Lemma 5.4.2 *Suppose A is an ideal in R, a commutative ring with identity for multiplication. The quotient ring R/A is a field iff A is a maximal ideal in R.*

Proof. First note that R/A automatically inherits all the properties of a field from R except closure under inverse for nonzero elements – a property that R does not necessarily have. In particular, $[0] = A$ is the identity for addition in R/A and $[1] = 1 + A$ is the identity for multiplication.

Suppose that R/A is a field. If B is an ideal such that $A \subset B \subset R$ but $B \neq A$, then we need to show that $B = R$. Since $B \neq A$, there is an element $x \in B - A$. This means $[x] \neq [0]$ in R/A. Since R/A is a field, there exists $[y] \in R/A$ such that $[x]\,[y] = [1]$. This means $xy - 1 \in A$. Thus $1 = xy - u$ for some $u \in A$. But then, because B is an ideal containing x and u, it follows that $1 \in B$. Therefore, for any $r \in R$, we have $r = 1 \cdot r \in B$ and $B = R$. Thus A is a maximal ideal.

Now suppose that A is maximal. We need to show that R/A is a field. Suppose $[x] \neq [0]$ in R/A. We need to find $[x]^{-1}$. Look at the ideal B generated by A and x. That is $B = \{u + rx \mid u \in A, r \in R\} = \langle A, x \rangle$. Then $A \subset B \subset R$. We know that $A \neq B$. Since A is maximal, it follows that $B = R$. But then $1 \in B$. So $1 = u + rx$ for some $u \in A$, $r \in R$. This implies $[r]\,[x] = [1]$. So $[r] = [x]^{-1}$ and we are done. ▲

Exercise 5.4.5 *In the preceding proof, show that* $B = \{u + rx \mid u \in A, r \in R\}$ *is an ideal in R, assuming that A is an ideal in R.*

Example 1. Which ideals $\langle n \rangle$ in $\mathbb{Z}$ are maximal ideals? The answer is that the nonzero prime ideals in $\mathbb{Z}$ are the maximal ideals in $\mathbb{Z}$. We know this since we proved finite integral domains are fields in Proposition 5.3.1. ▲

Exercise 5.4.6 *Prove the nonzero prime ideals in* $\mathbb{Z}$ *are the maximal ideals in* $\mathbb{Z}$ *directly by showing* $\mathbb{Z}/n\mathbb{Z}$ *is a field iff n is prime.*

Example 2. What are the maximal ideals in $\mathbb{Z}_{12}$?

First we show that all ideals A in $\mathbb{Z}_{12}$ are principal. To prove this, consider the corresponding ideal $\widetilde{A}$ in $\mathbb{Z}$, which is $\widetilde{A} = \{m \in \mathbb{Z} \mid [m] \in A\}$. Here we use the coset notation $[m] = m + 12\mathbb{Z}$. We leave it as an exercise to show that $\widetilde{A}$ is an ideal in $\mathbb{Z}$. Now, we know any ideal in $\mathbb{Z}$ is principal. Thus $\widetilde{A} = \langle n \rangle = n\mathbb{Z}$ for some $n \in \mathbb{Z}$. But then $A = [n]\,\mathbb{Z}_{12}$. Why?

Next suppose that $[u]$ is a unit in $\mathbb{Z}_{12}$. Then we can show that $[u]\,[n]\,\mathbb{Z}_{12} = [n]\,\mathbb{Z}_{12}$. For the fact that $[r] = [u]^{-1}$ exists in $\mathbb{Z}_{12}$ implies $[n]\,\mathbb{Z}_{12} = [u]^{-1}\,[u]\,[n]\,\mathbb{Z}_{12} \subset [un]\,\mathbb{Z}_{12}$. There is no problem seeing the reverse inclusion $[un]\,\mathbb{Z}_{12} \subset [n]\,\mathbb{Z}_{12}$. This is a general fact about principal ideals, by the way. ▲

Exercise 5.4.7

(a) Show that if A is an ideal in $\mathbb{Z}_{12}$, then $\widetilde{A} = \{m \in \mathbb{Z} \mid [m] \in A\}$ is an ideal in $\mathbb{Z}$.
(b) Answer the "why?" at the end of the end of the first paragraph of Example 2.

The units in $\mathbb{Z}_{12}$ are $[1], [5], [7], [11]$. Of course the principal ideals generated by units are all of the ring. Moreover, elements a and ua, for a unit u, generate the same ideal. Now we can list all the ideals in $\mathbb{Z}_{12}$. They are (dropping the [ ]):

$$\begin{aligned}
&\langle 0 \rangle = \{0\}, \quad \langle 1 \rangle = \langle 5 \rangle = \langle 7 \rangle = \langle 11 \rangle = \mathbb{Z}_{12},\\
&\langle 2 \rangle = \langle 10 \rangle = 2\mathbb{Z}_{12} = \{0, 2, 4, 6, 8, 10 \pmod{12}\},\\
&\langle 3 \rangle = \langle 9 \rangle = 3\mathbb{Z}_{12} = \{0, 3, 6, 9 \pmod{12}\},\\
&\langle 4 \rangle = \langle 8 \rangle = 4\mathbb{Z}_{12} = \{0, 4, 8 \pmod{12}\},\\
&\langle 6 \rangle = 6\mathbb{Z}_{12} = \{0, 6 \pmod{12}\}.
\end{aligned}$$

The poset diagram for the ideals in $\mathbb{Z}_{12}$ under the relation $\subset$ is in Figure 5.3. It follows from Figure 5.3 that the maximal ideals in $\mathbb{Z}_{12}$ are $\langle 2 \rangle$ and $\langle 3 \rangle$. Note the connection with the poset diagram for the divisors of 12 under the relation $\mid$ of divisibility.

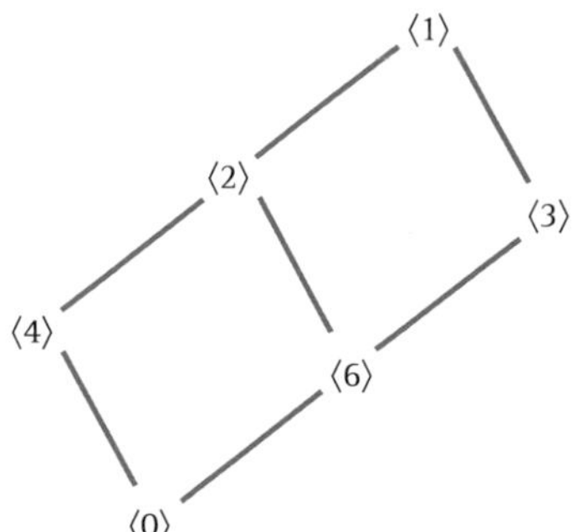

Figure 5.3 Poset diagram of the ideals in $\mathbb{Z}_{12}$

Exercise 5.4.8 *Explain the equalities for the ideals of $\mathbb{Z}_{12}$: for example, why is it that $\langle 8 \rangle = \langle 4 \rangle$?*

Exercise 5.4.9 *Show that every ideal in the ring $\mathbb{Z}_n$ is principal.*

Exercise 5.4.10 *Suppose that R is an integral domain. Show that, for $a, b \in R$, we have $aR = \langle a \rangle = \langle b \rangle = bR$ iff $a = ub$, where u is a unit in R.*

Exercise 5.4.11 *Find all the maximal ideals in* $\mathbb{Z}_{18}$. *Can you do the analogous exercise for the general case* $\mathbb{Z}_n$?

Exercise 5.4.12 *Draw the poset diagram for ideals in* $\mathbb{Z}_{30}$. *Which ideals are maximal?*

Our second method for the construction of rings is the ring analog of direct sum of groups.

Definition 5.4.5 *To create the* **direct sum** $R \oplus S$ *of two rings* R *and* S, *we start with the Cartesian product* $R \times S$ *and define the ring operations componentwise. That is, for* (a, b) *and* $(r, s) \in R \times S$, *we define* $(a, b) + (r, s) = (a + r, b + s)$ *and* $(a, b)(r, s) = (ar, bs)$.

Exercise 5.4.13 *Show that the preceding definition makes* $R \oplus S$ *a ring.*

Exercise 5.4.14 *Is* $\mathbb{Z}_2 \oplus \mathbb{Z}_2$ *a field, an integral domain? Same question for* $\mathbb{Z}_2 \oplus \mathbb{Z}_3$.

Exercise 5.4.15 *Find the characteristics of the following rings:* $\mathbb{Z}_2 \oplus \mathbb{Z}_2$, $\mathbb{Z}_2 \oplus \mathbb{Z}_4$, $\mathbb{Z}_2 \oplus \mathbb{Z}_3$.

Exercise 5.4.16 *Find a subring of* $R = \mathbb{Z} \oplus \mathbb{Z}$ *that is not an ideal.*

Hint. *Look at* $S = \{(a, b) \mid a + b \text{ is even}\}$.

The definitions of sum and product of ideals in the following exercise are basic to Dedekind's approach to arithmetic in rings of algebraic integers like $\mathbb{Z}\left[e^{\frac{2\pi i}{n}}\right]$, when – in general – not every ideal is principal.

Exercise 5.4.17 *If* A *and* B *are ideals in a commutative ring* R *with identity for multiplication, define the* **sum** $A + B = \{a + b | a \in A, b \in B\}$ *and the product* $AB = \{\sum_{i=1}^{n} a_i b_i \mid a_i \in A, b_i \in B\}$.

(a) Show that $A + B$ *and* AB *are ideals of* R.
(b) Show that $A + B = R$ *implies* $AB = A \cap B$.

The moral of the preceding exercise is that one can do arithmetic with ideals.

Exercise 5.4.18 *Suppose* $R = \mathbb{Z}$. *If* $A = \langle a \rangle$ *and* $B = \langle b \rangle$, *show that* $A + B = \langle \gcd(a, b) \rangle$. *If* $A + B = \langle 1 \rangle$, *show that* $AB = \langle ab \rangle = A \cap B$.

Exercise 5.4.19 *Suppose that* R, S, T, V *are rings. Show that if* $R \cong T$ *and* $S \cong V$, *then* $R \oplus S \cong T \oplus V$. *Here* $\cong$ *means that the rings are isomorphic.*

Exercise 5.4.20 *Suppose that* F *is a field. Find all the ideals in* F.

Exercise 5.4.21 *Suppose that* R *and* S *are rings. What are the ideals in the direct sum* $R \oplus S$?

Exercise 5.4.22 *Find a prime ideal in* $\mathbb{Z} \oplus \mathbb{Z}$ *which is not maximal.*

5.5 Polynomial Rings

Suppose that F is a field. We consider the ring $F[x]$ of polynomials in one indeterminate, x, and coefficients in F.

Beware: Do not confuse polynomials and functions. For example, in $\mathbb{Z}_3[x]$ the polynomials $f(x)=x^2+x+1$ and $g(x)=x^4+x+1$ represent the same function even though the two polynomials are different. To see this, plug in the elements of $\mathbb{Z}_3$.

$f(0)=1$	$f(1)=0$	$f(-1)=1$
$g(0)=1$	$g(1)=0$	$g(-1)=1$

This sort of thing has to happen, since the number of functions $T:\mathbb{Z}_3\to\mathbb{Z}_3$ is $3^3=27$, while the ring $\mathbb{Z}_3[x]$ is infinite.

You probably did Exercise 5.2.7 showing that $R[x]$ is a commutative ring with multiplicative identity if R is a commutative ring with multiplicative identity. If not, do that exercise now! The bad part is the associative law for multiplication.

If $R=\mathbb{Z}_3$, we add and multiply as in the following examples.

$(x^2+2x+2)+(x^3+x+2)=x^3+x^2+1$ (since $3\equiv 0 \pmod 3$ and $4\equiv 1 \pmod 3$);

$(x^2+2x+1)\cdot(x^3+2x)=x^5+2x^4+x^2+2x$,

using the same facts about congruences mod 3 as we used for addition. We learned to do this in the dim dark past by making the following table, with all numbers computed mod 3.

$$\begin{array}{ccccccccc}
 & & & & x^2 & + & 2x & + & 1 \\
 & & & & & & x^3 & + & 2x \\
\hline
 & & & & 2x^3 & + & 4x^2 & + & 2x \\
x^5 & + & 2x^4 & + & x^3 & & & & \\
\hline
x^5 & + & 2x^4 & + & 0 & + & x^2 & + & 2x
\end{array}$$

Recall that the units R^* of a ring R are the invertible elements for multiplication in R. When $R=\mathbb{Z}$, the only units are 1 and -1. When the ring is $\mathbb{Z}[x]$, it turns out the units are the same as for $\mathbb{Z}$, as we saw in Section 5.2. We get the analogous result when F is a field, that is:

$$(F[x])^*=F^*=F-\{0\}. \tag{5.8}$$

The proof is the same as that in Section 5.2 for $\mathbb{Z}[x]$.

Exercise 5.5.1 *Prove equation (5.8).*

Now we want to imitate what we said about the integers in Section 1.5. We will have analogs of primes, the division algorithm, the Euclidean algorithm, and the fundamental theorem of arithmetic for the ring $F[x]$, where F is any field. Pretty amazing!

Assumption. For the rest of this section **F is a field** and all polynomials are in $F[x]$.

We want to define the polynomial analog of prime. Before that, we should define polynomial divisors: we say that a polynomial $g(x)$ **divides** a polynomial $f(x)$ if there is some polynomial $h(x)$ such that $f(x) = g(x)h(x)$. Then we also say that $g(x)$ is a **divisor** of $f(x)$. This is quite analogous to divisibility of integers.

Definition 5.5.1 *A polynomial $f(x)$ of degree > 0 is* ***irreducible*** *iff $f(x) = g(x)h(x)$ for $g, f \in F[x]$ implies either g or h has degree 0. Otherwise the polynomial is* ***reducible.***

Now we want to get rid of the units in a factorization as we did in $\mathbb{Z}$ by allowing only positive non-unit integers to be called primes – assuming they cannot be factored non-trivially. To get rid of units in $F[x]$, we look at monic polynomials.

Definition 5.5.2 *A* ***monic polynomial*** *is a polynomial with leading coefficient (i.e., coefficient of the highest power of x) equal to 1.*

So a monic irreducible polynomial of degree > 0 is the analog of a prime in $F[x]$.

Examples: Irreducible Polynomials of Low Degree in $\mathbb{Z}_2[x]$
Degree 1 polynomials: $x, x + 1$. Both are irreducible.
Degree 2 polynomials: x^2, $x^2 + 1 = (x + 1)^2$, $x^2 + x = x(x + 1)$, $x^2 + x + 1$. The first three polynomials are clearly reducible. What about $x^2 + x + 1$? Does x or $x + 1$ divide $x^2 + x + 1$? The answer is: No! For we have $x^2 + x + 1 = x(x + 1) + 1$. This means that if we had $x^2 + x + 1 = xq(x)$, then x would divide $1 = xq(x) - x(x + 1)$. But that is impossible, as $0 = \deg(1) = \deg(x\{q(x) - x - 1\}) \geq 1$. A similar argument shows that $x + 1$ cannot divide $x^2 + x + 1$.

This means that $x^2 + x + 1$ **is the only irreducible polynomial of degree 2 in** $\mathbb{Z}_2[x]$.
Degree 3 polynomials with nonzero constant term: $x^3 + 1$, $x^3 + x + 1$, $x^3 + x^2 + 1$, $x^3 + x^2 + x + 1$. Which of these polynomials are irreducible? To answer this question more rapidly, it helps to know that $x - a$ divides a polynomial $f(x)$ iff $f(a) = 0$. Here we are assuming $a \in \mathbb{Z}_2$ and $f(x) \in \mathbb{Z}_2[x]$. This is Corollary 5.5.1 below.

The polynomial $f(x)$ of degree 3 will be reducible iff it has a factorization $f(x) = g(x)h(x)$ with $g(x), h(x) \in \mathbb{Z}_2[x]$, such that $\deg g \neq 0$ and $\deg h \neq 0$. But then $3 = \deg g + \deg h$ implies that either g or h has degree 1. This means that f is reducible iff $f(a) = 0$ for some $a \in \mathbb{Z}_2$. We prove the general version of this result in our proof of Proposition 5.5.1.

So we need to test $x^3 + 1$, $x^3 + x + 1$, $x^3 + x^2 + 1$, $x^3 + x^2 + x + 1$ for roots in $\mathbb{Z}_2$. However, the only possible root is 1, since we have already eliminated the polynomials with 0 as a root. The polynomials with an even number of terms will have 1 as a root in $\mathbb{Z}_2[x]$.

This implies that **there are only two irreducible degree 3 polynomials in** $\mathbb{Z}_2[x]$: $x^3 + x + 1$ **and** $x^3 + x^2 + 1$. ▲

Exercise 5.5.2 *Find the degree 4 irreducible polynomials in $\mathbb{Z}_2[x]$.*

Exercise 5.5.3 *Find the monic irreducible polynomials of degrees 1 and 2 in $\mathbb{Z}_3[x]$.*

In order to do the same things for rings of polynomials $F[x]$, when F is a field, that we did for the ring $\mathbb{Z}$ of integers, we will need a division algorithm. The division algorithm works just as it did in high school or wherever it was introduced. In fact, it really works the same way it did for the integers in elementary school (as in Section 1.5). It is important, however, that F is a field because we must divide by coefficients of the quotient polynomial.

Example 1. In $\mathbb{Z}_2[x]$, we have the following computation.

$$
\begin{array}{r|ll}
 & x^3 & +1 \\
\hline
x^2+x+1 & x^5+x^4+x^3 & +x^2+x+1 \\
 & x^5+x^4+x^3 & \\
\cline{2-3}
 & & x^2+x+1 \\
 & & x^2+x+1 \\
\cline{3-3}
 & & 0
\end{array}
$$

As a result, we have $x^5+x^4+x^3+x^2+x+1=(x^2+x+1)\,(x^3+1)$. The remainder is 0. ▲

Example 2. In $\mathbb{Z}_3[x]$, we have the following computation.

$$
\begin{array}{r|rl}
 & 2x & +1 \\
\hline
2x+1 & x^2+x & +2 \\
 & x^2+2x & \\
\cline{2-3}
 & 2x & +2 \\
 & 2x & +1 \\
\cline{2-3}
 & & 1
\end{array}
$$

This says $x^2+x+2=(2x+1)(2x+1)+1$. The remainder is 1. Note that we are definitely using the fact that $\mathbb{Z}_3$ is a field and $2^{-1}\equiv 2 \pmod 3$. That is, $2\cdot 2\equiv 1 \pmod 3$. ▲

Theorem 5.5.1 ***(The Division Algorithm for Polynomial Rings).*** *Suppose that F is a field. Given $f(x)$ and $g(x)\in F[x]$ with $g(x)$ not the zero polynomial, there are polynomials $q(x)$ (the quotient) and $r(x)$ (the remainder) in $F[x]$ such that $f(x)=g(x)q(x)+r(x)$ and $\deg r<\deg g$ or r is the zero polynomial.*

Sketch of Proof (Induction on deg f). If $\deg g=0$, the result is trivial, as g is then a unit in the ring $F[x]$. So we assume $\deg g>0$ from now on. The result is clear if $\deg f<\deg g$ as then we can take $q=0$ and $r=f$.

Induction step. Now assume the theorem true if $\deg f\leq m-1$. Suppose that f has degree m and

$$f(x)=b_m x^m+\cdots \quad \text{and} \quad g(x)=a_n x^n+\cdots, \quad \text{with} \quad a_n\neq 0 \text{ and } b_m\neq 0.$$

We may assume $m\geq n$ or we can take $r=f$ as we have said already.

Then we start the process by choosing the first term of $q(x)$ to be $a_n^{-1}b_m x^{m-n}$ so that

$$h(x)=f(x)-a_n^{-1}b_m x^{m-n}g(x)$$

has degree less than $\deg f=n$. That is

$$
\begin{array}{r|ll}
 & a_n^{-1}b_m x^{m-n} & + \cdots \\
\hline
g(x) = a_n x^n + \cdots & f(x) = b_m x^m + \cdots & \\
 & b_m x^m + \cdots & \\
\hline
 & 0 & h = \text{lower degree polynomial than } f
\end{array}
$$

This gets the induction going. The induction hypothesis allows us to divide h by g and we are done. ▲

Exercise 5.5.4 *Fill in the details of the preceding proof.*

Exercise 5.5.5 *Prove the uniqueness of the polynomials q and r in the division algorithm.*

Corollary 5.5.1 *Suppose that F is a field, $f(x) \in F[x]$, and $a \in F$. Then $f(a) = 0 \iff f(x) = (x - a)q(x)$ for some $q(x) \in F[x]$.*

Proof. $\Longrightarrow$ By the division algorithm, $f(x) = (x - a)q(x) + r(x)$, where $\deg r < 1$ or $r = 0$. It follows that r must be a constant in F. Thus $f(a) = (a - a)q(a) + r = r$. If $f(a) = 0$, then r is the 0 polynomial and $f(x) = (x - a)q(x)$.
$\Longleftarrow$ This should be clear. ▲

Corollary 5.5.2 *Suppose $f \in F[x]$ and $\deg f = n$. Then f has at most n roots in F counting* ***multiplicity****. This means that we count a not just once but $k > 1$ times if $(x - a)^k$ exactly divides $f(x)$ (meaning that $(x - a)^k$ divides $f(x)$ and $(x - a)^{k+1}$ does not divide $f(x)$).*

Proof. By the preceding corollary, $f(a) = 0$ implies $f(x) = (x - a)q(x)$ and $\deg f = n = 1 + \deg q$. Thus $\deg q = n - 1$. So we finish the proof by induction on the degree of f. ▲

The following corollary is the polynomial analog of Theorem 5.4.2 on ideals in $\mathbb{Z}$. The proof is essentially the same.

Corollary 5.5.3 *Every ideal in $F[x]$ is principal, when F is a field.*

Proof. Let A be an ideal in $F[x]$. If $A = \{0\} = \langle 0 \rangle$, we are done. Otherwise, let $f(x)$ be an element of A of minimal degree. Then we claim $A = \langle f \rangle$. To prove this, suppose $h \in A$. The division algorithm says there exist $q, r \in F[x]$ such that $h = qf + r$, with $\deg r < \deg f$ or $r = 0$. Thus $r = h - qf \in A$ since $h, f \in A$. This contradicts the minimality of the degree of f unless $r = 0$. Then $h \in \langle f \rangle$ and $A = \langle f \rangle$. ▲

The following corollary is an analog of the result from Section 5.3 that $\mathbb{Z}_n$ is a field if and only if n is a prime.

Corollary 5.5.4 ***(Irreducible Polynomials Give Rise to Fields).*** *The following are equivalent in $F[x]$, when F=field:*

(1) $\langle f(x) \rangle$ is a maximal ideal;
(2) $f(x) \in F[x]$ is irreducible;
(3) $F[x]/\langle f(x) \rangle$ =field.

Proof. We know from Lemma 5.4.2 that (1)$\Longleftrightarrow$(3). So we will show (1)$\Longleftrightarrow$(2).
(1)$\Longrightarrow$(2) Assume $\langle f\rangle$ is a maximal ideal. If $f=g\cdot h$, for $g,h\in F[x]$, such that neither g nor h is a unit, then $\langle f\rangle\subset\langle g\rangle\subset\langle 1\rangle=F[x]$ and $\langle f\rangle\subset\langle h\rangle\subset\langle 1\rangle=F[x]$ and none of the inclusions are equality. Why? Recall Exercise 5.4.10 that said $\langle a\rangle=\langle b\rangle\Longleftrightarrow b=au$, for some unit u in $F[x]$. This contradicts the maximality of $\langle f\rangle$.
(2)$\Longrightarrow$(1) Suppose f is irreducible. We know by the preceding corollary that every ideal in $F[x]$ is principal. So any ideal containing $\langle f\rangle$ must have the form $\langle g\rangle$, for some $g\in F[x]$. If $\langle f\rangle\subset\langle g\rangle\subset F[x]$, then $f=g\cdot h$ for some $h\in F[x]$. But the irreducibility of f says that either g or h is a unit. If g is a unit then $\langle g\rangle=F[x]$. If h is a unit, then $\langle f\rangle=\langle g\rangle$. Thus $\langle f\rangle$ is maximal. ▲

The following proposition has already been useful in searching for low degree irreducible polynomials.

Proposition 5.5.1 ***(Irreducibility Test for Low Degree Polynomials).*** *Suppose F is a field, $f(x)\in F[x]$ and $\deg f=2$ or 3. Then f is not irreducible iff $\exists c\in F$ such that $f(c)=0$.*

Proof. $\Longrightarrow$ We have a non-trivial factorization of f iff $f=gh$, for $g,h\in F[x]$, where either g or h has degree 1. This is true since $\deg f=\deg g+\deg h$ and $2=1+1$ or $3=2+1=1+2$ are the only possibilities for partitions of 2 or 3 into sums of two integers. It follows that we can take one of the factors, say $g(x)$, to be monic and linear: that is, $g(x)=x-c$ for some $c\in F$. But then $f(x)=(x-c)h(x)$ implies $f(c)=0$.
$\Longleftarrow$ If $f(c)=0$, then $x-c$ divides $f(x)$, by Corollary 5.5.1. ▲

Example. The preceding test fails for $\mathbb{Z}_6[x]$ since, for example, $f(x)=(2x+1)^2$ has no roots in $\mathbb{Z}_6$. Why is this not a contradiction to the proposition? ▲

Exercise 5.5.6 *In $\mathbb{Z}_3[x]$ show that the polynomials $f(x)=x^{13}-x$ and $g(x)=x^7-x$ determine the same function mapping $\mathbb{Z}_3$ into $\mathbb{Z}_3$.*

Exercise 5.5.7 *In $\mathbb{Z}_7[x]$ find the quotient and remainder upon dividing $f(x)=5x^4+3x^3+1$ by $g(x)=3x^2+2x+1$.*

Exercise 5.5.8 *Find all degree 3 monic irreducible polynomials in $\mathbb{Z}_3[x]$.*

Integral domains with a division algorithm are called **Euclidean domains.** Thus the ring of polynomials over a field F is a Euclidean domain. As such it has similar properties to $\mathbb{Z}$. One defines the greatest common divisor $d=\gcd(f,g)$ for $f,g\in F[x]$, to be the unique monic polynomial which divides both f and g such that any common divisor h of f and g must divide d. Again there is a Euclidean algorithm to compute d. One has the analog of Bézout's identity in (Theorem 1.5.2) which states that $d=uf+vg$ for some $u,v\in F[x]$, and the Euclidean algorithm can be used to find u and v.

Exercise 5.5.9 *Prove the preceding statements about* $\gcd(f, g)$ *for* $f, g \in F[x]$ *by imitating the proofs that worked in* $\mathbb{Z}$. *Here we must make a slight change in the algorithm to insure that the final result is a monic polynomial. That is, we must multiply by the reciprocal of the lead coefficient of the last nonzero remainder to make it monic.*

Exercise 5.5.10 *If F is a field, show that the ideal* $\langle a(x), b(x)\rangle$ *in* $F[x]$ *is* $\langle \gcd(a(x), b(x))\rangle$.

Exercise 5.5.11

(a) Show that if an ideal A of ring R contains an element of the unit group R^*, *then* $A = R$.
(b) Show that the only ideals of a field F are $\{0\}$ *and F itself.*

Exercise 5.5.12 *Find* $\gcd(x^3 + 2x^2 + 2x + 1, x^3 + 2x^2 + x + 2)$ *in* $\mathbb{Z}_3[x]$.

5.6 Quotients of Polynomial Rings

Now we have enough information to use quotients of polynomial rings in order to obtain finite fields. We already considered a field of order 9 in Section 5.3.

Example. A field with eight elements is $\mathbb{F}_8 = \mathbb{Z}_2[x]/\langle x^3 + x + 1\rangle$.

To see this, you just have to recall a few of the facts that we proved in preceding sections. First, we know from Section 5.5 that $x^3 + x + 1$ is irreducible in $\mathbb{Z}_2[x]$. Second, Corollary 5.5.4 says that $\mathbb{Z}_2[x]/\langle x^3 + x + 1\rangle$ is a field.

How do we know that our field has 2^3 elements? The answer is that a coset $[f]$ in $\mathbb{Z}_2[x]/\langle x^3 + x + 1\rangle$ is represented by the remainder of f upon division by $x^3 + x + 1$. The remainder has degree ≤ 2 and thus has the form $ax^2 + bx + c$, where $a, b, c \in \mathbb{Z}_2$. Moreover two polynomials g, h of degree ≤ 2 cannot be congruent modulo $x^3 + x + 1$ unless they are equal. Why? Congruent means the difference $g - h$ is a multiple of $x^3 + x + 1$. The only way a polynomial of degree ≤ 2, such as $g - h$, can be a multiple of a degree 3 polynomial is if $g - h$ is really the 0 polynomial. Thus g must equal h.

The preceding is analogous to what happens in $\mathbb{Z}_{163}$. The elements $[m]$ in $\mathbb{Z}_{163}$ are represented by $[r]$, where r is the remainder of m upon division by 163.

We can set $\theta = [x]$ in $\mathbb{Z}_2[x]/\langle x^3 + x + 1\rangle$. This means θ is a root of $x^3 + x + 1 = 0$. We are saying that $\mathbb{F}_8 = \{a\theta^2 + b\theta + c \mid a, b, c \in \mathbb{Z}_2\}$. Moreover, we can view $\mathbb{F}_8$ as a vector space over the field $\mathbb{Z}_2$. A basis for this vector space is $\{1, \theta, \theta^2\}$. See Section 7.1 for more information on vector spaces over fields.

If we express the elements of $\mathbb{F}_8$ in the form $a\theta^2 + b\theta + c$, for $a, b, c \in \mathbb{Z}_2$, it is easy to add the elements but hard to multiply. Thus it is useful to show that the multiplicative group of $\mathbb{F}_8$ is cyclic – a fact that can be proved for any finite field. See Exercises 6.3.10 and 7.5.9. It turns out the generator in this case is θ. It is not true that if our finite field is $K = \mathbb{Z}_p[x]/\langle k(x)\rangle$, where p is prime and $k(x)$ irreducible, then a root of $k(x)$ generates K^*. Only certain **primitive polynomials** $k(x)$ over the base field $\mathbb{Z}_p$ will have this property that a root of $k(x)$ is a generator of $(\mathbb{Z}_p[x]/\langle k(x)\rangle)^*$.

Next we create a table of powers of $\theta = [x]$ in $\mathbb{Z}_2[x]/\langle x^3 + x + 1\rangle$. We know that $\theta^3 + \theta + 1 = 0$. This implies that $\theta^3 = -\theta - 1 = \theta + 1$ since $-1 = +1$ in $\mathbb{Z}_2$. Then we note that $\theta(a_0 + a_1\theta + a_2\theta^2) = a_0\theta + a_1\theta^2 + a_2\theta^3 = a_0\theta + a_1\theta^2 + a_2(\theta + 1) = a_2 + (a_0 + a_2)\theta + a_1\theta^2$.

So multiplication by θ sends the coefficients (a_0, a_1, a_2) to $(a_2, a_0 + a_2, a_1)$. This is what is called a "**feedback shift register.**" So now it is easy to make a table of powers of θ

(see Table 5.1). The jth row will list the coefficients (a_0, a_1, a_2) of $\theta^j = a_0 + a_1\theta + a_2\theta^2$. To go from the jth row to the $(j+1)$st row, send (a_0, a_1, a_2) to $(a_2, a_0 + a_2, a_1)$.

Table 5.1 *Powers of θ for $\mathbb{F}_8$*

$\theta^j = a_0 + a_1\theta + a_2\theta^2$	a_0	a_1	a_2
θ	0	1	0
θ^2	0	0	1
θ^3	1	1	0
θ^4	0	1	1
θ^5	1	1	1
θ^6	1	0	1
$\theta^7 = 1$	1	0	0

This shows that the multiplicative group of units $\mathbb{F}_8^*$ is a cyclic group of order 7 generated by θ. We call $x^3 + x + 1$ a primitive polynomial in $\mathbb{Z}_2[x]$ for this reason. Figure 5.4 shows a picture of the feedback shift register corresponding to this polynomial. You can use primitive polynomials to construct other feedback shift registers. It is a finite state machine that will cycle through $2^n - 1$ states if $f(x)$ is a primitive polynomial of degree n in $\mathbb{Z}_2[x]$. The states are really the rows of the table of powers of θ for θ a root of $f(x)$, and this is really the multiplicative group of the finite field $\mathbb{Z}_2[x]/\langle f(x)\rangle$. The successive states of the registers are given in Table 5.1 for the example under consideration. ▲

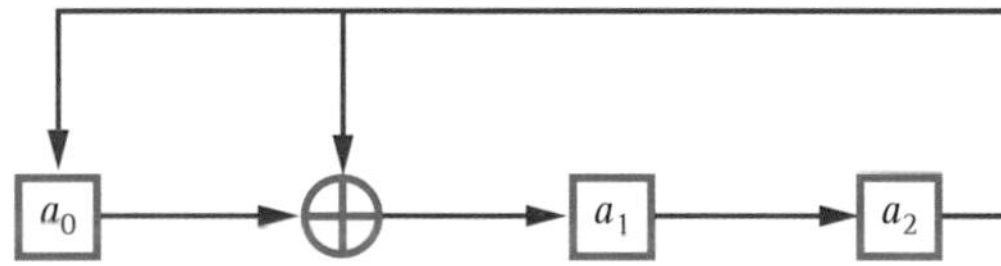

Figure 5.4 A feedback shift register diagram corresponding to the finite field $\mathbb{Z}_2[x]/\langle x^3 + x + 1\rangle$ and the multiplication table given in the text

Feedback shift registers are of interest in generating pseudo-random numbers and in cryptography. There are applications in digital broadcasting, communications, and error-correcting codes. See the first two sections of Chapter 8.

Exercise 5.6.1

(a) Show that $x^2 - 2$ is an irreducible polynomial in $\mathbb{Z}_5[x]$.

(b) Show that the factor ring $\mathbb{Z}_5[x]/\langle x^2 - 2\rangle$ is a field with 25 elements.

(c) Show that the field in part (b) can be identified with $\mathbb{Z}_5\left[\sqrt{2}\right] = \left\{a + b\sqrt{2} \,\middle|\, a, b \in \mathbb{Z}_5\right\}$, the smallest field containing $\mathbb{Z}_5$ and $\sqrt{2}$. Note that this is legal because the equation $a^2 = 2$ has no solution $a \in \mathbb{Z}_5$. Here we add and multiply as we would in $\mathbb{R}$, except that everything is mod 5.

(d) What is the characteristic of the field with 25 elements in parts (b) and (c)?

Exercise 5.6.2 *Identify $\mathbb{F}_{25} \cong \mathbb{Z}_5[x]/\langle x^2 - 2\rangle$ as the set $\mathbb{Z}_5\left[\sqrt{2}\right]$ as in the preceding exercise. Set $\theta = \sqrt{2}$. Find the table of powers $\theta^j = a_0 + a_1\theta$, where $\theta^2 = 2$, in a similar manner to the table we created for $\mathbb{Z}_2[x]/\langle x^3 + x + 1\rangle$. Do these powers give the whole unit group in $\mathbb{Z}_5[x]/\langle x^2 - 2\rangle$? This would say that the polynomial $x^2 - 2$ is primitive in $\mathbb{Z}_5[x]$? What is* *the feedback shift register diagram as in Figure 5.4 that corresponds to this polynomial? How many states does it cycle through before it repeats?*

Hint. *The order of the unit group $\mathbb{F}_{25}^*$ is 24 while the order of 2 is 4 and thus the order of $\sqrt{2}$ is 8. The symbol for multiplication by c in a feedback shift register is a circle with c inside it on the arrow going into the register.*

Exercise 5.6.3

(a) Show that the ideal $\langle x\rangle$ in $\mathbb{Z}_3[x]$ is maximal.
(b) Prove that $\mathbb{Z}_3$ is isomorphic to $\mathbb{Z}_3[x]/\langle x\rangle$.

Exercise 5.6.4

(a) Show that $\mathbb{Z}_3[x]/\langle x^2+x+2\rangle$ is a field $\mathbb{F}_9$ with nine elements which can be viewed as the field $\mathbb{Z}_3[\theta]=\{a+b\theta|a,b\in\mathbb{Z}_3\}$, where $\theta^2+\theta+2=0$.
(b) Imitating Table 5.1 in the example above, compute the powers of θ from part (a).
(c) Is the multiplicative group $\mathbb{F}_9^*$ in the field of part (a) cyclic?
(d) Draw the corresponding feedback shift register diagram as in Figure 5.4.

Exercise 5.6.5 *Find an infinite set of polynomials $f(x)\in\mathbb{Z}_3[x]$ such that $f(a)=0$ for all $a\in\mathbb{Z}_3$.*

Exercise 5.6.6 *Suppose a is a nonzero element of a finite field F with n elements. Show that $a^{n-1}=1$.*

Exercise 5.6.7

(a) Consider $\mathbb{Z}_5[i]=\{a+bi\mid a,b\in\mathbb{Z}_5\}$, where we view i as some entity not contained in $\mathbb{Z}_5$ such that $i^2=-1$. Show that this ring is not a field.
(b) Consider $\mathbb{Z}_7[i]=\{a+bi\mid a,b\in\mathbb{Z}_7\}$, where we view i as some entity not contained in $\mathbb{Z}_7$ such that $i^2=-1$. Show that this ring is a field.
(c) Can you develop a more general version of this problem for $\mathbb{Z}_p[i]$ where p is an odd prime according to whether p is congruent to 1 or 3 $\pmod 4$? Here you will need to use a fact from number theory. The congruence $c^2\equiv-1 \pmod p$ has a solution c iff $p\equiv 1 \pmod 4$.

Hint for (c). *The fact that the group $\mathbb{Z}_p^*$ is cyclic is useful for the proof of that number theory fact. We prove this later (in Exercise 6.3.10 as well as Theorem 7.5.4). Once we know $\mathbb{Z}_p^*$ is cyclic, then the main theorem on cyclic groups from Section 2.5 tells us that $\mathbb{Z}_p^*$ can only have a subgroup of order 4 like the subgroup generated by i if 4 divides $p-1$.*

Exercise 5.6.8 *We use the notation R[a] for ring R and element a of some larger ring S to mean that R[a] is the* ***smallest subring of*** *S containing R and a – also known as the* ***ring generated by*** *a* ***over*** *R. Show that R[a] consists of all polynomials $f(a)$, with $f(x)\in R[x]$. Explain why in the preceding exercise you only need polynomials of degree ≤ 1.*

Exercise 5.6.9 *Consider the quotient $\mathbb{Z}_2[x]/\langle x^4+1\rangle$. Is it a field? Why?*

Exercise 5.6.10 *Consider the quotient* $\mathbb{Z}_3[x]/\langle x^4 + 1\rangle$. *Is it a field? Why?*

Exercise 5.6.11 *Describe the following rings. Are they fields?*

(a) $\mathbb{Z}_7[x]/\langle 2\rangle$;
(b) $\mathbb{Z}_7[x]/\langle x+1\rangle$;
(c) $\mathbb{Z}_7[x]/\langle x^2\rangle$.

Exercise 5.6.12 *Construct a field of with* 49 *elements.*

6

Rings: There's More

6.1 Ring Homomorphisms

We have already discussed ring isomorphisms. Now you must agree that we need to consider the ring analog of group homomorphism from Section 3.5.

> **Definition 6.1.1** *Suppose that R and S are rings. Then a function $T: R \to S$ is a* **ring homomorphism** *iff T preserves both ring operations: that is,*
>
> $$T(a+b) = T(a) + T(b) \quad \text{and} \quad T(ab) = T(a)T(b), \quad \text{for all } a, b \in R.$$
>
> *If, in addition, T is 1–1 and onto, we say that T is a* **ring isomorphism** *and write $R \cong S$. If the ring homomorphism $T: R \to R$ is 1–1 and onto, it is called a* **ring automorphism**.

As we said earlier, we will usually identify isomorphic rings, just as we can identify isomorphic groups.

Example. $\pi: \mathbb{Z} \to \mathbb{Z}_n$ defined by $\pi(a) = a \pmod{n}$ is a ring homomorphism and is onto but not 1–1. This example is easily generalized to $\pi: R \to R/A$, where A is any ideal in a ring R and $\pi(x) = x + A = [x]$. ▲

Application: Test for Divisibility by 3. Any integer has a decimal expansion which we write $n = a_k a_{k-1} \cdots a_1 a_0$, for $a_j \in \{0,1,2,3,4,5,6,7,8,9\}$, where this means that $n = a_k 10^k + a_{k-1} 10^{k-1} + \cdots + a_1 10 + a_0$. Then we see that 3 divides $n = a_k a_{k-1} \cdots a_1 a_0$ iff 3 divides $a_k + a_{k-1} + \cdots + a_1 + a_0$, the sum of the digits of n.

Proof. Look at the homomorphism $\pi: \mathbb{Z} \to \mathbb{Z}_3$ defined by $\pi(a) = a \pmod 3$. Then, since $\pi(10) \equiv 1 \pmod 3$, we have

$$\begin{aligned} \pi(n) &= \pi\left(a_k 10^k + a_{k-1} 10^{k-1} + \cdots + a_1 10 + a_0\right) \\ &= \pi(a_k)\,\pi(10)^k + \pi(a_{k-1})\,\pi(10)^{k-1} + \cdots + \pi(a_1)\,\pi(10) + \pi(a_0). \\ &\equiv a_k + a_{k-1} + \cdots + a_1 + a_0 \pmod 3. \end{aligned}$$

▲

Example. Does 3 divide 314 159 265 358 979 323 846? We compute the sum of the digits to be 103 and then the sum of those digits is 4, which is not divisible by 3. So the answer is "No." ▲

Exercise 6.1.1 *Does* 3 *divide* 271 828 182 845 904 523 536*?*

Exercise 6.1.2 *Create a similar test to see whether* 11 *divides a number? Then use your theorem to see whether 11 divides the numbers in the preceding example and exercise.*

The following theorem gives the basic facts about ring homomorphisms – facts analogous to those for group homomorphisms in Section 3.5. In particular, we define the kernel of a ring homomorphism and find an analog of the first isomorphism theorem.

Theorem 6.1.1 ***(Properties of Ring Homomorphisms).*** *Suppose that S and S' are rings and $T: S \to S'$ is a ring homomorphism. Let 0 be the identity for addition in S and 1 the identity for multiplication in S, if S has an identity for multiplication. Let $0'$ be the identity for addition in S'. Then we have the following facts.*

(1) (a) $T(0)=0'$.
(b) The image $T(S)$ is a subring of S'. Here $T(S)=\{T(s) \mid s \in S\}$. If S has an identity for multiplication and $T(1) \neq T(0)$, then $T(1)$ is the identity for multiplication in the image ring $T(S)$.
(2) If S is a field, then $T(1) \neq T(0)$ implies that the image $T(S)$ is a field.
(3) Define the **kernel** *of T to be $\ker T = T^{-1}(0') = \{x \in S \mid T(x)=0'\}$. Then $\ker T$ is an ideal in S. Moreover T is 1–1 iff $\ker T = \{0\}$.*
(4) ***(First Isomorphism Theorem)*** $S/\ker T \cong T(S)$.

Proof.

(1a) follows from the corresponding fact about groups in Section 3.5, since S and S' are groups under addition.

(1b) The image $T(S)$ is an additive subgroup of S' from results of Section 3.5. To finish the proof of this part, we must think a little since S, S' are unlikely to be groups under multiplication. To see that $T(S)$ is closed under multiplication, note that $T(a)T(b)= T(ab) \in T(S)$, for all $a, b \in S$. As a subset of S', the image $T(S)$ is a ring, by the subring test. Then, if S has an identity for multiplication and $T(1) \neq T(0)$, for $a \in S$, we have

$$T(1)T(a) = T(1 \cdot a) = T(a),$$
$$T(a)T(1) = T(a \cdot 1) = T(a).$$

It follows that $T(a)$ is the identity for multiplication in $T(S)$.

(2) If S is a field, from part (1) we know that $T(1)$ is the identity for multiplication in $T(S)$. Since $S^* = S - \{0\}$ is a group under multiplication, we can use results from Section 3.5 to see that $T(S^*) = T(S)^*$ is a group under multiplication.

(3) We know that $\ker T$ is an additive subgroup of S by results of Section 3.5. To show $\ker T$ is an ideal we also need to show that if $a \in \ker T$ and $s \in S$, then sa and as are in $\ker T$. This is easy since $T(sa) = T(s)T(a) = T(s)0 = 0$ implies $sa \in \ker T$. The same sort of argument works for as.

(4) We imitate the proof of the first isomorphism theorem for groups in Section 3.5. As before, we define a map $\widetilde{T}: S/\ker T \to T(S)$ by setting $\widetilde{T}([a]) = \widetilde{T}(a + \ker T) = T(a)$. Then we need to show that $\widetilde{T}$ is a ring isomorphism.

$\widetilde{T}$ is **well defined** since if $[a]=a+\ker T=[b]$, then $b=a+u$, where $u\in\ker T$. This implies

$$\begin{aligned}\widetilde{T}([b])&=T(b)=T(a+u)=T(a)+T(u)\\&=T(a)+0=T(a)=\widetilde{T}([a]).\end{aligned}$$

$\widetilde{T}$ is **1–1** since $\widetilde{T}([a])=\widetilde{T}([b])$ implies $T(a)=T(b)$ and thus $a-b\in\ker T$ and $[a]=[b]$. $\widetilde{T}$ is **onto** since any element of $T(S)$ has the form $T(a)$ for some $a\in S$. Thus $\widetilde{T}([a])=T(a)$.

$\widetilde{T}$ **preserves both ring operations** since, for any $a,b\in S$, we have the following, using the definition of addition and multiplication in the quotient $S/\ker T$, the definition of $\widetilde{T}$, and the fact that T is a ring homomorphism:

$$\begin{aligned}\widetilde{T}([a]+[b])&=\widetilde{T}([a+b])=T(a+b)=T(a)+T(b)=\widetilde{T}([a])+\widetilde{T}([b]),\\\widetilde{T}([a]\cdot[b])&=\widetilde{T}([a\cdot b])=T(a\cdot b)=T(a)\cdot T(b)=\widetilde{T}([a])\cdot\widetilde{T}([b]).\end{aligned}$$ ▲

Example. Define $\mathbb{Z}_3[i]=\{x+iy\mid x,y\in\mathbb{Z}_3\}$, where $i^2=-1$. Here, for $x,y,u,v\in\mathbb{Z}_3$, we define $(x+iy)+(u+iv)=x+u+i(y+v)$ and $(x+iy)\cdot(u+iv)=xu-yv+i(xv+yu)$. We can use the first isomorphism theorem to show that

$$\mathbb{Z}_3[i]\cong\mathbb{Z}_3[x]/\langle x^2+1\rangle.$$

Note that the right-hand side is a field because x^2+1 is irreducible in $\mathbb{Z}_3[x]$ since the fact that -1 is not a square mod 3 means x^2+1 has no roots in $\mathbb{Z}_3$. Here we use Corollary 5.5.4. The left-hand side can be shown directly to be a field by proving that it is possible to find the multiplicative inverse of any nonzero element.

First we define a ring homomorphism $T\colon\mathbb{Z}_3[x]\to\mathbb{Z}_3[i]$ by $T(f(x))=f(i)$ for any polynomial $f(x)\in\mathbb{Z}_3[x]$. The map T is well defined, preserves the ring operations, and is onto. For example, it is easily seen that

$$T(f+g)=(f+g)(i)=f(i)+g(i)=T(f)+T(g)$$

and

$$T(f\cdot g)=(f\cdot g)(i)=f(i)\cdot g(i)=T(f)\cdot T(g).$$

We claim $\ker T=\langle x^2+1\rangle$. To see this, note first that $\langle x^2+1\rangle\subset\ker T$, since $i^2+1=0$. To prove $\ker T\subset\langle x^2+1\rangle$, let $g(x)\in\ker T$. By the division algorithm, we have $g(x)=(x^2+1)\,q(x)+r(x)$, where $\deg r<2$ or $r=0$. Since $g(i)=0$, we see that $r(i)=0$. But if $r\neq0$, $\deg r=0$ or 1, and we have $r(x)=ax+b$, with $a,b\in\mathbb{Z}_3$. But then $ai+b=0$. If $a\neq0$, this would mean $i=-b/a\in\mathbb{Z}_3$, a contradiction to the fact that -1 is not a square in $\mathbb{Z}_3$. Thus r must be the 0-polynomial and $\ker T\subset\langle x^2+1\rangle$ to complete the proof that $\ker T=\langle x^2+1\rangle$.

It follows then from the first ring isomorphism theorem (which was part (4) of Theorem 6.1.1) that $\mathbb{Z}_3[i]\cong\mathbb{Z}_3[x]/\langle x^2+1\rangle$. ▲

Exercise 6.1.3 *Suppose that R,S are rings, A is a subring of R. Let $T\colon R\to S$ be a ring homomorphism.*

(a) Show that for all $r\in R$ and all $n\in\mathbb{Z}^+$, we have $T(nr)=nT(r)$ and $T(r^n)=T(r)^n$. Here $n\cdot r=\underbrace{r+\cdots+r}_{n}$.

(b) Show that if A is an ideal of R and $T(R)=S$, then $T(A)$ is an ideal of S.

Exercise 6.1.4 *Under the same hypotheses as in the preceding exercise, prove that if B is an ideal in S, then we have the following facts.*

(a) $T^{-1}(B)$ *is an ideal of R. Here the* ***inverse image*** *of B under T is*

$$T^{-1}(B) = \{a \in A \mid T(a) \in B\}.$$

Do not assume the inverse function of T exists.

(b) If T is a ring isomorphism of R onto S, then the inverse function T^{-1} *is a ring isomorphism of S onto R.*

Exercise 6.1.5 *Show that isomorphism of rings has the properties of an equivalence relation.*

Exercise 6.1.6

(a) If we try to define $T:\mathbb{Z}_5 \to \mathbb{Z}_{10}$ *by setting* $T(x) = 5x$ *we do not really have a well-defined function. Explain.*

(b) Show that $T:\mathbb{Z}_4 \to \mathbb{Z}_{12}$ *given by* $T(x) = 3x$ *is well defined but does not preserve multiplication.*

(c) Show that every homomorphism $T:\mathbb{Z}_n \to \mathbb{Z}_n$ *has the form* $T(x) = ax$ *for some fixed a in* $\mathbb{Z}_n$ *with* $a^2 = a \pmod{n}$. *Find a value of a that works when* $n = 12$.

Exercise 6.1.7

(a) Show that the factor ring $\mathbb{R}[x]/\langle x^2+1\rangle$ *is isomorphic to the ring of complex numbers* $\mathbb{C}$. *Here* $\mathbb{R}[x]$ *is the ring of polynomials in one indeterminate x and real coefficients.*

(b) Show that complex conjugation $\phi(a+ib) = a - ib$, *for* a, b *in* $\mathbb{R}$ *and* $i^2 = -1$, *defines a ring automorphism of* $\mathbb{C}$.

(c) Show that $\mathbb{C}$ *is not isomorphic to* $\mathbb{R}$.

(d) Show that $\mathbb{C}$ *is isomorphic to the ring* $\left\{ \begin{pmatrix} a & b \\ -b & a \end{pmatrix} \middle| \, a, b \in \mathbb{R} \right\}$, *where the operations are the usual matrix addition and multiplication.*

Exercise 6.1.8

(a) Show that the only ring automorphisms of $\mathbb{Q}$ *are the map* $T(x) = 0$, $\forall x \in \mathbb{Q}$, *and the identity* $I(x) = x$, $\forall x \in \mathbb{Q}$. ***Hint.*** *First look at the map on* $\mathbb{Z}$.

(b) Show that the only ring automorphism of $\mathbb{R}$ *is the identity map.* ***Hint.*** *First, recall that the positive reals are squares of nonzero reals and vice versa. Then recall that* $a < b$ $\iff b - a > 0$. *Use this to show that* $a < b$ *implies* $\phi(a) < \phi(b)$. *Then suppose that* $\exists a$ *such that* $\phi(a) \neq a$. *Consider the two cases that* $a < \phi(a)$ *and* $\phi(a) < a$. *There is a rational number between a and* $\phi(a)$. *Use the fact that* ϕ *must be the identity on the rationals to get a contradiction.*

Finally we have an exercise which is the finite characteristic analog of Exercise 5.3.12.

Exercise 6.1.9 *Suppose that F is a field of prime characteristic p. Show that F contains a subfield isomorphic to* $\mathbb{Z}_p$.

Hint. *Define a mapping* $T:\mathbb{Z} \to F$ *by setting* $T(n) = \underbrace{1 + \cdots + 1}_{n}$. *Set* $T(0) = 0$ *and* $T(-n) = -T(n)$, *for* $n \in \mathbb{Z}^+$. *Show that T preserves the operations of addition and multiplication and induces a field isomorphism from* $\mathbb{Z}_p$ *onto a subfield of F.*

Exercise 6.1.10 ***(Second Isomorphism Theorem for Rings).*** *Suppose that R is a ring, A is a subring of R, and B is an ideal of R. Show that*

$$A+B=\{a+b \mid a\in A, b\in B\}$$

is a subring of R and $A\cap B$ is an ideal of R. Then show that

$$(A+B)/B\cong A/A\cap B.$$

There are also third and fourth isomorphism theorems. See Dummit and Foote [28, p. 246].

Exercise 6.1.11 *Find two ring automorphisms of $\mathbb{Z}_3[i]=\{x+iy \mid x,y\in\mathbb{Z}_3\}$, where $i^2=-1$, that fix elements of $\mathbb{Z}_3$.*

Exercise 6.1.12 *Show that the only ring automorphism of $\mathbb{Z}$ is the identity map $I(x)=x$, for all $x\in\mathbb{Z}$.*

Exercise 6.1.13 *Consider the ring $\mathbb{Z}_p^{2\times 2}$ of 2×2 matrices with entries in $\mathbb{Z}_p$ for prime p. The ring operations are componentwise addition and matrix multiplication as in Exercise 5.2.1. Define $T:\mathbb{Z}_p^{2\times 2}\to\mathbb{Z}_p$ by $T\begin{pmatrix} a & b \\ c & d\end{pmatrix}=a+d$. Is this a ring homomorphism? Why?*

Exercise 6.1.14 *We can define a mapping $T:\mathbb{Z}_{24}\to\mathbb{Z}_{12}$ by $T(x\,(\text{mod } 24))=x \text{ mod } 12$. Show that this map is a ring homomorphism. Find the kernel. Is it onto? What does the first ring isomorphism say?*

6.2 The Chinese Remainder Theorem

An example of the Chinese remainder theorem can be found in a manuscript by Sun-Tzu from the third century AD.

Theorem 6.2.1 ***(The Chinese Remainder Theorem for Rings).*** *Assume that the positive integers m, n satisfy $\gcd(m,n)=1$. The mapping $\widetilde{T}:\mathbb{Z}_{mn}\to\mathbb{Z}_m\oplus\mathbb{Z}_n$ defined by $\widetilde{T}([s])=(s\ (\text{mod } m), s\ (\text{mod } n))$ is a ring isomorphism, showing that $\mathbb{Z}_{mn}$ is isomorphic to the ring $\mathbb{Z}_m\oplus\mathbb{Z}_n$.*

Proof. First consider the mapping $T:\mathbb{Z}\to\mathbb{Z}_m\oplus\mathbb{Z}_n$ defined by $T(s)=(s\,(\text{mod } m), s\,(\text{mod } n))$. Then T is a ring homomorphism. Note that Exercise 3.6.3 showed that it is an additive group homomorphism. To see that it preserves multiplication, let $a,b\in\mathbb{Z}$. Then, using the definition of T and the definition of multiplication in $\mathbb{Z}_m\oplus\mathbb{Z}_n$, we have:

$$\begin{aligned} T(a\cdot b) &= (a\cdot b\ (\text{mod } m), a\cdot b\ (\text{mod } n)) \\ &= (a\ (\text{mod } m), a\ (\text{mod } n))\cdot(b\ (\text{mod } m), b\ (\text{mod } n)) = T(a)\cdot T(b). \end{aligned}$$

It follows from the first ring isomorphism theorem (which was part (4) of Theorem 6.1.1) that $\widetilde{T}:\mathbb{Z}/\ker T\to T(\mathbb{Z})$ defined by $\widetilde{T}(x+\ker T)=T(x)$, for $x\in\mathbb{Z}$, is an isomorphism. So we need to compute the kernel of T. This is, since $\gcd(m,n)=1$,

$$\begin{aligned} \ker T &= \{a\in\mathbb{Z} \mid a\equiv 0\ (\text{mod } m) \text{ and } a\equiv 0\ (\text{mod } n)\} \\ &= \{a\in\mathbb{Z} \mid m \text{ divides } a \text{ and } n \text{ divides } a\} = mn\mathbb{Z}. \end{aligned}$$

This map $\widetilde{T}$ is 1–1. Thus the image of $\widetilde{T}$ must have mn elements. It follows that $\widetilde{T}$ must map $\mathbb{Z}_{mn}$ onto $\mathbb{Z}_m \oplus \mathbb{Z}_n$ because $\mathbb{Z}_m \oplus \mathbb{Z}_n$ has mn elements. ▲

In particular, the Chinese Remainder theorem says that if $\gcd(m, n) = 1$, there is a solution $x \in \mathbb{Z}$ to the simultaneous congruences:

$$x \equiv a \pmod{m},$$
$$x \equiv b \pmod{n}.$$

When the theorem is discussed in elementary number theory books, only the onto-ness of the function is emphasized. Many examples like the following are given. This result and its generalizations have many applications: for example, to rapid and high-precision computer arithmetic. See the next section or Dornhoff and Hohn [25], Knuth [54], Richards [90], Rosen [91], or Terras [116].

Example. Solve the following simultaneous linear congruences for x:

$$3x \equiv 1 \pmod{5},$$
$$2x \equiv 3 \pmod{7}.$$

The first congruence has the solution $x \equiv 2 \pmod{5}$, as one can find by trial and error. Then plug $x = 2 + 5u$ into the second congruence to get

$$2x = 2(2 + 5u) \equiv 3 \pmod{7} \quad \text{and thus} \quad 4 + 10u \equiv 3 \pmod{7}.$$

This becomes $3u \equiv 6 \pmod{7}$. One immediately sees a solution $u \equiv 2 \pmod{7}$. This means that $u = 2 + 7t$. Plug this back into our formula for x and get $x = 2 + 5(2 + 7t) = 12 + 35t$. This means $x \equiv 12 \pmod{35}$. You should check that x solves the original simultaneous congruences. ▲

There are many ways to understand the Chinese remainder theorem. The first step is to extend it to an arbitrary number of relatively prime moduli. If the positive integers m_i satisfy $\gcd(m_1, \ldots, m_r) = 1$, and $m = m_1 m_2 \cdots m_r$, then the rings $\mathbb{Z}_m$ and $\mathbb{Z}_{m_1} \oplus \cdots \oplus \mathbb{Z}_{m_r}$ are isomorphic under the mapping $f(x \bmod m) = (x \bmod m_1, \ldots, x \bmod m_r)$. We leave the proof of this as an exercise.

Exercise 6.2.1 *Show that if we assume that the positive integers m_i satisfy $\gcd(m_1, \ldots, m_r) = 1$, and $m = m_1 m_2 \cdots m_r$, then the rings $\mathbb{Z}_m$ and $\mathbb{Z}_{m_1} \oplus \cdots \oplus \mathbb{Z}_{m_r}$ are isomorphic under the mapping $f(x \bmod m) = (x \bmod m_1, \ldots, x \bmod m_r)$. Recall Exercise 3.1.6.*

We want to look at the case $r = 2$ again. To create the isomorphism between $\mathbb{Z}_{15}$ and $\mathbb{Z}_3 \oplus \mathbb{Z}_5$, for example, you can make a big table with its rows and columns indexed by the positive integers, as in Table 6.1.

Next we fill in the blanks in the upper left 3×5 part of the table by moving the first number on the diagonal of the big table which is left out of that upper 3×5 part of the table. That number is 4. Move it up three rows. Similarly we move 5 up three rows. The next number 6 must be moved up three rows and then moved left five columns. Equivalently

Table 6.1

	1	2	3	4	5	1	2	3	4	5	1	2	3	4	5
1	1														
2		2													
3			3												
1				4											
2					5										
3						6									
1							7								
2								8							
3									9						
1										10					
2											11				
3												12			
1													13		
2														14	
3															15

put 6 diagonally down from 5 next to the smaller grid below and then move it left five columns. This produces the following table.

	1	2	3	4	5
1	1			4	
2		2			5
3	6		3		

Continue in this way to complete the table which embodies the Chinese remainder theorem for the case $\mathbb{Z}_3 \oplus \mathbb{Z}_5$.

	1	2	3	4	5
1	1	7	13	4	10
2	11	2	8	14	5
3	6	12	3	9	15

Since the isomorphism preserves addition and multiplication we can compute mod 15 by computing mod 3 and mod 5. This is not so impressive but it would work better if we took a lot of primes like

$$m = 2^5 \cdot 3^3 \cdot 5^2 \cdot 7^2 \cdot 11 \cdot 13 \cdot 17 \cdot 19 \cdot 23 \cdot 29 \cdot 31 \cdot 37 \cdot 41 \cdot 43 \cdot 47,$$

which is $> 3 \cdot 10^{21}$. Then computing $\bmod m$ would be the same as computing $\bmod m_i$, for $m_1 = 2^5, m_2 = 3^3, \ldots, m_{15} = 47$. See Dornhoff and Hohn [25, p. 238] for more information.

In another visualization, we see the additive group $\mathbb{Z}_{15}$ as a discrete circle by taking the Cayley graph $X(\mathbb{Z}_{15}, \{\pm 1 \pmod{15}\})$. See Figure 6.1. Now we can view the same group as a two-dimensional product of the circles $\mathbb{Z}_3$ and $\mathbb{Z}_5$. This is a torus or doughnut graph. It is also the Cayley graph $X(\mathbb{Z}_{15}, \{5, 6, 9, 10 \pmod{15}\})$ in Figure 6.2.

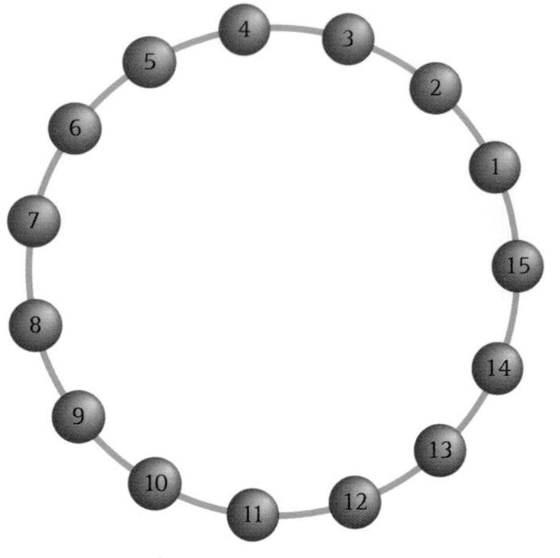

Figure 6.1 The Cayley graph $X(\mathbb{Z}_{15}, \{\pm 1 \pmod{15}\})$

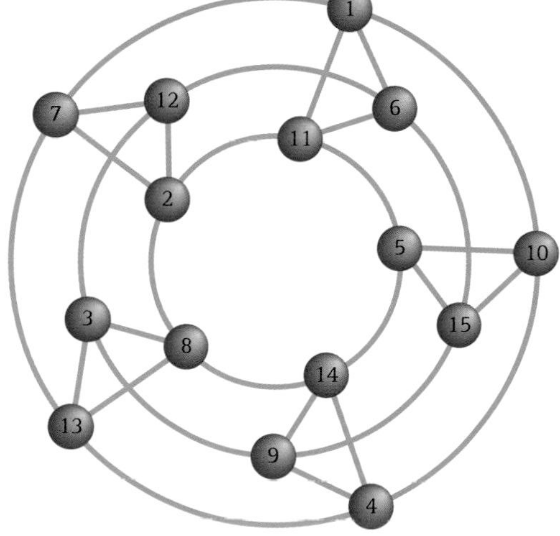

Figure 6.2 The Cayley graph $X(\mathbb{Z}_{15}, \{5, 6, 9, 10 \pmod{15}\})$

Exercise 6.2.2 *Draw analogous figures to Figures 6.1 and 6.2 for* $\mathbb{Z}_{35}$.

As we noted above, the most important thing for a number theorist is the onto-ness of the isomorphism $\widetilde{T}: \mathbb{Z}_{mn} \to \mathbb{Z}_m \oplus \mathbb{Z}_n$ for $\gcd(m, n) = 1$, defined by $\widetilde{T}([s]) = T(s) = (s \pmod{m}, s \pmod{n})$. We want to discuss an old method to give an explicit formula for the solution of the simultaneous congruences that expresses this onto-ness. For example, suppose that $m = 3$ and $n = 5$, and we want to solve

$$\left.\begin{aligned} x &\equiv a \pmod{3}, \\ x &\equiv b \pmod{5}. \end{aligned}\right\} \tag{6.1}$$

Then we first solve two sets of simultaneous congruences:

$$5u \equiv 1 \pmod{3} \quad \text{and} \quad 3v \equiv 1 \pmod{5}.$$

Then set $x = 5au + 3bv \pmod{15}$. It is easily checked that x does solve the problem of formula (6.1). The method is preferable if you want a formula for the answer.

Exercise 6.2.3 *Find an analogous procedure to that of the last paragraph to solve the simultaneous congruences:*

$$x \equiv a \pmod{3},$$
$$x \equiv b \pmod{5},$$
$$x \equiv c \pmod{7}.$$

What is the solution when $a = 2, b = 4, c = 1$?

Hint. *Let* $x = 35ta + 21ub + 15vc$, *where* $35t \equiv 1 (\text{mod } 3), 21u \equiv 1 (\text{mod } 5)$, *and* $15v \equiv 1 (\text{mod } 7)$.

Exercise 6.2.4 *Use the Chinese remainder theorem to finish the proof of equation (2.4) from Section 2.3 for the Euler phi-function.*

There are also applications of the Chinese remainder theorem in RSA cryptography (as we have seen), secret sharing, the fast Fourier transform, and proving the Gödel incompleteness theorem in logic.

There are many puzzles related to the Chinese remainder theorem. Some are ancient. The following is a puzzle found in Ore [84, pp. 118ff].

> An old woman goes to market and a horse steps on her basket and crushes the eggs. The rider offers to pay for the damage and asks her how many eggs she had brought. She does not remember the exact number, but when she had taken them out 2 at a time there was 1 egg left. The same happened when she picked them out 3, 4, 5, and 6 at a time, but when she took them out 7 at a time they came out even. What is the smallest number of eggs she could have had?

Exercise 6.2.5 *Solve the preceding puzzle.*

Exercise 6.2.6 *Suppose the* 2×2 *integer matrix* M *has determinant* d *such that* $\gcd(d, n) = 1$. *Show that we can solve the simultaneous congruences represented by the vector congruence* $Mv = a \pmod{n}$.

Hint. *Use the inverse matrix* $(\text{mod } n)$.

Exercise 6.2.7 *Generalize the Chinese remainder theorem to a commutative ring* R *with identity for multiplication and ideals* A *and* B *such that* $A + B = R$, $A \neq \{0\}, B \neq \{0\}$. *Show that*

$$R/ (A \cap B) \cong (R/A) \oplus (R/B) .$$

Exercise 6.2.8 *Show that* $\mathbb{Z}_3[x]/ \langle x^2 - 1 \rangle \cong \mathbb{Z}_3 \oplus \mathbb{Z}_3$.

Exercise 6.2.9 *Suppose* F *is a field. Consider two rings. The first ring is* $R = F^{2\times 2}$ *under componentwise addition and matrix multiplication. The second ring is* $S = F^{2\times 2}$ *under componentwise addition and componentwise multiplication. Are these two rings isomorphic? Give a reason for your answer.*

Exercise 6.2.10 *Nineteen bears have only 14 berry baskets. The first 13 berry baskets have an equal number of berries and the 14th basket has 3 berries. What is the least number of berries present in total if we know that should they be able to put the berries into 19 baskets, then each basket would have an equal number of berries?*

Exercise 6.2.11 *Suppose you have some beads in a jar and you know that when you take them out three at a time you have two left, but when you take them out five at a time you have four left, and finally when you take them out seven at a time you have six left. How many beads are in the jar?*

Exercise 6.2.12 *Suppose that* F *is a field and* $f(x), g(x) \in F[x]$ *with* $\gcd(f(x), g(x)) = 1$. *Show that* $F[x] / \langle (fg)(x) \rangle \cong (F[x] / \langle f(x) \rangle) \oplus (F[x] / \langle g(x) \rangle)$.

Exercise 6.2.13 *Show that* $F[x] / \langle x^2 - 1 \rangle \cong F \oplus F$, *for any field* F.

6.3 More Stories about *F*[*x*] Including Comparisons with $\mathbb{Z}$

Suppose that F is a field and $F[x]$ is the ring of polynomials in one indeterminate with coefficients in F. We have made a table of comparisons between the ring of integers $\mathbb{Z}$ and the ring $F[x]$ (see Table 6.2). Most of the facts about $F[x]$ stated in the table were proved in Section 5.5. We will make the rest exercises.

Exercise 6.3.1

(a) Compute $h(x) = \gcd(x^2 + 1, x^4 + x^3 + x^2 + 1)$ *in* $\mathbb{Z}_2[x]$. *Find polynomials* $u(x)$ *and* $v(x)$ *in* $\mathbb{Z}_2[x]$ *such that* $h(x) = u(x)\,(x^2 + 1) + v(x)\,(x^4 + x^3 + x^2 + 1)$.
(b) Factor $x^3 + 1$ *as a product of monic irreducible polynomials in* $\mathbb{Z}_7[x]$.
(c) Write $x^4 + x^3 + x^2 + 1$ *as a product of monic irreducible polynomials in* $\mathbb{Z}_2[x]$.

Exercise 6.3.2 *Prove the analog of Euclid's Lemma 1.5.1 for* $F[x]$, *where* F *is a field. Then extend the result to the case that a monic irreducible polynomial divides a product of* n *polynomials.*

Exercise 6.3.3 *Assume that* F *is a field. Prove that every nonzero polynomial* $f(x) \in F[x]$ *factors as*

$$f(x) = \text{unit} \cdot p_1(x) \cdots p_r(x),$$

where a unit is an element of $F - \{0\}$ *and, for each* i, $p_i(x)$ *is a monic irreducible with* $\deg p_i \geq 1$. *Then prove that this factorization is unique up to order.*

Exercise 6.3.4 *Suppose* F *is a field. Show that the ring of polynomials* $F[x]$ *is isomorphic to the ring* V *consisting of infinite sequences* $\{x_n\}_{n \geq 0}$, *where* $x_n \in F$, *for all* n *and all but a finite number of* x_n *are* 0. *Addition and multiplication in* V *are given by the formulas:*

$$\{x_n\} + \{y_n\} = \{x_n + y_n\},$$
$$\{x_n\} \cdot \{y_n\} = \{z_n\},$$
$$\text{with} \quad z_n = \sum_{\substack{j+k=n \\ 0 \leq j,k \leq n}} x_j y_k$$

Table 6.2 *Comparing $\mathbb{Z}$ and $F[x]$*

Property	$\mathbb{Z}$	$F[x]$
infinite ring	yes	yes
integral domain	yes	yes
unit group	$\{1, -1\}$	$F^* = F - \{0\}$
division algorithm	$n = mq + r, 0 < r \leq m$	$f(x) = g(x)q(x) + r(x)$, $r = 0$ or $\deg r < \deg g$
divisibility	$m \mid n \Longleftrightarrow n = mq$, for some $q \in \mathbb{Z}$	$g(x) \mid f(x) \Longleftrightarrow f(x) = g(x)q(x)$, for some $q(x) \in F[x]$
prime	$p > 1$ s.t. $p = a \cdot b \Longrightarrow$ either a or b is a unit in $\mathbb{Z}$	$f(x)$ monic irreducible, $\deg f(x) \geq 1$ $f = g \cdot h$ implies either g or h is a unit in $F[x]$
unique factorization into primes	$n \neq 0 \Longrightarrow n = (\pm 1)\, p_1 p_2 \cdots p_r$, p_i prime, factorization unique up to order	$f(x) \neq 0 \Longrightarrow$ $f(x) = \text{unit} \cdot p_1(x) \cdots p_r(x)$, $p_i(x)$ monic irreducible $\deg p_i \geq 1$ factorization unique up to order
every ideal A principal	$A = \langle n \rangle$, n least positive element of A, if $A \neq \{0\}$	$A = \langle f \rangle$, f element of A of least degree if $A \neq \{0\}$
maximal ideal	$A = \langle p \rangle$, p prime	$A = \langle f(x) \rangle$, $f(x)$ irreducible $\deg (f) \geq 1$
$R/A =$ field when A maximal	$\mathbb{Z}/p\mathbb{Z} =$ field when p is prime	$F[x]/\langle f(x) \rangle =$ field when f irreducible, $\deg f \geq 1$
Euclidean algorithm for $\gcd(a, b) = na + mb$ (Bézout's identity)	yes, Euclidean algorithm works	yes, Euclidean algorithm works
Euclid's lemma	p prime, $p \mid ab \Longrightarrow p \mid a$ or $p \mid b$	$f(x)$ irreducible, $f(x) \mid a(x)b(x)$ $\Longrightarrow f(x) \mid a(x)$ or $f(x) \mid b(x)$

The slightly frightening moral of the preceding exercise is that you do not really need the powers of the indeterminate x when computing with polynomials in $F[x]$. Fraleigh [32, pp. 240–241] addresses the matter, saying: "Why carry x around when you do not even need it? Mathematicians have simply become used to it, that is why." He also notes that replacing x with the sequence

$$\{0, 1, 0, 0, 0, \ldots\}$$

would be pretty annoying to many – including x. Then he asks us not to fuss too much about what x really is. So we call it an "indeterminate." Moreover we resist calling x a variable – for as we noted earlier – a polynomial is not a function – particularly over a finite field.

Next we want to consider our favorite fact about $F[x]$ and its quotients – a fact that allows us to construct all finite fields. Recall that f irreducible implies $\deg f > 0$.

Proposition 6.3.1 ***(A Field with p^n Elements, p Prime).*** *If p is prime and $f(x)$ is an irreducible polynomial in $\mathbb{Z}_p[x]$, $\deg f \geq 1$, then the quotient $\mathbb{F}_{p^n} = \mathbb{Z}_p[x]/\langle f(x) \rangle$ is a field with p^n elements, where $n = \deg f$.*

Proof. Since f is irreducible, the ideal $\langle f\rangle$ is maximal and thus $F[x]/\langle f(x)\rangle$ is a field by Corollary 5.5.4. To see that this field has p^n elements, we just need to see that the elements of $\mathbb{Z}_p[x]/\langle f(x)\rangle$ are represented by the remainders of $h\in\mathbb{Z}_p[x]$ upon division by f. These remainders have the form $r(x)=a_{n-1}x^{n-1}+\cdots+a_1x+a_0$, where $a_j\in\mathbb{Z}_p$. Moreover, the cosets in $\mathbb{Z}_p[x]/\langle f(x)\rangle$ of two distinct remainders cannot be the same. Otherwise f would divide the difference of the remainders. But $\deg f$ is greater than the degree of the difference of two remainders. Contradiction. How many such remainders are there? There are p possibilities for each a_j. Thus there are p^n remainders and thus p^n elements of $\mathbb{Z}_p[x]/\langle f(x)\rangle$. ▲

We have already considered the following example in Section 5.3, but it will not hurt too much to reconsider it, given our accumulated knowledge.

Example: A field with nine elements. $\mathbb{Z}_3[x]/\langle x^2+1\rangle=\mathbb{F}_9$.

If we view $\mathbb{Z}_3$ as an analog of the real numbers, this field may be viewed a a finite analog of the complex numbers. From the preceding proposition, we know that $\mathbb{Z}_3[x]/\langle x^2+1\rangle$ is a field with nine elements, once we know that x^2+1 is irreducible in $\mathbb{Z}_3[x]$. From Proposition 5.5.1, we know that x^2+1 is irreducible iff it has no roots in $\mathbb{Z}_3$. Consider $f(x)=x^2+1$ and plug in the elements of $\mathbb{Z}_3$. You get $f(0)=1, f(1)=f(-1)=2$. Thus x^2+1 is irreducible in $\mathbb{Z}_3[x]$.

The field $\mathbb{Z}_3[x]/\langle x^2+1\rangle$ is isomorphic to $\mathbb{Z}_3[i]=\{a+bi \mid a,b\in\mathbb{Z}_3\}$, which is the smallest field containing $\mathbb{Z}_3$ and i such that $i^2=-1$. Perhaps we should use some other letter than i here. We are not talking about the complex number i. The letter i just stands for something not in $\mathbb{Z}_3$ such that $i^2=-1$. We showed in the preceding paragraph that no element of $\mathbb{Z}_3$ satisfies the equation for i. To prove that $\mathbb{Z}_3[x]/\langle x^2+1\rangle\cong\mathbb{Z}_3[i]$, you can define an onto map $T:\mathbb{Z}_3[x]\to\mathbb{Z}_3[i]$ by $T(f(x))=f(i)$. Then $\ker T=\langle x^2+1\rangle$ since if $g\in\ker T$, we have $g(x)=q(x)(x^2+1)+r(x)$, where $\deg r(x)<2$ or $r=0$. So $0=g(i)=r(i)$ means $r=0$, and then $g\in\langle x^2+1\rangle$. Conversely any element of $\langle x^2+1\rangle$ must be in $\ker T$. Thus, the first isomorphism theorem says $\mathbb{Z}_3[x]/\langle x^2+1\rangle$ is isomorphic to $\mathbb{Z}_3[i]$. Here we really identify the coset $[x]$ in $\mathbb{Z}_3[x]/\langle x^2+1\rangle$ with i. That is quite analogous to what happened when $\mathbb{C}$ was created as $\mathbb{R}[i]$ in Section 5.4. ▲

The Quadratic Formula for $\mathbb{Z}_p$, $p>2$

Next we consider $\mathbb{Z}_p$, for p prime, to be an analog of the real numbers $\mathbb{R}$ and we ask: **is there an analog of the quadratic formula?** So consider the quadratic equation $ar^2+br+c=0$ for $a,\ b,\ c\in\mathbb{Z}_p$ and $a\neq 0$. Now we do the $\mathbb{Z}_p$ analog of completing the square, as long as $p\neq 2$. We can divide by a since $\mathbb{Z}_p$ is a field and obtain

$$r^2+\frac{b}{a}r=\frac{-c}{a}.$$

Then for $p\neq 2$ we can add $\left(\frac{b}{2a}\right)^2$ to both sides and get

$$r^2+\frac{b}{a}r+\left(\frac{b}{2a}\right)^2=\frac{-c}{a}+\left(\frac{b}{2a}\right)^2.$$

Now the left-hand side is a square and we have

$$\left(r+\frac{b}{2a}\right)^2=\frac{-c}{a}+\left(\frac{b}{2a}\right)^2=\frac{-4ac+b^2}{4a^2}.$$

Define the **discriminant** $D=b^2-4ac$ and obtain

$$\left(r+\frac{b}{2a}\right)^2=\frac{D}{(2a)^2}.$$

So we take square roots of both sides and note that we may have to go to a larger field than $\mathbb{Z}_p$ to find $\sqrt{D}$. This gives:

$$r=\frac{-b\pm\sqrt{D}}{2a}.$$

This is the "same" quadratic formula that you may be familiar with from high school. We have two cases (again assuming $p>2$):

(1) If $\sqrt{D}\in\mathbb{Z}_p$ then $r\in\mathbb{Z}_p$.
(2) If $\sqrt{D}\notin\mathbb{Z}_p$, we can view r as an element of the field

$$\mathbb{Z}_p[\sqrt{D}]\cong\mathbb{Z}_p[x]/\langle x^2-D\rangle=\mathbb{F}_{p^2},$$

which is our analog of the complex numbers.

For the real numbers we also had two cases:

Case 1. $D=b^2-4ac\geq 0\Longrightarrow$ roots real.
Case 2. $D=b^2-4ac<0\Longrightarrow$ roots complex and not real.

Exercise 6.3.5 *Explain the preceding cases for the real numbers $\mathbb{R}$ and then produce an analogous result for the rational numbers $\mathbb{Q}$.*

You may be wondering about the case $p=2$. When $p=2$, the quadratic formula does not make sense as $1/2$ makes no sense in $\mathbb{Z}_2$.

There are many other facts that you learned in high school or college that are just as true in "any" field. For example, in Section 7.1 we recall a bit of linear algebra, just to make sure that you believe it works for any field as well as for the field of real numbers.

Exercise 6.3.6

(a) *Find all roots of $f(x)=3x^2+x+4$ in $\mathbb{Z}_7$ by the process of substituting all elements of $\mathbb{Z}_7$.*
(b) *Find all roots of the polynomial $f(x)$ in part (a) using the quadratic formula for $\mathbb{Z}_7$. Do your answers agree? Should they?*
(c) *Same as (a) for $g(x)=x^2+x+4$ over $\mathbb{Z}_5$.*
(d) *Same as (b) for $g(x)$ in part (c).*
(e) *Same as (a) for $h(x)=x^2-3x+2$.*
(f) *Same as (b) for $h(x)=x^2-3x+2$.*

Exercise 6.3.7 *Suppose that* $D \in \mathbb{Z}^+$ *is not a square: that is,* $D \neq n^2$*, for any* $n \in \mathbb{Z}$*. Set* $\mathbb{Q}\left[\sqrt{D}\right] = \left\{x + y\sqrt{D} \,\middle|\, x, y \in \mathbb{Q}\right\}$. *Show that* $\mathbb{Q}[\sqrt{D}]$ *is a field.*

Exercise 6.3.8 *Using the definition in the preceding exercise, show that the mapping* $T : \mathbb{Q}\left[\sqrt{7}\right] \to \mathbb{Q}\left[\sqrt{11}\right]$ *defined by* $T\left(x + y\sqrt{7}\right) = x + y\sqrt{11}$*, for all* $x, y \in \mathbb{Q}$*, is not a ring isomorphism.*

Exercise 6.3.9 *Show that, using the notation of the preceding exercises – with* $\mathbb{Q}$ *replaced by* $\mathbb{R}$ *– we have* $\mathbb{R}\left[\sqrt{-1}\right] = \mathbb{R}\left[\sqrt{-7}\right] = \mathbb{C}$.

The following exercise is an important theorem. Do not skip it!

Exercise 6.3.10 *Show that, for prime p, the multiplicative group* $\mathbb{Z}_p^*$ *is cyclic.*

Hint. *Recall Exercise 3.6.17 which states that if G is an Abelian group, and* $g, h \in G$*, there is an element of G of order the least common multiple* $\mathrm{lcm}[|g|, |h|]$*. This implies that if g has maximal order r in G then* $x^r = e$*, the identity, for all* $x \in G$*. Then use the fact that a polynomial of degree d over a field F has at most d roots in F.*

Exercise 6.3.11 ***(Euler's Criterion).*** *Show that if p is a prime* > 2*, and* $a \in \mathbb{Z}_p^*$*, then* $a = b^2$ *for some* $b \in \mathbb{Z}_p^*$ *iff* $a^{\frac{p-1}{2}} \equiv 1 \pmod{p}$.

Exercise 6.3.12 *Suppose that* $\mathbb{F}_p$ *is the finite field with a prime number p of elements. Suppose that A and B are non-squares in* $\mathbb{F}_p$*. Show that* $\mathbb{F}_p[\sqrt{A}] = \mathbb{F}_p[\sqrt{B}]$*, again using the notation of the previous exercises, with* $\mathbb{R}$ *and* $\mathbb{Q}$ *replaced with* $\mathbb{F}_p$.

Hint. $A = C^2 B$ *for some* $C \in \mathbb{F}_p$.

Exercise 6.3.13

(a) Assume p is prime. Show that there are $(p-1)/2$ *irreducible polynomials of the form* $f(x) = x^2 - b$ *in* $\mathbb{Z}_p[x]$.
(b) Show that for every prime p, there exists a field with p^2 *elements.*

There is actually a formula for the number of irreducible polynomials of degree d over $\mathbb{Z}_p$ or any finite field. See Dornhoff and Hohn [25, p. 377].

6.4 Field of Fractions or Quotients

You might argue that this section could or should have appeared in Chapter 1. The idea of the field of fractions (or quotients) generalizes the idea from elementary school that created the rational numbers $\mathbb{Q}$ from the integers $\mathbb{Z}$. It also generalizes the construction of the field of rational functions

$$F(x) = \left\{ \frac{f(x)}{g(x)} \,\middle|\, f(x), g(x) \in F[x], g(x) \neq 0 \right\}$$

from the ring of polynomials $F[x]$ over a field F. Recall that we need to identify the fractions: $\frac{1}{3}=\frac{2}{6}=\frac{-2}{-6}$ or $\frac{3}{4}=\frac{75}{100}$. You need to recall how to add and multiply them too: $\frac{2}{3}+\frac{5}{7}=\frac{14+15}{21}=\frac{29}{21}$, $\frac{2}{3}\cdot\frac{5}{7}=\frac{10}{21}$. The same sort of things happen with fractions of polynomials. You need to identify things like

$$\frac{1}{x}=\frac{x^2+1}{x^3+x}.$$

You also need to add them

$$\frac{1}{x}+\frac{1}{x-1}=\frac{x-1}{x^2-x}+\frac{x}{x^2-x}=\frac{2x-1}{x^2-x}.$$

Once you remember this, you should be able to generalize the idea to any integral domain. Thus you would produce the following definition, state the next theorem, and do the ensuing exercises. These things are worked out in detail in many references such as [9], sometimes just for $\mathbb{Z}$.

Definition 6.4.1 *Suppose that D is an integral domain. Then we can construct the field of fractions or quotients F by first creating a set S whose elements are the symbols $\frac{a}{b}$, where $a, b \in D$ and $b \neq 0$. An equivalence relation on S is given by saying $\frac{a}{b} \sim \frac{c}{d}$ iff $ad = bc$. Then define F to be the set of equivalence classes $\left[\frac{a}{b}\right]$ of S. Addition is defined by*

$$\left[\frac{a}{b}\right]+\left[\frac{c}{d}\right]=\left[\frac{ad+bc}{bd}\right],$$

and multiplication is defined by

$$\left[\frac{a}{b}\right]\left[\frac{c}{d}\right]=\left[\frac{ac}{bd}\right],$$

for $a, b, c, d \in D$. In these definitions of addition and multiplication, we always assume $bd \neq 0$.

Theorem 6.4.1 *The preceding definition creates a field F which contains a subring isomorphic to D.*

We consign the proof of this theorem to the following exercises.

Exercise 6.4.1 *Prove the following statements.*

(a) The relation $\sim$ in Definition 6.4.1 is indeed an equivalence relation.
(b) The definitions of addition and multiplication of equivalence classes for the relation in part (a) are independent of representative.

Exercise 6.4.2 *Prove the following claims.*

(a) The set F satisfies the field axioms.
(b) The subring isomorphic to D consists of classes $\left[\frac{a}{1}\right]$, $a \in D$.

The earliest known use of fractions (according to the all-wise internet) goes back to 2800 BC in India (the Indus valley).

Exercise 6.4.3 *Is is possible to create a field containing a subring isomorphic to $\mathbb{Z}_6$ or more generally any ring with zero divisors?*

Exercise 6.4.4 *Suppose an integral domain D is a subring of the field E. Show that the field of fractions of D is isomorphic to a subfield of E.*

Exercise 6.4.5 *What is the field of fractions of $\mathbb{Z}_5$?*

Exercise 6.4.6 *Show that the field of fractions of an integral domain is unique up to isomorphism.*

Suppose that, instead of $\mathbb{Z}$, we start with $F[x_1,\ldots,x_n]$, where F is a field. Here $F[x_1,\ldots,x_n]$ denotes the ring of polynomials in n indeterminates. This ring was considered in Exercise 5.2.15. Then the field of fractions for $F[x_1,\ldots,x_n]$ is the **field of rational functions in several indeterminates:**

$$f(x_1,\ldots,x_n)=\frac{g(x_1,\ldots,x_n)}{h(x_1,\ldots,x_n)},$$
$$\text{where } g(x_1,\ldots,x_n) \text{ and } h(x_1,\ldots,x_n)\in F[x_1,\ldots,x_n], \text{ with } h\neq 0.$$

We use the notation $F(x_1,\ldots,x_n)$ for the corresponding field of rational functions.

We should perhaps mention a big theorem that intertwines group theory and polynomials. As for polynomials, elements σ of the symmetric group S_n act on rational functions $f(x_1,\ldots,x_n)$ by $(\sigma f)(x_1,\ldots,x_n)=f\left(x_{\sigma(1)},\ldots, x_{\sigma(n)}\right)$. We defined the **elementary symmetric polynomials** in Section 3.7:

$$\begin{aligned}
s_1(x_1,\ldots,x_n)&=x_1+x_2+\cdots+x_n,\\
s_2(x_1,\ldots,x_n)&=x_1x_2+x_1x_3+\cdots+x_{n-1}x_n=\sum_{1\le i<j\le n} x_ix_j,\\
&\ \ \vdots\\
s_k(x_1,\ldots,x_n)&=\sum_{1\le i_1<i_2<\cdots<i_k\le n} x_{i_1}x_{i_2}\cdots x_{i_k},\\
&\ \ \vdots\\
s_n&=x_1x_2\cdots x_n.
\end{aligned}\tag{6.2}$$

The symmetric group acts on the rational functions in $F(x_1,\ldots,x_n)$. Then one can show that symmetric rational functions are rational functions of the elementary symmetric polynomials. Moreover, one finds that S_n is the Galois group of $F(x_1,\ldots,x_n)$ over $F(s_1,\ldots,s_n)$. We will not consider such Galois groups here, but see Herstein [42] or Dummit and Foote [28]. With a little more effort, one can prove the fundamental theorem on symmetric polynomials – which says that any symmetric polynomial can be expressed as a polynomial in the elementary symmetric polynomials. (It is a starred exercise in Herstein [42].) Thus, for example, when $n=2$, we have $x_1^2+x_2^2=(x_1+x_2)^2-2x_1x_2=s_1^2-2s_2$.

Exercise 6.4.7 *Write the following symmetric polynomials in the indeterminates x_1 and x_2 as polynomials in the elementary symmetric polynomials $s_1=x_1+x_2$ and $s_2=x_1x_2$.*

(a) $x_1^3+x_2^3$,
(b) $x_1^4+x_2^4$.

It is also possible to create rings with smaller sets of denominators than the set of nonzero elements. Rings of the sort created in the following exercise have been very useful in number theory. See the book by Samuel [97, Chapter V].

Exercise 6.4.8 ***(The Localization of $\mathbb{Z}$ at a Prime p).*** *If p is a prime, let $\mathbb{Z}_{(p)}$ denote the subset of $\mathbb{Q}$ consisting of fractions $\frac{m}{n}$, with $m, n \in \mathbb{Z}$, $\gcd(m, n) = 1$, such that p does not divide n. Show that $\mathbb{Z}_{(p)}$ is a subring of $\mathbb{Q}$. Then show that the nonzero ideals of $\mathbb{Z}_{(p)}$ have the form (p^n), $n = 1, 2, 3, \ldots$*

There are other sorts of rings of fractions that have been useful in number theory. Instead of considering fractions such as those in the preceding exercise, one can consider $\frac{m}{n}$, with $m, n \in \mathbb{Z}$, nonzero n, $\gcd(m, n) = 1$, such that p *does* divide n – denominators in the complement of the set of denominators in the localization. All you need for the possible denominators is that they form a closed set under multiplication (and, of course, do not contain 0). See Ribenboim [89, Chapter 12]. S-integers in rings of algebraic integers like $\mathbb{Z}\left[e^{2\pi i/n}\right]$ appear in recent work on the Stark conjectures. A somewhat expository paper is that of Stark [111]. The object of such creations as the S-integers seems to be to kill off a number of prime ideals in order to go from unique factorization into prime ideals in the ring of integers of an algebraic number field to unique factorization into prime numbers – simplifying many computations.

Exercise 6.4.9 *We defined an ordered integral domain in Section 1.3. Suppose that F is the field of fractions of an ordered integral domain D. Show that if we define $\frac{a}{b}$ positive to mean that ab is positive in D, this turns F into an ordered field as in Definition 5.3.7.*

Exercise 6.4.10 *Show that in an ordered integral domain D any nonzero square b^2 must be positive. Next suppose we create the field of fractions F of D as in Definition 6.4.1, and we know that F is an ordered field. Show that then $\frac{a}{b} > 0$ for $a, b \in D$ implies $ab > 0$.*

Exercise 6.4.11 *Consider the integral domain $\mathbb{Z}\left[\sqrt{5}\right] = \left\{a + b\sqrt{5} \mid a, b \in \mathbb{Z}\right\}$. What is the field of fractions for $\mathbb{Z}\left[\sqrt{5}\right]$?*

7

Vector Spaces and Finite Fields

7.1 Matrices and Vector Spaces over Arbitrary Fields and Rings like $\mathbb{Z}$

We want to restrict ourselves to a consideration of the mere basics of linear algebra. We will make much of this subject an exercise – assuming that you know most of this from calculus. You can find solutions in Dornhoff and Hohn [25], for example. Or you could look at whatever book you used for this part of your calculus course and ask what remains true if we replace $\mathbb{R}$ by $\mathbb{Z}_p$ or some other field F. Most of the earlier chapters on Gaussian elimination, dimension, determinants work as before. So, for example, you might take the book [115] by Strang (or the book with the same title by various other authors) and convince yourself that all of the results of the early chapters work for arbitrary fields.

The favorite calculations from linear algebra involve Gaussian elimination. It turns out that this is not due to Gauss at all but appears in a Chinese book – parts of which were written as early as 150 bc. **Gaussian elimination** allows one to put a matrix $A \in F^{m\times n}$ into echelon form using elementary row operations.

The **elementary row operations over the field** F are:

(1) permute row i and row j;
(2) multiply row i by a nonzero element of F;
(3) replace row i by row i plus an element of F times row j.

A matrix in **row echelon form** means that it has the following properties.

(1) Rows with at least one nonzero element are above the rows of all 0s. The first nonzero entry in a nonzero row is called a **pivot**. Below each pivot is a column of 0s.
(2) Each pivot is to the right of the pivot in the row above.

Using only elementary row operations over F, you can put any matrix with entries in F into row-echelon form.

You can even put any matrix in $F^{m\times n}$ into **row-reduced echelon form** by using elementary row operation (2) to normalize all pivots to be 1 and then by putting 0s above all pivots.

Example. Suppose the field is $\mathbb{Z}_3$ and the matrix is

$$\begin{pmatrix} 2 & 1 & 0 & 2 \\ 1 & 0 & 0 & 2 \\ 1 & 0 & 0 & 0 \end{pmatrix}.$$

Assuming we remember to compute mod 3, we can replace (row 2) by (row $2 - 2 \cdot$ row 1) and do the same for (row 3) to get

$$\begin{pmatrix} 2 & 1 & 0 & 2 \\ 0 & 1 & 0 & 1 \\ 0 & 1 & 0 & 2 \end{pmatrix}.$$

Finally replace (row 3) by (row 3 − row 2) to get

$$\begin{pmatrix} 2 & 1 & 0 & 2 \\ 0 & 1 & 0 & 1 \\ 0 & 0 & 0 & 1 \end{pmatrix}.$$

This matrix is in row-echelon form.

The row-reduced echelon form of the matrix is

$$\begin{pmatrix} 1 & 0 & 0 & 0 \\ 0 & 1 & 0 & 0 \\ 0 & 0 & 0 & 1 \end{pmatrix}.$$

Of course, one also has the analogous elementary column operations. These are really the elementary row operations applied to the transposed matrix. ▲

Exercise 7.1.1 *Over the field $\mathbb{Z}_2$ put the following matrix (which we shall see again in the section on error-correcting codes)*

$$\begin{pmatrix} 1 & 1 & 0 & 1 & 0 & 0 & 0 \\ 0 & 1 & 1 & 0 & 1 & 0 & 0 \\ 0 & 0 & 1 & 1 & 0 & 1 & 0 \\ 0 & 0 & 0 & 1 & 1 & 0 & 1 \end{pmatrix}$$

into row-echelon form and row-reduced echelon form.

Since the elementary row operations on the matrix A of a homogeneous linear equation $Ax = 0$ do not change the solution set, we get the theorem saying that a homogeneous system of m linear equations in n unknowns over a field F has a non-trivial solution if $m < n$. To see this, you just need to see that if $Ax = 0$, and you perform the elementary row operations on A, to produce a new matrix A', you will still find that $A'x = 0$. Moreover, you can reverse all the row operations to bring A' back to A with the same sort of operations. This means that $A'x = 0$ implies $Ax = 0$. This same argument is also clear from the elementary matrix version of the row operations. For the elementary row operations each multiply A on the left by an invertible matrix U with elements in the field F – as is discussed in the next paragraph.

We consider an example of the matrix version of the system of equations $Ax = 0$, with $A \in F^{m\times n}, x \in F^n$. Each elementary row operation on A corresponds to finding an $m \times m$ non-singular matrix U with entries in the field F and replacing A by UA. Here, as usual, we define matrix multiplication in the same way as we did over $\mathbb{R}$ with formula (1.3) in Section 1.8 and the statement following it. Here, as usual, a non-singular matrix is a square matrix with nonzero determinant – or equivalently a matrix in $F^{m\times m}$ with an inverse for multiplication in $F^{m\times m}$ – see Exercise 7.3.13. For example, the first operation we did to the matrix in the example above was to replace (row 2) by (row 2 − 2 · row 1). This is achieved by multiplying the matrices below – remembering that the coefficients are in $\mathbb{Z}_3$:

$$\begin{pmatrix} 1 & 0 & 0 \\ 1 & 1 & 0 \\ 0 & 0 & 1 \end{pmatrix} \begin{pmatrix} 2 & 1 & 0 & 2 \\ 1 & 0 & 0 & 2 \\ 1 & 0 & 0 & 0 \end{pmatrix} = \begin{pmatrix} 2 & 1 & 0 & 2 \\ 0 & 1 & 0 & 1 \\ 1 & 0 & 0 & 0 \end{pmatrix}.$$

The second operation to replace (row 3) with (row 3 − 2 · row 1) is achieved by multiplying the matrices below:

$$\begin{pmatrix} 1 & 0 & 0 \\ 0 & 1 & 0 \\ 1 & 0 & 1 \end{pmatrix} \begin{pmatrix} 2 & 1 & 0 & 2 \\ 0 & 1 & 0 & 1 \\ 1 & 0 & 0 & 0 \end{pmatrix} = \begin{pmatrix} 2 & 1 & 0 & 2 \\ 0 & 1 & 0 & 1 \\ 0 & 1 & 0 & 2 \end{pmatrix}.$$

The third operation to replace (row 3) by (row 3 − row 2) is achieved by multiplying the following matrices:

$$\begin{pmatrix} 1 & 0 & 0 \\ 0 & 1 & 0 \\ 0 & 2 & 1 \end{pmatrix} \begin{pmatrix} 2 & 1 & 0 & 2 \\ 0 & 1 & 0 & 1 \\ 0 & 1 & 0 & 2 \end{pmatrix} = \begin{pmatrix} 2 & 1 & 0 & 2 \\ 0 & 1 & 0 & 1 \\ 0 & 0 & 0 & 1 \end{pmatrix}.$$

To put 1s on the diagonal we multiply:

$$\begin{pmatrix} 2 & 0 & 0 \\ 0 & 1 & 0 \\ 0 & 0 & 1 \end{pmatrix} \begin{pmatrix} 2 & 1 & 0 & 2 \\ 0 & 1 & 0 & 1 \\ 0 & 0 & 0 & 1 \end{pmatrix} = \begin{pmatrix} 1 & 2 & 0 & 1 \\ 0 & 1 & 0 & 1 \\ 0 & 0 & 0 & 1 \end{pmatrix}.$$

Then to put zeros in the non-diagonal entry of the second row, we multiply

$$\begin{pmatrix} 1 & 0 & 0 \\ 0 & 1 & 2 \\ 0 & 0 & 1 \end{pmatrix} \begin{pmatrix} 1 & 2 & 0 & 1 \\ 0 & 1 & 0 & 1 \\ 0 & 0 & 0 & 1 \end{pmatrix} = \begin{pmatrix} 1 & 2 & 0 & 1 \\ 0 & 1 & 0 & 0 \\ 0 & 0 & 0 & 1 \end{pmatrix}.$$

Similarly one can produce zeros in the non-diagonal entries of the first row to get to the row reduced-echelon form matrix.

$$\begin{pmatrix} 1 & 0 & 0 & 0 \\ 0 & 1 & 0 & 0 \\ 0 & 0 & 0 & 1 \end{pmatrix}.$$

Exercise 7.1.2 *Write down the 3×3 matrix that is needed to produce the result of the last sentence.*

Exercise 7.1.3 *Write down the product of all the 3×3 matrices that we used to go from*

$$\begin{pmatrix} 2 & 1 & 0 & 2 \\ 1 & 0 & 0 & 2 \\ 1 & 0 & 0 & 0 \end{pmatrix} \quad \text{to} \quad \begin{pmatrix} 1 & 0 & 0 & 0 \\ 0 & 1 & 0 & 0 \\ 0 & 0 & 0 & 1 \end{pmatrix},$$

Example. Suppose the field is $\mathbb{Z}_3$ and the matrix is that of the previous example

$$A = \begin{pmatrix} 2 & 1 & 0 & 2 \\ 1 & 0 & 0 & 2 \\ 1 & 0 & 0 & 0 \end{pmatrix}.$$

Then the corresponding system of equations is $Ax = 0$, ${}^{\mathrm{T}}x = (x_1, x_2.x_3)$:

$$\begin{cases} 2x_1 + x_2 + 2x_4 = 0 \\ x_1 + 2x_4 = 0 \\ x_1 = 0. \end{cases}$$

An equivalent system is that corresponding to the row-reduced echelon form matrix

$$\begin{pmatrix} 1 & 0 & 0 & 0 \\ 0 & 1 & 0 & 0 \\ 0 & 0 & 0 & 1 \end{pmatrix},$$

which is

$$\begin{cases} x_1 = 0 \\ x_2 = 0 \\ x_4 = 0. \end{cases}$$

The result is that $x_1 = x_2 = x_4 = 0$ and the other coordinate x_3 is arbitrary (i.e., free to be whatever it wants to be in $\mathbb{Z}_3$). Of course it was pretty obvious at the beginning that the equations did not involve x_3. Moreover it might have been better to interchange row 3 and row 1 first. Gaussian elimination always involves choices. ▲

Exercise 7.1.4 *Over the field $\mathbb{Z}_2$ write down the homogeneous system of equations $Gx = 0$ corresponding to the following matrix*

$$G = \begin{pmatrix} 1 & 1 & 0 & 1 & 0 & 0 & 0 \\ 0 & 1 & 1 & 0 & 1 & 0 & 0 \\ 0 & 0 & 1 & 1 & 0 & 1 & 0 \\ 0 & 0 & 0 & 1 & 1 & 0 & 1 \end{pmatrix}.$$

From Exercise 7.1.1, the row-reduced echelon form of G is $G' = (I_4\ A)$, where I_4 is the 4×4 identity matrix. If

$$H = \begin{pmatrix} -A \\ I_3 \end{pmatrix},$$

solve $xH = 0$, for $x \in \mathbb{Z}_2^7$. How does the set of such vectors x compare with the set of vectors $y = uG$, for $u \in \mathbb{Z}_2^4$?

The three elementary row operations are obtained by multiplying the matrix on the left by three types of **elementary matrices** for which we give 3×3 examples with entries that are assumed to be in the field $F = \mathbb{Z}_7$:

1. a **permutation matrix** U, which is a square matrix such that each row and each column has one entry equal to 1 while the rest of the entries are 0

$$\begin{pmatrix} 1 & 0 & 0 \\ 0 & 0 & 1 \\ 0 & 1 & 0 \end{pmatrix},$$ to interchange (rows 2) and (row 3);

2. a matrix having all entries 0 except for the diagonal entries, all of which are 1 except for the ii entry which is some $\alpha \in F^*$

$$\begin{pmatrix} 1 & 0 & 0 \\ 0 & 5 & 0 \\ 0 & 0 & 1 \end{pmatrix}, \text{ to multiply (row 2) by 5 (mod 7);}$$

3. a matrix having all entries 0 except for the diagonal entries, all of which are 1, and the ij entry which is some $\alpha \in F$

$$\begin{pmatrix} 1 & 0 & 0 \\ 5 & 1 & 0 \\ 0 & 0 & 1 \end{pmatrix}, \text{ to replace (row 2) by ((row 2) + 5(row 1)).}$$

The elementary matrices corresponding to the elementary row operations on $F^{n\times n}$ can be shown to generate the **general linear group** $GL(n, F)$ consisting of all invertible $n \times n$ matrices with entries in the field F.

Exercise 7.1.5 *Give some examples to show that if you multiply a matrix $A \in F^{3\times 3}$ on the left by an elementary matrix $U \in F^{3\times 3}$, it produces elementary row operations on A. Show also that the elementary matrices in your examples are invertible for matrix multiplication.*

Exercise 7.1.6 *Show that $n \times n$ elementary matrices generate the general linear group $GL(n, F)$, for any field F.*

Exercise 7.1.7 *Show that, if $U \in GL(n, F)$, the map from $A \in F^{n\times m}$ to UA gives a group action (as in Definition 3.7.1) of the general linear group $GL(n, F)$ on $F^{n\times m}$, for any field F. Give representatives of the orbits of this group action when $n = 3$ and $m = 2$.*

Exercise 7.1.8 *Suppose $U \in GL(n, F)$ and $A \in F^{n\times m}$. Why is it that the solutions $x \in F^m$ of $Ax = 0$ are the same as the solutions to $UAx = 0$?*

Question. Does Gaussian elimination work over Euclidean domains D like $\mathbb{Z}$ and $F[x]$? The answer is that it does work well. In fact, you could even allow D to be a **principal ideal domain** – meaning that D is an integral domain such that every ideal in D is a principal ideal.

The **elementary row operations over Euclidean domains** D are:

1. permute row i and row j;
2. replace row i by a unit in D^* times row i;
3. replace row i by row i plus an element of D times row j.

Again these elementary row operations on a matrix $A \in D^{m\times n}$ correspond to multiplication of A on the left by matrices in the general linear group $GL(n, D)$ which consists of $n \times n$ matrices U with elements in D such that the determinant $\det(U)$ is a unit in D. Here, as usual, the group operation is matrix multiplication. For then the formula for the inverse of a matrix from Exercise 7.3.12 below implies that D^{-1} has entries in D. Similarly one can define column operations over D. These will be necessary if one wants to diagonalize the matrix A.

Elementary row and column operations over $\mathbb{Z}$, for example, allow us to put a matrix of integers into the **Smith normal form,** meaning a matrix such that all entries are 0 except those on the diagonal and such that if d_i is the ith diagonal entry then d_i divides d_{i+1} for all i given by

$$\begin{pmatrix} 1 & & & & & & & & \\ & \ddots & & & & & & & \\ & & 1 & & & & 0 & & \\ & & & d_{s+1} & & & & & \\ & & & & \ddots & & & & \\ & 0 & & & & d_{s+m} & & & \\ & & & & & & 0 & & \\ & & & & & & & \ddots & \\ & & & & & & & & 0 \end{pmatrix}, \text{ with } d_i > 1. \tag{7.1}$$

The prime power divisors of the diagonal entries are called **elementary divisors** and are unique up to multiplication by units in D. This result in turn can be used to prove the fundamental theorem of finitely generated Abelian groups stated below. Here **finitely generated group** G just means that G has a finite generating set. The Smith normal form is also useful in computations of algebraic number theory and algebraic topology.

Example. Put $A = \begin{pmatrix} 1 & 2 \\ 3 & 4 \end{pmatrix}$ into Smith normal form over $\mathbb{Z}$.

We replace (row 2) by ((row 2) $- 3 \cdot$ (row 1)). This gives

$$\begin{pmatrix} 1 & 2 \\ 3 & 4 \end{pmatrix} \to \begin{pmatrix} 1 & 2 \\ 0 & -2 \end{pmatrix}.$$

Then replace (row 1) by ((row 1) + (row 2)) and (row 2) by $-$(row 2) to get

$$\begin{pmatrix} 1 & 0 \\ 0 & 2 \end{pmatrix}.$$

So the elementary divisors of A are 1 and 2. ▲

We did not have to use column operations in the preceding example, but it was pretty simple. In general, you need column operations. Doing this calculation can be quite difficult – especially if you replace $\mathbb{Z}$ with the ring of polynomials $F[x]$, F a field. Luckily now software exists to help. I use Scientific Workplace both to type this book and to compute Smith normal forms for matrices over $\mathbb{Z}$ or $F[x]$. Now I can just put the mouse on the matrix and then hit tools, matrices, Smith normal form, and the answer pops out:

$$\begin{pmatrix} 1 & 2 \\ 3 & 4 \end{pmatrix}, \text{ Smith normal form: } \begin{pmatrix} 1 & 0 \\ 0 & 2 \end{pmatrix}.$$

What a change from the time I taught a course in which everyone got a different answer for the elementary divisors of 3×3 matrices over a polynomial ring.

To put a matrix with integer entries into Smith normal form, your first goal is to put the greatest common divisor of the entries in the upper left-hand corner of the new matrix. Start by putting the smallest entry in absolute value in the upper left-hand position. Then multiply by -1 if necessary to make this guy positive. You can slowly achieve your goal by replacing any other entry in the first row or column by its remainder upon division by the first entry. Again take the smallest remainder obtained and put it in the favorite upper left position. At some point all the entries in the first row and column except for the first must be 0. An induction has begun.

Theorem 7.1.1 ***(The Fundamental Theorem of Finitely Generated Abelian Groups).*** *Any finitely generated Abelian group G is isomorphic to the direct sum* $\mathbb{Z}_{d_1} \oplus \cdots \oplus \mathbb{Z}_{d_m} \oplus \mathbb{Z}^n$. *Here* $\mathbb{Z}_{d_i}$ *and* $\mathbb{Z}$ *are viewed as additive groups.*

This is a central theorem – in both group theory and algebraic number theory, as well as algebraic topology. You can find a proof in the spirit of our discussion in Schreier and Sperner [101, Chapter IV, Section 20]. More modern books have less computational versions. See Dummit and Foote [28] for example.

Let us now give a sketch of a proof of the fundamental theorem of Abelian groups following Schreier and Sperner – feeling free to cheat if necessary. The idea is that you have a presentation of your group G with a finite set of generators. Fix a generating set $\{g_1, \ldots, g_r\}$. Any element $g \in G$ has an expression of the form $g = g_1^{v_1} \cdots g_r^{v_r}$, for some $v_i \in \mathbb{Z}$. We say the generating set $\{g_1, \ldots, g_r\}$ is a **basis** of G if $g_1^{v_1} \cdots g_r^{v_r} = g_1^{w_1} \cdots g_r^{w_r}$ implies $g_i^{v_i} = g_i^{w_i}$, for all $i = 1, \ldots, r$. For the generating set to be a basis, it suffices to show that the representation of the identity element is unique.

To show there exists a basis, first define the relation vectors $v = (v_1, \ldots, v_r) \in \mathbb{Z}^r$ to be those corresponding to relations $g^{v_1} \cdots g_r^{v_r} = e$, the identity of G. Note that the set R of relation vectors forms a subgroup of the group $\mathbb{Z}^r$ under componentwise addition. In fact, modern authors would call R a submodule of the $\mathbb{Z}$-module $\mathbb{Z}^r$. This means that R is closed under scalar multiplication by some integer $n \in \mathbb{Z}$ as well as addition. We will say a bit about modules after we discuss vector spaces at the end of this section.

Our relation vectors R form a matrix with r columns and infinitely many rows. We need to go all Smith normal form on this matrix. Certainly it is legal to perform elementary row operations on the matrix – except for the infinite number of rows part.[1] Well, induction should work. But we need to be able to do elementary column operations too. How is this possible? For this, you must be willing to replace or renumber the generators. To interchange columns i and j you must interchange generators g_i and g_j. To multiply column i by -1, you must replace generator g_i by g_i^{-1}. To replace (column j) by ((column j) $-$ n(column i)), for $n \in \mathbb{Z}$, you need to replace generator g_i by generator $g_i g_j^{-n}$.

Performing the elementary row and column operations leads to a matrix in Smith normal form – in which the bottom infinite number of rows are 0 and a diagonal matrix above most of these 0 rows looks like the matrix (7.1) above. The basis relations are the nonzero relation vectors.

What does this mean? If e denotes the identity of G, the basic relations are $g_1 = e, \ldots, g_s = e, g_{s+1}^{d_1} = e, \ldots, g_{s+m}^{d_m} = e$. This implies that we can leave out the first s generators as they are all the identity. Then, renumbering the generators, we are looking at a group with generators $g_1, \ldots, g_{m+n}$, where the first m have finite orders. The last n generators have infinite order. The result is that we have the fundamental theorem of Abelian groups.

[1] However, that is really a red herring by a theorem about submodules of finitely generated free modules being finitely generated. See Hungerford [47, Chapter 2, Section 1].

Another application of the Smith normal form replaces the ring $\mathbb{Z}$ of integers with the polynomial ring $F[x]$, for a field F and indeterminate x. If two $n \times n$ matrices A, B have entries in F, one can use the Smith normal form of $A - xI$ and $B - xI$ (over the polynomial ring $F[x]$) to decide whether A and B are **similar** matrices, meaning that $B = UAU^{-1}$ for some invertible matrix $U \in F^{n \times n}$. The Smith normal form is due to H. J. Smith (1826–1883). If the Smith normal form of $xI - A$ has diagonal entries: $1, \ldots, 1, f_j(x), f_{j+1}(x), \ldots, f_n(x)$, the **rational canonical form** of A will be the matrix with companion matrices of the non-trivial polynomials $f_i(x)$ along the diagonal. If F is a field and

$$f(x) = x^n - a_{n-1}x^{n-1} - \cdots - a_1 x - a_0$$

is a polynomial in $F[x]$, define the **companion matrix**

$$C_f(x) = \begin{pmatrix} 0 & 0 & 0 & \cdots & 0 & a_0 \\ 1 & 0 & 0 & \cdots & 0 & a_1 \\ 0 & 1 & 0 & \cdots & 0 & a_2 \\ \vdots & \vdots & \vdots & \ddots & \vdots & \vdots \\ 0 & 0 & 0 & \cdots & 1 & a_{n-1} \end{pmatrix}. \tag{7.2}$$

Some authors transpose the companion matrix.

A favorite reference for me on this subject is the Schaum's outline by Ayres [4]. As a student I found it very frustrating that the texts I read would never tell me how to compute the rational canonical form of a given matrix. Instead they would tell me a few tricks that worked in special cases along with a general proof that was not computational. Of course you could argue that now you do not need to know how Scientific Workplace computes the rational canonical form of a matrix. Other references for canonical forms are Dornhoff and Hohn [25] and Dummit and Foote [28].

The Jordan form of a matrix is slightly different. The aim is to produce a matrix as close to diagonal as possible. You need to consider a field F such that all elements of $F[x]$ factor completely into linear factors – or to assume that all the eigenvalues of the matrix under consideration lie in F. Then you find that the **Jordan form** of matrix M with entries in F is a matrix that is block diagonal with Jordan blocks along the diagonal. A **Jordan block** is a matrix of the form

$$\begin{pmatrix} \lambda & 1 & & & 0 \\ & \lambda & & & \\ & & \ddots & 1 & \\ & & & \lambda & 1 \\ 0 & & & & \lambda \end{pmatrix}.$$

Again the matrix M is similar to its Jordan form matrix. For the details, see Dummit and Foote [28].

It will be useful to try an example.

Example. Find the Smith normal form of the following matrix over $\mathbb{Q}[x]$:

$$xI - A = \begin{pmatrix} x-1 & -2 \\ -3 & x-4 \end{pmatrix}.$$

$$\begin{pmatrix} x-1 & -2 \\ -3 & x-4 \end{pmatrix} \to \begin{pmatrix} -3 & x-4 \\ x-1 & -2 \end{pmatrix} \to \begin{pmatrix} 1 & \frac{-1}{3}(x-4) \\ x-1 & -2 \end{pmatrix}$$
$$\to \begin{pmatrix} 1 & \frac{-1}{3}(x-4) \\ 0 & -2+\frac{1}{3}(x-4)(x-1) \end{pmatrix} \to \begin{pmatrix} 1 & \frac{-1}{3}(x-4) \\ 0 & x^2-5x-2 \end{pmatrix}$$
$$\to \begin{pmatrix} 1 & 0 \\ 0 & x^2-5x-2 \end{pmatrix}.$$

It follows that the matrix A is similar to the **companion matrix** of $f(x)=x^2-5x-2$, which is the matrix

$$C(f)=\begin{pmatrix} 0 & 1 \\ 2 & 5 \end{pmatrix}.$$

This happens because the Smith normal form of $Ix-C(f)$ is the same as that of $Ix-A$. This is the basic reasoning behind the computational theory of the **rational canonical form** of a matrix. Again Scientific Workplace does the computation for me and gets the transpose of my result – probably because it has transposed the companion matrices, which some people will:

$$\begin{pmatrix} 1 & 2 \\ 3 & 4 \end{pmatrix} = \begin{pmatrix} 1 & 1 \\ 0 & 3 \end{pmatrix} \begin{pmatrix} 0 & 2 \\ 1 & 5 \end{pmatrix} \begin{pmatrix} 1 & -\frac{1}{3} \\ 0 & \frac{1}{3} \end{pmatrix}.$$ ▲

Exercise 7.1.9 *Put the following matrix of integers into its Smith normal form using only elementary row operations over* $\mathbb{Z}$:

$$\begin{pmatrix} 1 & 2 & 3 \\ 4 & 5 & 6 \\ 7 & 8 & 9 \end{pmatrix}.$$

Exercise 7.1.10 *Put the following two matrices in their Smith normal forms using only elementary row operations over* $\mathbb{Z}_3[x]$:

$$\begin{pmatrix} x-1 & 1 \\ 0 & x-1 \end{pmatrix}, \begin{pmatrix} x-1 & 0 \\ 1 & x-2 \end{pmatrix}.$$

Now we move to some topics that lead to the abstraction of matrices as linear functions (also called linear maps or transformations) in the next section. Before defining linear functions, we need to define the abstract version of F^n: namely, the vector space over F.

Definition 7.1.1 *A non-empty set V (the **vectors**) is a **vector space** over a field F (the **scalars**) if V is an Abelian group under addition and there is a function from $F \times V$ into V sending $(\alpha, v) \in F \times V$ to $\alpha \cdot v = \alpha v$ (multiplication by scalars) such that, $\forall v, w \in V$ and $\forall \alpha, \beta \in F$, we have the following four properties:*

1. $\alpha(v+w)=\alpha v+\alpha w$;
2. $(\alpha+\beta)v=\alpha v+\beta v$;
3. $\alpha(\beta v)=(\alpha\beta)v$;
4. $1v=v$.

Note that we do not put arrows over our vectors. We are content to differentiate vectors from scalars by using Greek letters for scalars when possible.

Example. The plane $\mathbb{R}^2 = \{(x, y) \mid x, y \in \mathbb{R}\}$ is a vector space over the field $\mathbb{R}$. Here addition is componentwise, that is,

$$(x, y) + (u, v) = (x + u, y + v),$$

and multiplication by scalars is also componentwise, that is,

$$\alpha(x, y) = (\alpha x, \alpha y),$$

for all $x, y, \alpha \in \mathbb{R}$.

If you replace $\mathbb{R}$ by any field F, you make F^2 into a vector space with analogous definitions of addition and multiplication by scalars. Similarly you obtain a vector space F^n over F, by considering the set of vectors $(x_1, \dots, x_n)$, $x_i \in F$, $i = 1, \dots, n$, with componentwise addition and scalar multiplication:

$$(x_1, \dots, x_n) + (y_1, \dots, y_n) = (x_1 + y_1, \dots, x_n + y_n),$$
$$\alpha\,(x_1, \dots, x_n) = (\alpha x_1, \dots, \alpha x_n), \quad \text{for } x_i, y_i \in F, i = 1, \dots, n \text{ and } \alpha \in F.$$ ▲

Next we need some definitions in order to know what we mean by dimension of a vector space – span of a set of vectors and linearly independent set of vectors.

Definition 7.1.2 *The **span** of a set S of vectors in the vector space V over the field F is the set of finite F-linear combinations of vectors from S: that is,*

$$\mathrm{Span}(S) = \left\{ \sum_{i=1}^{k} \alpha_i v_i \;\middle|\; \alpha_i \in F, v_i \in S, \forall i, \text{ with } k \text{ any positive integer} \right\}.$$

By definition, a **finite-dimensional vector space** has a finite spanning set. If the vector space V does not have a finite set S of elements such that any element of V is a finite linear combination of elements from S, then V is **infinite dimensional**. We will mostly consider finite-dimensional vector spaces here.

Definition 7.1.3 *S is a **set of linearly independent vectors** if*

$$\sum_{i=1}^{n} \alpha_i v_i = 0$$

for some scalars $\alpha_i \in F$ and vectors $v_i \in S$ implies all $\alpha_i = 0$.

Definition 7.1.4 *A finite subset B of the finite-dimensional vector space V over the field F is a **basis** of V if it has the following two properties:*

1. *B spans V, that is, $V = \mathrm{Span}(B)$.*
2. *B is a set of linearly independent vectors.*

*The **dimension** of a finite-dimensional vector space V over the field F is the size of a basis B of V.*

To show that the idea of dimension makes sense, one must prove that any finite-dimensional vector space has a basis. We do this in the next section. Then one must show that any two bases of a vector space have the same number of elements.

Exercise 7.1.11 *Prove that any two bases A and B of a finite-dimensional vector space V over a field F must have the same number of elements.*

Hint. *See Birkhoff and MacLane [9], Dornhoff and Hohn [25], Schreier and Sperner [101], or Strang [115]. Schreier and Sperner argue that if* $A=\{a_1,\ldots,a_p\}$ *spans V and* $B=\{b_1,\ldots,b_q\}$ *is linearly independent, then* $q\le p$ *(using induction on q). In particular, they prove that we can replace q of the vectors in A with the vectors in B and still span V. They start with* $q=0$*, which is a very silly case – but nevertheless legal and sensible to begin with. Assuming the result for the subset* $\{b_1,\ldots,b_{q-1}\}$ *of B, they argue that, by renumbering the set A, we can replace the first* $q-1$ *vectors in A with the first* $q-1$ *vectors in B and still span V. Then we know that* $b_q=\sum_{i=1}^{p}\beta_i a_i$*, for some* $\beta_i\in F$*. Not all* β_i *for* $i>q-1$ *can vanish, since the set B consists of linearly independent vectors and* $a_1=b_1,\ldots,a_{q-1}=b_{q-1}$*. In particular, the sum must have terms in it beyond the first* $q-1$ *terms. This means that indeed* $p\ge q$ *and some* $\beta_i\neq 0$*, say for* $i=q$*. Then we can replace* a_q *with* b_q *and still span V.*

Example. The plane $\mathbb{R}^2$ is two-dimensional with basis $\{(1,0),(0,1)\}$. This is the standard basis. Another basis of $\mathbb{R}^2$ is $\{(1,1),(2,0)\}$. ▲

Exercise 7.1.12 *Prove the preceding statements and then the analog replacing* $\mathbb{R}$ *by* $\mathbb{Z}_3$*.*

Exercise 7.1.13 *Show that, for a matrix in row-echelon form, the set of nonzero rows is a linearly independent set.*

Notation. We will view elements of $F^n=\{(x_1,\ldots,x_n)|\,x_i\in F\}$ as row vectors (mostly). This means we need to write vM if $M\in F^{n\times m}$. The function $v\longmapsto vM$ composes badly with another matrix function since the function part, M, is on the right rather than the left. That is why I would really prefer to write column vectors. But they take up a lot of space on a page.

Exercise 7.1.14 *Show that a basis for* F^n *is the set of n vectors with* $(n-1)$ *zero entries and one entry of 1:*

$$(1,0,0,\ldots,0),(0,1,0,\ldots,0),(0,0,1,\ldots,0),\ldots,(0,0,0,\ldots,1).$$

You could surely make the following definition yourself, having seen the definitions of subgroup and subring.

Definition 7.1.5 *A* ***subspace*** *W of a vector space V over the field F is a non-empty subset* $W\subset V$ *which is a vector space under the same operations as those of V.*

Example 1. The plane $\mathbb{R}^2$ has as subspace $W=\{(x,0)\,|\,x\in\mathbb{R}\}$, the real line. Similarly $\mathbb{Z}_3^2$ has as a subspace $W=\{(x,0)\mid x\in\mathbb{Z}_3\}$. ▲

Example 2. A subfield F of a field E is a vector subspace of E, considering E and F as vector spaces over any subfield of F. ▲

Exercise 7.1.15 *If S is any subset of the vector space V over the field F, prove that* $\mathrm{Span}(S)$ *is indeed a vector subspace of V.*

Exercise 7.1.16 *Prove that elementary row operations do not change the span of the set of rows of a matrix.*

Your linear algebra text may define the row rank of a matrix as the number of pivots of the matrix. Recall that the pivots are the first entries in the nonzero rows of the row echelon form of the matrix. Now we want to define the row rank using the idea of dimension.

Definition 7.1.6 *The **(row) rank** of a matrix $A \in F^{m\times n}$ is the dimension of the span of the set of row vectors of A.*

Thanks to the following exercises, we just say "rank" rather than the row rank of a matrix.

Exercise 7.1.17 *Show that the row rank of a matrix $A \in F^{m\times n}$ is the same as the number of pivots, that is, the first nonzero entries of the nonzero rows of the row echelon form of A.*

Exercise 7.1.18 *Prove that the row rank of a matrix is the same as the column rank (which is the dimension of the span of the set of columns of the matrix).*

Exercise 7.1.19 *Show that if F is a field then the set of polynomials $F[x]$ in one indeterminate and coefficients in F forms a vector space over F with the usual addition of polynomials and multiplication by constants in F. Then show that $F[x]$ is an infinite-dimensional vector space over F.*

Since we have already discussed replacing the field F in a matrix with a ring like $\mathbb{Z}$ or $F[x]$, it makes sense to think about what happens to the definition of vector space when the field F of scalars is replaced with a ring R of scalars. Well, algebraists have a word for that – an **R-module** M or, maybe, a left R-module M if R is not commutative. It is really an exercise to define it – just copy the definition of vector space. Of course, you need to leave out the requirement that $1v = v$, for all $v \in M$, if the ring does not have an identity 1 for multiplication – and you need to add the word "left" if R is not commutative. Similarly you should be able to define submodule, module homomorphism, quotient module, direct product of modules. Here we are mostly interested in $R = \mathbb{Z}$ or $F[x]$ – both commutative and, even better, Euclidean, meaning they have a division algorithm. Thus you can compute such things as the Smith normal form of a matrix using the division algorithm.

A $\mathbb{Z}$-module is really an Abelian group. So we are not motivated to take up the subject of modules, which is usually left to graduate algebra courses. The $\mathbb{Z}$-module that you get from $\mathbb{Z}^n$ under componentwise addition and the scalar multiplication $\gamma(a_1, \ldots, a_n) = (\gamma a_1, \ldots, \gamma a_n)$, for γ and $a_i \in \mathbb{Z}$, has a special name – the **free $\mathbb{Z}$-module of rank n.** The **rank** of a free module is analogous to the dimension of a vector space. Moreover, one has the result that a submodule M of a free $\mathbb{Z}$-module N of rank n is also free and has rank $\leq n$.

This was useful in our sketch of a proof of the fundamental theorem of Abelian groups. Many algebra books cover this subject: for example, Dummit and Foote [28], Hungerford [47], and Lang [65]. Some of these texts may be a bit hard to read because they are directed at graduate students and usually refuse to assume their rings are Euclidean integral domains like $\mathbb{Z}$ – meaning that there is a division algorithm. Instead a weaker condition is assumed – that all ideals are principal – when proving something implying the fundamental theorem of Abelian groups.

7.2 Linear Functions or Mappings

Modern mathematicians – Mr. Bourbaki we are thinking of you – try not to write equations with subscripts and subsubscripts. This means they really want to eliminate matrices from polite conversation. This leads to the following definition.

Definition 7.2.1 *If V and W are both vector spaces over the field F, then we say that a function $T: V \to W$ is a* ***linear function,*** *also called a* ***linear mapping*** *or* ***linear transformation,*** *if T has the following two properties for all $v, w \in V$ and all $\alpha \in F$:*

1. $T(v+w) = T(v) + T(w)$,
2. $T(\alpha v) = \alpha T(w)$.

Here we will usually call a linear mapping T a linear map for short. Elsewhere – particularly for infinite-dimensional vector spaces – it may be called a linear operator. Strangely, it is not called a vector space homomorphism.

Definition 7.2.2 *If V and W are both vector spaces over the field F, then a linear mapping $T: V \to W$ is a* ***vector space isomorphism*** *over F iff T is 1–1 and onto. Then we write $V \cong W$ and say V is* ***isomorphic*** *to W.*

Example 1. If $B = \{b_1, \ldots, b_m\}$ is a basis of the vector space V over the field F, then $V \cong F^m$. The mapping M_B is defined by writing $v \in V$ in the form $v = \sum_{i=1}^{m} \alpha_i b_i$, with $\alpha_i \in F$, and then setting

$$M_B(v) = (\alpha_1, \ldots, \alpha_m). \tag{7.3}$$

We leave it as an exercise to check that M_B is indeed a vector space isomorphism. ▲

Exercise 7.2.1 *Check that M_B defined above is a vector space isomorphism $M_B : V \to F^m$.*

Exercise 7.2.2 *Show that if $m \neq n$, then F^n is not isomorphic to F^m.*

Exercise 7.2.3 *Show that if the vector space*

$$V = C^\infty(\mathbb{R}) = \left\{ f: \mathbb{R} \to \mathbb{R} \mid f^{(n)}(x) \text{ exists } \forall x \in \mathbb{R}, \forall n \geq 1 \right\}$$

the derivative mapping $Lf = f', f \in V$, is linear.

Example 2. Consider the linear mapping $T:\mathbb{Z}_2^4 \to \mathbb{Z}_2^7$ defined by mapping a row vector $v \in \mathbb{Z}_2^4$ to the row vector $T(v) = vG$, where

$$G = \begin{pmatrix} 1 & 1 & 0 & 1 & 0 & 0 & 0 \\ 0 & 1 & 1 & 0 & 1 & 0 & 0 \\ 0 & 0 & 1 & 1 & 0 & 1 & 0 \\ 0 & 0 & 0 & 1 & 1 & 0 & 1 \end{pmatrix}.$$

▲

Exercise 7.2.4 *Show that if we define T as in Example 2, the image space $T(\mathbb{Z}_2^4)$ is a vector subspace of $\mathbb{Z}_2^7$. Find a basis for $T(\mathbb{Z}_2^4)$. What is the dimension of the image space?*

Now we address the problem of showing that a basis of a finite-dimensional vector space exists.

Methods to find a Basis of a Finite-Dimensional Vector Space *V* over a Field *F*

Method 1. If you have a finite spanning set S of vectors in V, you keep deleting any vectors that can be written as a (finite) linear combination of other vectors from S.

Method 2. Start with one nonzero vector or any non-empty set of linearly independent vectors in V and keep adding vectors from V that are not in the span of the vectors you already have.

Exercise 7.2.5 *Prove that the two methods to find a basis of V actually work.*

Given that V and W are both (finite-dimensional) vector spaces over the field F, then the **matrix of a linear mapping** $T: V \to W$ with respect to the (ordered) bases $B = \{b_1, \ldots, b_m\}$ of V and $C = \{c_1, \ldots, c_n\}$ of W is the $n \times m$ array of scalars $\mu_{ij} \in F$ defined by:

$$Mat_{C,B}(T) = (\mu_{ij})_{\substack{1 \le i \le n, \\ 1 \le j \le m}} \quad \text{where} \quad T(b_j) = \sum_{i=1}^{n} \mu_{ij} c_i, \quad \text{for } j = 1, \ldots, m. \tag{7.4}$$

We have defined the matrix of a linear transformation this way so that the composition of linear transformations corresponds to product of matrices.

Exercise 7.2.6 *Suppose V, W are vector spaces over the field F and $T: V \to W$ is a linear map. If B, C are (ordered) bases of V, W, show that, using the definition of M_B in equation (7.3) from Example 1 above – as well as the definition of $Mat(T)_{C,B}$ in (7.4) – we have*

$${}^{\mathrm{T}}M_C(T(v)) = Mat(T)_{C,B}\, {}^{\mathrm{T}}M_B(v),$$

*where if $A = (\alpha_{ij})_{1 \le i \le n, 1 \le j \le m}$, the **transpose** of A denoted ${}^{\mathrm{T}}A$ is the matrix $(\alpha_{ji})_{1 \le j \le m, 1 \le i \le n}$, where the rows and columns are interchanged.*

Hint. *By the linearity of T, we have for (ordered) bases $B = \{b_1, \ldots, b_m\}$ of V and $C = \{c_1, \ldots, c_n\}$ of W:*

$$T\left(\sum_{j=1}^{m} \alpha_j b_j\right) = \sum_{j=1}^{m} \alpha_j T(b_j) = \sum_{j=1}^{m} \alpha_j \sum_{i=1}^{n} \mu_{ij} c_i.$$

Interchange the sums over i and j are you will have done the exercise.

Exercise 7.2.7 *Suppose V, W, U are vector spaces over the field F and $T: V \to W$, $S: W \to U$ are both linear maps. Show that the composition $S \circ T: V \to U$ is also a linear map. Then show that if B, C, D are (ordered) bases of V, W, U respectively, then*

$$Mat_{D,B}(S \circ T) = Mat_{D,C}(S)\, Mat_{C,B}(T)\,, \tag{7.5}$$

where the product on the right is the usual matrix multiplication.

Given that V and W are both vector spaces over the field F, and $T: V \to W$ is a linear map, the **image of** T, $T(V)$, is a vector subspace of W. The space $T(V)$ is also called the **range** of T. Then the **rank of the linear map** T is defined to be the dimension of the image $L(V) = \{Lv \mid v \in V\}$.

Exercise 7.2.8

(a) Show that, assuming V and W are both vector spaces over the field F and $T: V \to W$ is a linear transformation, then the image $T(V)$ is indeed a vector subspace of W.
(b) Show that the rank of a linear transformation L is the same as the rank of a matrix of L using bases B of V and C of W.

Exercise 7.2.9 *Consider the field $\mathbb{Z}_3[i]$, where $i^2 + 1 = 0$.*

(a) Show that $\mathbb{Z}_3[i]$ is a vector space over the field $\mathbb{Z}_3$.
(b) Define the map $F: \mathbb{Z}_3[i] \to \mathbb{Z}_3[i]$ by $F(z) = z^3$. Show that F is linear. Then find a matrix of F using the basis $\{1, i\}$ for $\mathbb{Z}_3[i]$ as a vector space over the field $\mathbb{Z}_3$.

Exercise 7.2.10 *Suppose that V and W are vector spaces and $T: V \to W$ is a 1–1 linear transformation. Show that if B is a basis of V, then $T(B)$ is a basis of $T(V)$. Conclude that if V and W are isomorphic finite-dimensional vector spaces, then they have the same dimension. Moreover, show that if $V \cong W$ and V is finite dimensional, so is W.*

There is another vector space associated to a linear transformation T – the kernel of T.

Definition 7.2.3 *Suppose that V, W are vector spaces and $T: V \to W$ is a linear transformation. The* ***kernel*** *of T, also called the* ***nullspace*** *of T, is*

$$\ker T = \{v \in V \mid Tv = 0\}.$$

The dimension of $\ker T$ is often called the ***nullity*** *of T (or of a matrix corresponding to T in the case that V and W are finite dimensional).*

Theorem 7.2.1 *Suppose that V and W are both (finite-dimensional) vector spaces over the field F, and $T: V \to W$ is a linear map. If $\ker T = \{v \in V \mid Tv = 0\}$ then*

$$\dim \ker T + \dim T(V) = \dim V.$$

Proof. Take a basis $B = \{b_1, \ldots, b_m\}$ of $\ker T$ and extend it to a basis $C = \{b_1, \ldots, b_m, b_{m+1}, \ldots, b_n\}$ of V. Then we claim that $\{T(b_{m+1}), \ldots, T(b_n)\}$ is a basis for $T(V)$. ▲

Exercise 7.2.11 *Fill in the details in the preceding proof.*

Exercise 7.2.12 *Prove that if V is a finite-dimensional vector space over the field F, then a linear mapping $T: V \to V$ is 1–1 iff it is onto.*

Exercise 7.2.13 *Suppose that $B = \{b_1, \ldots, b_n\}$ and $C = \{c_1, \ldots, c_n\}$ are two (ordered) bases of the vector space V over the field F. Let $1_V(x) = x$, $\forall x \in V$ be the identity map (which is certainly linear).*

(a) *Show that if $M = \mathrm{Mat}_{C,B}(1_V)$, then M is invertible.*
(b) *Show that for any linear map $L: V \to V$, we have*

$$\mathrm{Mat}_{C,B}(1_V)\mathrm{Mat}_{B,B}(L) = \mathrm{Mat}_{C,C}(L)\mathrm{Mat}_{C,B}(1_V).$$

Hint for part (b). *Use equation (7.5) to see that both sides of the equality are $\mathrm{Mat}_{C.B}(L)$.*

Definition 7.2.4 *If A and B are $n \times n$ matrices with entries in the field F, we say that A and B are* **similar** *iff there is an invertible $n \times n$ matrix U such that $B = U^{-1}AU$.*

Exercise 7.2.14 *Here we ask for proofs of some useful facts about similarity.*

(a) *Show that similarity is an equivalence relation on $F^{n\times n}$.*
(b) *Show that two $n \times n$ matrices A and B over the field F are similar iff they are the matrices of the same linear transformation $T: F^n \to F^n$ with respect to two bases of F^n.*

As we said in Section 7.1, the Smith normal form allows one to obtain canonical forms of matrices (such as the Jordan form or rational canonical form) so that any matrix will be similar to only one matrix of a given canonical form. See Dornhoff and Hohn [25] or Dummit and Foote [28] for more details. This can be useful despite the reluctance of some applied books to consider the Jordan form of a matrix.

For example, if one needs to do Fourier analysis on the general linear group $G = GL(n, \mathbb{Z}_p)$ of invertible 2×2 matrices over the field $\mathbb{Z}_p$, p prime, one must know the conjugacy classes $\{g\} = \{x^{-1}gx \mid x \in G\}$. That is, one needs to know the similarity classes.

These **similarity (conjugacy) classes in** $GL(2, \mathbb{Z}_p)$ for odd primes p are

central	$\left\{\begin{pmatrix} r & 0 \\ 0 & r \end{pmatrix}\right\}$,	where $r \neq 0$;
parabolic	$\left\{\begin{pmatrix} r & 1 \\ 0 & r \end{pmatrix}\right\}$,	where $r \neq 0$;
hyperbolic	$\left\{\begin{pmatrix} r & 0 \\ 0 & s \end{pmatrix}\right\}$,	where $rs \neq 0$ and $r \neq s$;
elliptic	$\left\{\begin{pmatrix} r & s\delta \\ s & r \end{pmatrix}\right\}$,	where δ is not a square in $\mathbb{Z}_p$ and $s \neq 0$.

See Terras [116, p. 366] for more information.

Recall that the characteristic p of a finite field F is a prime number p which is the order of 1 in the additive group of F. Exercise 6.1.9 showed that if F is a finite field of characteristic p, then F contains $\mathbb{Z}_p$ as a subfield. The next proposition says a little more.

Proposition 7.2.1 *A finite field F of characteristic p (necessarily a prime) is a vector space over $\mathbb{Z}_p$.*

Proof. Look at the additive subgroup H of F which is generated by 1. Then $T:\mathbb{Z}\to H\subset F$ defined by $T(n)=n\cdot 1$ is a ring homomorphism mapping $\mathbb{Z}$ onto H. By the definition of characteristic, we know that $\ker T=p\mathbb{Z}$. By the first isomorphism theorem, we know that H is isomorphic to $\mathbb{Z}/p\mathbb{Z}=\mathbb{Z}_p$. This implies that we may view $\mathbb{Z}_p$ as a subfield of F. But then scalar multiplication by $a\in\mathbb{Z}_p$ makes sense for elements $v\in F$, implying that F is indeed a vector space over $\mathbb{Z}_p$. ▲

Corollary 7.2.1 *A finite field F of characteristic p is a vector space over $\mathbb{Z}_p$ which is necessarily finite dimensional. If the dimension of F over $\mathbb{Z}_p$ is n, then $F\cong\mathbb{Z}_p^n$. This implies that F has p^n elements.*

Exercise 7.2.15 *Show that there is no integral domain with exactly six elements. More generally, show that there is no integral domain with pq elements if p and q are distinct primes.*

Notation. When p is a prime, we write $\mathbb{F}_{p^n}$ for the field with p^n elements since we will be able to show that there is only one such field up to isomorphism fixing elements of $\mathbb{F}_p$. However, many texts write $GF(p^n)$ instead of $\mathbb{F}_{p^n}$ and they call the field a **Galois field.** See Gallian [33, Chapter 22]. The name honors Galois who first introduced the idea in 1830. However, Gauss had certainly considered congruences and $\mathbb{F}_p$ much earlier in 1799 with his book *Disquisitiones Arithmeticae*. In notation, as usual, we are following Bourbaki and using blackboard bold ($\mathbb{F}$) rather than the usual font for F to free up the letter F for other uses. However, we might note that $\mathbb{Z}_p$ has another meaning in number theory (the ring of p-adic integers, which will not be considered in this text). There are never enough letters!

Example: The Quaternions. Consider a four-dimensional vector space $\mathbb{H}$ over $\mathbb{R}$ with basis $1, i, j, k$. We define multiplication by first defining how to multiply the basis vectors as in the multiplication table for the quaternion group in Section 3.6. That is, $i^2=j^2=k^2=ijk=-1$. Then assume that the multiplication satisfies the usual associative and distributive laws plus $(\alpha v)\cdot w=\alpha(v\cdot w)=v\cdot(\alpha w)$, for all $\alpha\in\mathbb{R}$ and $v, w\in\mathbb{H}$. This gives a non-commutative ring – also called a division algebra – as in Definition 7.2.5 below. It turns out that you can divide by nonzero elements. That is because we have an analog of complex conjugate:

$$\overline{\alpha_1+\alpha_2 i+\alpha_3 j+\alpha_4 k}=\alpha_1-\alpha_2 i-\alpha_3 j-\alpha_4 k.$$

Then if $v=\alpha_1+\alpha_2 i+\alpha_3 j+\alpha_4 k$, for $\alpha_n\in\mathbb{R}$, we have the **norm** of v given by $Nv=v\overline{v}=\alpha_1^2+\alpha_2^2+\alpha_3^2+\alpha_4^2$. This means that when $v\neq 0$, we have $v^{-1}=\frac{1}{v\overline{v}}\overline{v}\in\mathbb{H}$.

We told some of the story of Hamilton's discovery of the quaternions in Section 3.6 where we introduced the quaternion group. The quaternions have proved useful in physics and number theory. The construction has been generalized, replacing $\mathbb{R}$ by other fields. Finite quaternions turn out not to be so interesting as they are full matrix algebras like $\mathbb{R}^{n\times n}$. ▲

Exercise 7.2.16 *Suppose we multiply two quaternions*

$$(\alpha_1+\alpha_2 i+\alpha_3 j+\alpha_4 k)(\beta_1+\beta_2 i+\beta_3 j+\beta_4 k)=\gamma_1+\gamma_2 i+\gamma_3 j+\gamma_4 k. \tag{7.6}$$

Here the α_r, β_r, and γ_r are in $\mathbb{R}$. Show that $\gamma_1 = \alpha_1\beta_1 - \alpha_2\beta_2 - \alpha_3\beta_3 - \alpha_4\beta_4$. Obtain similar equations for the rest of the γ_r, $r = 2, 3, 4$.

Exercise 7.2.17 *Show that, in the quaternions $\mathbb{H}$, we have $\overline{x \cdot y} = \overline{y} \cdot \overline{x}$.*

Exercise 7.2.18 *Use what we know about quaternions to prove that we have the multiplicative property of the norm, which says $NvNw = N(vw)$. This gives* ***Lagrange's identity****, which states that if the relationship of the γs to the αs and βs is as in equation (7.6), then*

$$\left(\alpha_1^2 + \alpha_2^2 + \alpha_3^2 + \alpha_4^2\right)\left(\beta_1^2 + \beta_2^2 + \beta_3^2 + \beta_4^2\right) = \gamma_1^2 + \gamma_2^2 + \gamma_3^2 + \gamma_4^2.$$

The following definition generalizes the quaternions.

Definition 7.2.5 *An (associative)* ***algebra*** *A over a field F is a finite-dimensional vector space over F with a multiplication operation that makes A a ring such that for all $\alpha \in F$ and $x, y \in A$ we have $\alpha(xy) = (\alpha x)y = x(\alpha y)$. The algebra is called a* ***division algebra*** *if, in addition, A has an identity 1 for multiplication and, for every $a \in A$, there is an inverse $a^{-1} \in A$ such that $aa^{-1} = a^{-1}a = 1$.*

In 1878 Frobenius proved that the only associative division algebras over $\mathbb{R}$ are $\mathbb{R}$, $\mathbb{C}$, and $\mathbb{H}$ (the quaternions).

Another example of an algebra over a field F is $A = F^{n\times n}$, consisting of all $n \times n$ matrices over the field F (under the usual componentwise addition and matrix multiplication) with componentwise multiplication by scalars in F. This algebra is **simple**, meaning that it has no two-sided ideals except $\langle 0 \rangle$ and $\langle 1 \rangle = A = F^{n\times n}$.

Exercise 7.2.19 *Prove the last statement.*

In 1908 Wedderburn proved a converse to the simplicity of $F^{n\times n}$. A special case of Wedderburn's theorem says that any simple algebra over $\mathbb{C}$ is isomorphic to the algebra $\mathbb{C}^{n\times n}$. Wedderburn's general result says that if A is a simple algebra over any field F, then there is a division algebra D over F such that A is isomorphic to $D^{n\times n}$, for some integer $n > 0$.

Exercise 7.2.20 *Prove that any division algebra is simple.*

Another example of an algebra is the **group algebra** $F[G]$ over the field F associated to a finite group G. This algebra is defined as follows. Suppose that $G = \{g_1, \ldots, g_n\}$. Identify the elements of G with a basis for a vector space over F. Then as a vector space the algebra $F[G]$ consists of vectors $\sum_{i=1}^{n} \alpha_i g_i$, for $\alpha_i \in F$. Then we add and multiply in A as follows, with $\alpha_i, \beta_i \in F$,

$$\begin{aligned} \sum_{i=1}^{n} \alpha_i g_i + \sum_{i=1}^{n} \beta_i g_i &= \sum_{i=1}^{n} (\alpha_i + \beta_i)\, g_i, \\ \left(\sum_{i=1}^{n} \alpha_i g_i\right) \cdot \left(\sum_{j=1}^{n} \beta_j g_j\right) &= \sum_{k=1}^{n} \sum_{\substack{i,j \\ g_i g_j = g_k}} (\alpha_i \beta_j)\, g_k. \end{aligned}$$

If we identify g_1 with the identity of G and $\alpha \in F$ with αg_1, then we have multiplication by scalars as

$$\alpha \sum_{i=1}^{n} \beta_i g_i = \sum_{i=1}^{n} (\alpha\beta_i)\, g_i.$$

In short, our formulas say that multiplication of the basis elements of the group algebra $F[G]$ just comes from the multiplication of the group elements. Thus the quaternion algebra over $\mathbb{C}$ is just the group algebra $\mathbb{C}[Q]$, where Q is the quaternion group.

Fourier analysis on G is often carried out (thanks to Emmy Noether) via the study of the group algebra $F[G]$. See Dummit and Foote [28, Chapter 15]. This requires Wedderburn's theory of the structure of algebras like $F[G]$ as a direct sum of simple algebras over F. I personally prefer to avoid Wedderburn theory and use the direct approach to Fourier analysis on finite groups in [116].

7.3 Determinants

We have already referred to determinants numerous times. You should know the formula for 2×2 determinants:

$$\det \begin{pmatrix} a & b \\ c & d \end{pmatrix} = ad - bc$$

and the analogous formula in three dimensions which has six terms. What happens in n dimensions? One answer is to write a sum of $n!$ terms – one term for every element of the symmetric group. In short, our formula is:

$$\det \begin{pmatrix} a_{11} & \cdots & a_{1n} \\ \vdots & \ddots & \vdots \\ a_{n1} & \cdots & a_{nn} \end{pmatrix} = \sum_{\sigma \in S_n} \operatorname{sgn}(\sigma) a_{\sigma(1)1} a_{\sigma(2)2} \cdots a_{\sigma(n)n}. \tag{7.7}$$

This formula is not so good for evaluating determinants – thanks to the humongous number of terms – even for relatively small n such as 50. It is not so good for proving anything about determinants either. For that we prefer the following definition. We assume that F is any field. It does not have to be the real numbers as it was in calculus. For many things it could just be a commutative ring with identity.

Definition 7.3.1 *The* ***determinant*** *of a matrix is a function $d: F^{n\times n} \to F$ with the following three properties.*

1. *d is a* ***multilinear*** *function: that is, it is a linear function of each column, holding the other columns fixed;*
2. *d is* ***alternating:*** *that is, $d(A) = 0$ if two columns of A are equal.*
3. *$d(I) = 1$, if I is the identity matrix.*

From this definition, it is possible to deduce (7.7) and all the standard facts about determinants. We mostly follow Dornhoff and Hohn [25] for this discussion.

We will write our matrix $A = (A_1 \dots A_n) \in F^{n\times n}$, meaning that A_j denotes the jth column of A. We will write $a_{ij} \in F$ for the ij entry of A.

Exercise 7.3.1

(a) Show that if d is a determinant then $d(A_1, A_2, A_3, \dots, A_n) = -d(A_2, A_1, A_3, \dots, A_n)$: that is, switching two columns changes the sign of the determinant.
(b) Then show that if you permute columns using $\sigma \in S_n$, you get $d\left(A_{\sigma(1)}, \dots, A_{\sigma(n)}\right) = \operatorname{sgn}(\sigma)\, d(A_1, \dots, A_n)$.

Using the preceding exercise, we see that Definition 7.3.1 implies the following computation

$$\begin{aligned} d(AB) &= d\left(\sum_{j_1=1}^{n} A_{j_1} b_{j_1 1}, \dots, \sum_{j_n=1}^{n} A_{j_n} b_{j_n n}\right) \\ &= \sum_{j_1=1}^{n} \cdots \sum_{j_n=1}^{n} d\left(b_{j_1 1} A_{j_1}, \dots, b_{j_n n} A_{j_n}\right) \\ &= \sum_{j_1=1}^{n} \cdots \sum_{j_n=1}^{n} b_{j_1 1} \cdots b_{j_n n}\, d\left(A_{j_1}, \dots, A_{j_n}\right). \end{aligned}$$

Now we can see that, provided that the j_i are pairwise distinct, the n-tuple $(j_1, \dots, j_n) = (\sigma(1), \dots, \sigma(n))$ for some permutation $\sigma \in S_n$. Therefore the multiple sum over the j_i with n^n terms becomes a single sum over S_n with $n!$ terms – and we see that

$$d(AB) = \sum_{\sigma \in S_n} b_{\sigma(1)1} \cdots b_{\sigma(n)n} \operatorname{sgn}(\sigma) d(A). \tag{7.8}$$

Exercise 7.3.2 *How did the multiple sum over n^n terms involving the $(j_1, \dots, j_n)$ become a single sum over $\sigma \in S_n$ with only $n!$ terms? Explain. Then explain why $d\left(A_{\sigma(1)}, \dots, A_{\sigma(n)}\right) = \operatorname{sgn}(\sigma) d(A)$.*

Hint. *What property of determinants causes the terms to vanish in which $j_i = j_k$ for some $i \neq k$?*

Now set $A = I$ in equation (7.8) and we obtain the result we sought (using the fact that $d(I) = 1$)

$$d(B) = \det(B) = \sum_{\sigma \in S_n} \operatorname{sgn}(\sigma) b_{\sigma(1)1} \cdots b_{\sigma(n)n}.$$

The preceding argument shows that there is a unique function $d = \det$ having the properties in Definition 7.3.1. The preceding argument also shows that

$$\det(AB) = \det(A)\det(B).$$

Exercise 7.3.3 *Explain why $\det(AB) = \det(A)\det(B)$ follows from the preceding discussion.*

Exercise 7.3.4 *Show that, in the plane, a point* (x_1, x_2) *is on the line joining the point* (a_1, a_2) *to the point* (b_1, b_2) *iff*

$$\det\begin{pmatrix} x_1 & x_2 & 1 \\ a_1 & a_2 & 1 \\ b_1 & b_2 & 1 \end{pmatrix} = 0.$$

It is possible to use Definition 7.3.1 to prove the rest of the basic results that one knows about determinants from that calculus course: expansion by minors, Cramer's rule, the formula for A^{-1}, and the Laplace expansion. Such results can be found in the references such as Dornhoff and Hohn [25], Dummit and Foote [28], or Schreier and Sperner [101]. We will also give a few exercises with some of these results.

Exercise 7.3.5 *Show that the determinant of an upper triangular matrix is the product of the entries on the diagonal.*

Exercise 7.3.6 *Explain how elementary column operations affect determinants, deriving your results from Definition 7.3.1. Then show, by considering the matrix*

$$\begin{pmatrix} 1 & 2 & 0 \\ 2 & 1 & 4 \\ 3 & 1 & 1 \end{pmatrix} \in \mathbb{Z}_5^{3\times 3},$$

how Gaussian elimination can be used to compute a determinant.

Exercise 7.3.7 *As usual, if* $A = (a_{ij}) \in F^{n\times n}$, *we write the transpose of* A *as* ${}^{\mathrm{T}}A = (a_{ji})$. *Show that* $\det({}^{\mathrm{T}}A) = \det(A)$.

Exercise 7.3.8 *Suppose* $T\colon V \to V$ *is a linear mapping of a finite-dimensional vector space* V *and* B *is a basis of* V. *Using the notation (7.4) from Section 7.2, show that* $\det(\mathrm{Mat}_{B,B}(L))$ *is independent of the basis* B.

Other references for determinants are Birkhoff and Maclane [9], Dummit and Foote [28], Herstein [42, Chapter 6], and any linear algebra book. The modern way of doing these things is called "exterior algebra" or alternating multilinear algebra. We do not have time to cover this subject but if you dislike messy formulas with subscripts like formula (7.7), then exterior algebra is for you. It is exterior algebra or the algebra of differential forms that clarifies Stokes' theorem and the many other formulas of multivariable integral calculus that can be derived from the general version of Stokes' theorem. For example, the view of determinants given in Definition 7.3.1 is important in several-variables integral calculus for helping to understand why the Jacobian determinant appears in the change of variables formula for multiple integrals. References are Schreier and Sperner [101], Lang [63], or Courant and John [19].

We wish to investigate the connection between determinants and volume.

We take our field F to be $\mathbb{R}$, the field of real numbers. Consider a **parallelepiped** (also called a **parallelotope**) $P(A)$ in $\mathbb{R}^n$ spanned by the vectors $A_1, \ldots, A_n$ making up the columns of a square matrix $A \in \mathbb{R}^{n\times n}$

$$P(A) = \left\{ \sum_{i=1}^{n} x_i A_i \,\middle|\, x_i \in \mathbb{R}, 0 \le x_i \le 1 \right\}. \tag{7.9}$$

Check it out for $n=2$.

What are the defining properties of the volume of $P(A)$?

Definition 7.3.2 *A* ***volume function*** $v:\mathbb{R}^{n\times n}\to\mathbb{R}$ *is a function with the following properties:*

1. $v(A)\geq 0$, *for all* $A\in\mathbb{R}^{n\times n}$;
2. v *is an additive function of the ith column, holding the rest of the columns fixed, for all* $i=1,\ldots,n$;
3. *if we multiply the ith column by a scalar c the determinant is multiplied by* $|c|$;
4. $v(I)=1$;
5. $v(A)=0$ *if the vectors* A_j *are not linearly independent.*

Of course, in the case of property 5, one would not really consider the parallelepiped $P(A)$ to be n-dimensional and thus it is not always included in the definition.

Exercise 7.3.9 *Check that the preceding definition is a reasonable definition of the volume of a parallelogram for* $n=2$. *Draw some pictures of parallelograms.*

The determinant of the matrix $A=(A_1,\ldots,A_n)$ may be negative. What does this mean? It means that the vectors $A_1,\ldots,A_n$ are arranged so as to give a coordinate system which is "left handed" – that is, not the usual right-handed $x_1,\ldots,x_n$-coordinates in $\mathbb{R}^n$. The standard right-hand coordinate system is such that when you hold your right hand out with palm up, fingers curled from x-axis to y-axis, then the thumb points up in the direction of the z-axis. The left hand would have a thumb pointing down. This question of **orientation of the coordinates** has only two possible answers. The right-hand rule makes many appearances in physics.

Once you have this definition of the volume of a parallelepiped, you find that the volume of the parallelepiped $P(A)$ in (7.9) is $|\det(A)|$. And this begins to explain the change of variables formula for a multiple integral. See Lang [64] for more information on that.

Exercise 7.3.10 *State whether each of the following statements about two matrices* $A,B\in F^{n\times n}$, *where F is a field, is true or false and give reasons for your answers. Recall that we have defined a square matrix A to be non-singular iff* $\det(A)\neq 0$.

(a) *If the entries of A and B are all the same except for the upper left-hand corner entry, where* $b_{11}=-a_{11}$, *then* $\det B=-\det A$.

(b) *Suppose A is non-singular and B is singular. Then* $A+B$ *is singular.*

(c) *Suppose A is non-singular and B is non-singular. Then* $A+B$ *is non-singular.*

Exercise 7.3.11 *Suppose that A is an* $n\times n$ *matrix over a field F. Define the i,j* ***minor*** M_{ij} *of matrix* $A=(a_{ij})_{1\leq i,j\leq n}$ *to be the determinant of the* $(n-1)\times(n-1)$ *matrix obtained from A by crossing out the ith row and the jth column. Prove the formula for* $\det(A)$ *using* ***expansion by minors*** *of the jth column:*

$$\det A=\sum_{i=1}^{n}(-1)^{i+j}M_{ij}a_{ij}.$$

Use this formula to compute $\frac{\partial}{\partial a_{ij}}\det A$.

Exercise 7.3.12 *Suppose that A is an $n \times n$ matrix over a field F such that $\det A \neq 0$. Prove that the ij entry of A^{-1} can be expressed as:*

$$\frac{1}{\det A}(-1)^{i+j} M_{ji}.$$

Hint. *Use the preceding problem and the fact that the determinant of a matrix is 0 if two columns of the matrix are the same. If δ_{kj} is the **Kronecker delta**, that is, $\delta_{kj}=0$ if $k \neq j$ and $\delta_{kk}=1$, this gives*

$$\delta_{kj} \det A = \sum_{i=1}^{n} (-1)^{i+j} M_{ij} a_{ik}.$$

One can use the preceding exercise to prove **Cramer's rule** which gives a formula for the solution of n linear equations in n unknowns as a quotient of determinants. This formula – like many of the formulas involving determinants – is not so important for solving linear equations as for using the result theoretically to see some property of the solutions of the linear equations. Thus it would be a mistake to leave it out of a course on linear algebra, but this is not such a course so we will leave it out. Nevertheless we encourage the reader to look at the discussion of Cramer's rule in other texts: for example, Birkhoff and Maclane [9] or Dummit and Foote [28].

Determinants are often objects of extreme prejudice – even among mathematicians. Those who want to compute things really hate them. Theorists hate messy formulas with lots of subscripts. But for those who love matrix groups, determinants are our friends. They have appeared all over this text already. I am also reminded that I needed some of the messiest formulas for determinants when updating my book [119]. In particular, a formula derived from the Cauchy–Binet formula below appeared in a study of a generalized central limit theorem for positive matrices. It would be impossible for me to live without determinants and their messiness. Multivariate statisticians would also have a difficult time without knowing these results. See Horn and Johnson [45] or Schreier and Sperner [101, p. 112] or Wikipedia for the formulas stated below.

Let b and c denote ordered sets, each consisting of r numbers between 1 and $\min\{m,n\}$. Define $\Delta^{b_1,\ldots,b_r}_{c_1,\ldots,c_r}(X)$ to be the $r \times r$ subdeterminant of the $m \times n$ matrix X obtained by taking the rows from b and the columns from c. The **Cauchy–Binet formula** says that, for an $m \times k$ matrix L and an $k \times n$ matrix M, if $r \leq \min\{m,n,k\}$ and a is an ordered set of r numbers between 1 and m, while b is an ordered set of r numbers between 1 and n, we have

$$\Delta^{b}_{a}(LM) = \sum_{1 \leq c_1 < \cdots < c_r \leq k} \Delta^{c_1,\ldots,c_r}_{a_1,\ldots,a_r}(L)\, \Delta^{b_1,\ldots,b_r}_{c_1,\ldots,c_r}(M). \tag{7.10}$$

Here the sum is over all ordered sets of r numbers between 1 and k. In the special case that $r=1$, we are looking at the formula for matrix multiplication.

For the **Laplace expansion** of the determinant of an $n \times n$ matrix M, let a denote an ordered set consisting of r numbers between 1 and n. Define a' to be the complementary ordered set of $n-r=s$ numbers between 1 and n that are not included in a. The Laplace expansion of $\det(M)$ with respect to a is:

$$\det(M) = \sum_{1 \leq c_1 < \cdots < c_r \leq n} (-1)^{a_1+\cdots a_r+c_1+\cdots+c_r} \Delta^{c_1,\ldots,c_r}_{a_1,\ldots,a_r}(M)\, \Delta^{c'_1,\ldots,c'_s}_{a'_1,\ldots,a'_s}(M),$$

$$\text{where} \quad r+s=n. \tag{7.11}$$

This formula generalizes the formula from Exercise 7.3.11 for **expansion by minors** – the case $r = 1$.

We have defined a square matrix A to be **non-singular** iff $\det(A) \neq 0$. Applied to a linear transformation T of a vector space V into itself, this definition is equivalent to saying that T is invertible (i.e., 1–1 and onto). This means that T is a vector space automorphism. Otherwise T is called **singular**. Wikipedia gives around 20 equivalent conditions for the non-singularity of a square matrix A over a field.

Exercise 7.3.13 *Give at least 12 equivalent conditions for the non-singularity of a square matrix* $A \in F^{n \times n}$ *over the field* F.

One can define a non-singular matrices $A \in R^{n \times n}$, where R is a commutative ring with identity for multiplication, to be a matrix such that there is a multiplicative inverse $A^{-1} \in R^{n \times n}$ – equivalently $\det(A)$ is a unit in R. This allows the definition of a general linear group over such a ring. For example, the general linear group (also called the modular group) $GL(n, \mathbb{Z})$ consists of all $n \times n$ integer matrices A with $\det(A) = \pm 1$. For non-commutative rings, one does not have the standard concept of determinant.

7.4 Extension Fields: Algebraic versus Transcendental

If field F is a subfield of field E we say E is a extension field of F, as in Definition 5.3.6. This implies that E is a vector space over F which may be infinite dimensional.

Example

1. $\mathbb{C}$ is an extension field of $\mathbb{R}$ having dimension 2 as a vector space. A vector space basis can be taken to be $\{i, 1\}$.
2. $\mathbb{C}$ is an infinite-dimensional extension of $\mathbb{Q}$. This is a bit harder to understand. We will say more shortly.
3. Suppose that $f(x) \in F[x]$ is irreducible, then the quotient ring $E = F[x] / \langle f(x) \rangle$ is a field containing (an isomorphic copy of) F.

To see this, note that coset representatives for $F[x] / \langle f(x) \rangle$ are the polynomials of degree less than $n = \deg f$:

$$\sum_{j=0}^{n-1} a_j x^j, \quad \text{with} \quad a_j \in F,\ j = 0, \ldots, n-1.$$

This follows from the division algorithm in the same way that we got representatives for $\mathbb{Z}/n\mathbb{Z}$ from the remainders of division of an integer by n. In short, the proof is the same as that of Proposition 6.3.1 for which F was a finite field. One sees also that the equivalence classes $[1], [x], \ldots, [x^{n-1}]$ form a basis for the vector space $E = F[x] / \langle f(x) \rangle$ over F. Thus $n = \deg f = \dim_F E$. You can set $\theta = [x]$ and then you see that a vector space basis for E over F is $\{1, \theta, \theta^2, \ldots, \theta^{n-1}\}$. Moreover θ is a root of $f(\theta) = 0$. This generalizes the first example. The complex number i seems concrete to most of us, although it is called imaginary for good historical reasons. However, once we replace $\mathbb{R}$ with any field F and the polynomial $x^2 + 1$ with any irreducible polynomial over F, then this thing we call $\theta = [x]$ does indeed seem rather abstract or imaginary. However, we got accustomed to computation in $\mathbb{Z}/n\mathbb{Z}$

and thus we should be able to get accustomed to computation in $F[x] / \langle f(x) \rangle$. But here we have to think $f(\theta) \equiv 0$ rather than $n \equiv 0$. ▲

Definition 7.4.1 *The* **degree** *d of an extension $F \subset E$ of fields is the dimension of E as a vector space over F. The notation is $d = [E : F]$.*

Proposition 7.4.1 *Suppose we have three fields K, E, F and $F \subset E \subset K$. If E is a finite-degree extension of F and K is a finite-degree extension of E, then K is a finite-degree extension of F and*

$$[K : F] = [K : E] [E : F].$$

Proof. Suppose $\{v_i\}$ is a vector space basis of K over E and $\{w_j\}$ is a vector space basis of E over F. Then we claim that $\{v_i w_j\}$ is a vector space basis of K over F. We leave it as an exercise to prove this. ▲

Exercise 7.4.1 *Prove the claim in the proof of the preceding proposition.*

Suppose that E is a field extension of F. If $a \in E$, let $F(a)$ be the **smallest subfield of** E **containing a and** F. It is also called the field generated by a over F or the field obtained by adjoining a to F. Note the difference between $F(a)$ and $F[a]$. The latter is the smallest **subring** of E containing a and F. The notation is reminiscent of but not the same as that for the ring of polynomials $F[x]$ over F and the larger ring of rational functions $F(x)$ over F since in that situation x is an indeterminate, while a is not. In fact, when a is an element of a large field containing the field F, $F[a]$ and $F(a)$ may be the same entity, as in the following examples $F[i] = F(i)$.

Examples. Suppose that i is a root of $x^2 + 1 = 0$ in some extension field of a field F such that $x^2 + 1$ is irreducible.

1. $F = \mathbb{R}$: $\mathbb{R}(i) = \mathbb{C} = \{a + bi \mid a, b \in \mathbb{R}\} = \mathbb{R}[i]$.
2. $F = \mathbb{Q}$: $\mathbb{Q}(i) = \{a + bi \mid a, b \in \mathbb{Q}\} = \mathbb{Q}[i]$.
3. $F = \mathbb{F}_3$: $\mathbb{F}_3(i) = \{a + bi \mid a, b \in \mathbb{F}_3\} = \mathbb{F}_3[i]$. Here we can replace 3 with any prime p such that -1 is not a square mod p. ▲

Definition 7.4.2 *If K is an extension field of F and $a \in K$, we say that a is* **algebraic** *over F if $f(a) = 0$ for some polynomial $f(x) \in F[x]$. Otherwise we say that a is* **transcendental** *over F.*

Definition 7.4.3 *If K is an extension field of F such that every element of K is algebraic over F, we call K an* **algebraic extension field** *of F. Otherwise K is a* **transcendental extension field** *of F.*

Examples. In all three of the examples above, i is algebraic over $F = \mathbb{R}, \mathbb{Q}$, or $\mathbb{F}_3$. ▲

Showing that something is transcendental is hard. For example, both e and π are real numbers that are transcendental over the rationals. Thus $\mathbb{R}$ is a transcendental extension of $\mathbb{Q}$. We will not prove these things here. Liouville showed (in 1844) that certain real numbers (for example, 0.101 001 000 000 1.... – the decimal expansion with 1s separated by 1!, 2!, 3!, 4!, ... zeros) are transcendental over $\mathbb{Q}$. In 1873 Hermite showed that e is transcendental over $\mathbb{Q}$. Then in 1882 Lindemann showed π to be transcendental over $\mathbb{Q}$. A reference for the transcendence of e is Herstein [42]. References for more results on the subject are A. Baker [5] as well as S. J. Miller and R. Takloo-Bighash [78].

Exercise 7.4.2 *Assuming that e is known to be transcendental over $\mathbb{Q}$, state whether the following numbers are transcendental over $\mathbb{Q}$.*
(a) $e-1$; (b) $e^{2\pi i}$, where $i=\sqrt{-1}$; (c) $\sqrt{2}+\sqrt{3}$; (d) $-e$.

Exercise 7.4.3 *State whether each of the following statements is true or false and explain why.*

(a) The sum of two transcendental numbers over $\mathbb{Q}$ is transcendental over $\mathbb{Q}$.
(b) If a is algebraic over $\mathbb{Q}$, then so is $2a$.

Definition 7.4.4 *Suppose that K is an extension field of F and $a \in K$. If a is algebraic over F, then there is a monic polynomial in $F[x]$ of least degree such that $f(a)=0$. We call f the **minimal polynomial** of a over F.*

It is admissible to say "the" minimal polynomial by Exercise 7.4.4 below.

Example. The minimal polynomial of i over $\mathbb{Q}$ or $\mathbb{R}$ or $\mathbb{F}_3$ is x^2+1. ▲

Exercise 7.4.4 *Show that the minimal polynomial of a over F is unique.*

Hint. *Suppose f and g are both minimal polynomials for a over F. Use the division algorithm.*

It is tempting to think about finding the minimal polynomial of more examples over $\mathbb{Q}$ such as $\sqrt{2}+\sqrt{3}$ or $e^{2\pi i/n}$, where $n=3,4,5,\ldots$. However, that seems like a subject for another sort of text. One would need to discuss irreducibility tests. So we will avoid the subject and stick to finite fields for the most part. Testing for irreducibility of polynomials of low degree over finite fields can be done by the same sort of method that works to test whether an integer is prime – assuming that integer is not too big. Just divide by the irreducible polynomials of degree $\leq \lfloor n/2 \rfloor$. There are better methods for factoring polynomials over finite fields etc. See Lidl and Niederreiter [69]. Another reference is the handbook edited by Mullin and Panario [79] which includes an article on the construction of irreducibles among other fascinating topics. See other texts such as Birkhoff and MacLane [9], Fraleigh [32], Gallian [33], or Herstein [42] for more information on irreducibility tests for polynomials over the field of rational numbers.

Exercise 7.4.5 *Is x^4+1 irreducible over $\mathbb{F}_3$?*

Exercise 7.4.6 *Is x^4+1 irreducible over $\mathbb{F}_5$?*

Proposition 7.4.2 *Suppose that K is an extension field of F, $a \in K$, and a is algebraic over F. Let $f(x)$ be the minimal polynomial of a over F. Then f is irreducible and $F[x] / \langle f(x) \rangle$ is isomorphic to the field $F(a)$. Moreover the cosets of the quotient ring $F[x] / \langle f(x) \rangle$ are represented by the remainders of polynomials upon division by f. Then $F(a) = F[a]$.*

Proof. Everything has been proved in the preceding paragraphs (including exercises) except the irreducibility of f, the minimal polynomial of a over F. Otherwise we have $f = gh$, where $g, h \in F[x]$ and $0 < \deg g, \deg h < \deg f$. But then $0 = f(a) = g(a)h(a)$ implies either $g(a) = 0$ or $f(a) = 0$. But this contradicts the minimality of the degree of f with $f(a) = 0$. ▲

Exercise 7.4.7 *Show that if K is an extension field of F and there is a transcendental element $a \in K$ over F, then K is an infinite-dimensional vector space over F. In fact, show that then $F(a)$ is isomorphic to the field of fractions of the polynomial ring $F[x]$. This is a case in which $F(a) \neq F[a]$.*

Exercise 7.4.8 *Suppose that F is a finite field of characteristic p. Show that every element of F is algebraic over $\mathbb{F}_p$.*

Hint. *Look at the group F^* of nonzero elements of F and recall Lagrange's theorem.*

Algebraically Closed Fields

Some things in the later chapters of linear algebra texts (such as eigenvalues) do not work well for fields like $\mathbb{R}$ but instead require the field to be a larger field like $\mathbb{C}$ where all polynomials factor completely into a product of degree 1 polynomials – that is, fields containing all roots of $\det(M - xI) = 0$ for any square matrix M. For $\mathbb{C}$, we are referring to the **fundamental theorem of algebra** which says that $\mathbb{C}$ is an algebraically closed field. A field F is **"algebraically closed"** if all polynomials $f(x)$ in $F[x]$ factor completely into a product of degree 1 polynomials from $F[x]$. Equivalently, to say F is algebraically closed is to say that all roots of $f(x)$ in $F[x]$ lie in F. Our favorite fields $\mathbb{F}_p, \mathbb{Q}$, and $\mathbb{R}$ are not algebraically closed, but $\mathbb{C}$ is.

The history of proofs of the fundamental theorem of algebra is very interesting. The first proofs (given by d'Alembert in 1746 and C. F. Gauss in 1799) had flaws. Gauss later published three correct proofs. Some analysis is usually required and thus we will not prove the theorem here. My favorite proof uses Liouville's theorem from complex analysis. See Birkhoff and MacLane [9] for a topological proof. Another reference is Courant and Robbins [20].

Sadly finite fields are **never** algebraically closed. See the exercise below. However, there is an extension field of a finite field that is algebraically closed. Such a field E is called an **algebraic closure** of $\mathbb{F}_q$. We can say "the algebraic closure" because if we are given two algebraic closures of $\mathbb{F}_p$, there is a field isomorphism from one algebraic closure to the other fixing every element of $\mathbb{F}_p$. To show that an arbitrary field has an algebraic closure involves use of **Zorn's lemma**, which is equivalent to the axiom of choice, as well as something called transfinite induction. We have tried to avoid these axioms here and so will avoid thinking about general algebraic closures. The axiom of choice may sound completely reasonable. It can be phrased as a statement about the **Cartesian product** of an arbitrary number of sets. Given a family of non-empty sets S_i, indexed by a non-empty set I, the Cartesian product

$\prod_{i \in I} S_i$ consists of functions $f\colon I \to \bigcup_{i \in I} S_i$ such that $f(i) \in S_i$. The **axiom of choice** says that $\prod_{i \in I} S_i$ is non-empty. See Dummit and Foote [28], Fraleigh [32], or Hungerford [47] for more information. We should also note that the axiom of choice allows one to prove the famous Banach–Tarski paradox. This says that a solid sphere in $\mathbb{R}^3$ can be decomposed into five disjoint subsets, which can then be composed (by translation and rotation) into two identical copies of the original sphere. The catch is that the five sets are scattered all over the sphere and their volumes are undefined.

Exercise 7.5.17 in the next section says that no finite field is algebraically closed. In fact, there are irreducible polynomials of every degree over any finite field and there is a formula for the number of irreducible polynomials of given degree over $\mathbb{F}_q$. See Dornhoff and Hohn [25] or Lidl and Niederreiter [69].

When I am feeling finite, this make me very sad. Of course, you can keep adding roots of polynomials to $\mathbb{F}_p$. This may be done in various ways. For example, PlanetMath.org envisions taking

$$\mathbb{F}_{p^\infty} - \bigcup_{n=1}^{\infty} \mathbb{F}_{p^{n!}} \tag{7.12}$$

to get an algebraically closed field containing $\mathbb{F}_p$. The finite field $\mathbb{F}_{p^m}$ can be constructed as the splitting field of the polynomial $x^{p^m} - x$, as we shall see in Section 7.3.

Exercise 7.4.9

(a) Is a union of fields necessarily a field? Explain your answer.

(b) Show that $\mathbb{F}_{p^\infty} = \bigcup_{n=1}^{\infty} \mathbb{F}_{p^{n!}}$ is a field, using a fact from the next section that $\mathbb{F}_{p^{m!}} \subset \mathbb{F}_{p^{n!}}$ if $m \leq n$.

Exercise 7.4.10 *Consider the eigenvalues of the adjacency matrix of a finite undirected graph. Show that such eigenvalues must be algebraic numbers over $\mathbb{Q}$.*

Exercise 7.4.11 *Represent the field $\mathbb{Q}\left(e^{2\pi i/3}\right)$ as a quotient $\mathbb{Q}[x] / \langle f(x) \rangle$. Note that $\omega = e^{2\pi i/3}$ satisfies $\omega^3 = 1$ but $\omega^n \neq 1$, for $n = 1$ or 2. Thus ω is what is called a* ***primitive third root of unity****.*

Exercise 7.4.12 *Do the analog of the preceding exercise with $\mathbb{Q}$ replaced by $\mathbb{F}_2$. That is, find a quotient $\mathbb{F}_2[x] / \langle f(x) \rangle$ containing a primitive third root of unity.*

7.5 Subfields and Field Extensions of Finite Fields

Next suppose that E is a finite field of characteristic p (for prime p) with subfield F. As in the last section, we say that E is a field extension of F. Both E and F are extensions of $\mathbb{Z}_p$ by Proposition 7.2.1. We can view E as a vector space over F and then $E \cong F^r$, where r is the vector space dimension of E over F. If the dimension of F over $\mathbb{Z}_p$ is n, then $|F| = p^n$ and $|E| = p^{nr}$. All elements of E have to be algebraic over F. So we are in the situation of the theorems of the preceding section.

Notation. We write $\mathbb{F}_{p^k}$ for the finite field with p^k elements. We can use the word "the" (implying uniqueness up to isomorphism) by the results of this section. So we have two notations if $k = 1$, namely, $\mathbb{F}_p = \mathbb{Z}_p$. We can identify $\mathbb{F}_{p^k}$ with the quotient ring $\mathbb{F}_p[x] / \langle f(x) \rangle$

for some irreducible polynomial of degree k over $\mathbb{F}_p$ – assuming we know that such a polynomial exists. And we can think of the elements of $\mathbb{F}_{p^k}$ as polynomials of degree less than k with coefficients in $\mathbb{F}_p$. Usually we write $\theta = [x]$ in $\mathbb{F}_p[x] / \langle f(x) \rangle$ and then elements of $\mathbb{F}_{p^k}$ have the form

$$\sum_{i=0}^{k-1} a_i\theta^i, \quad a_i \in \mathbb{F}_p.$$

Of course there will usually be many irreducible polynomial of degree k over $\mathbb{F}_p$. But the resulting fields will be isomorphic.

Proposition 7.5.1 *We have* $\mathbb{F}_{p^k} \subset \mathbb{F}_{p^n} \iff k$ *divides* n.

Proof. $\implies \mathbb{F}_{p^k} \subset \mathbb{F}_{p^n}$ means $\left(p^k\right)^r = p^n$ where r is the dimension of $\mathbb{F}_{p^n}$ as a vector space over $\mathbb{F}_{p^k}$. It follows that $n = kr$ and thus k divides n.
$\Longleftarrow$ We postpone this proof until we have proved that $\mathbb{F}_{p^n}$ is the splitting field of $x^{p^n} - x$, meaning the field where this polynomial factors completely into degree 1 factors. ▲

The preceding proposition implies that any subfield F of $K = \mathbb{F}_{p^n}$ must have the form $F = \mathbb{F}_{p^k}$ such that k divides n. Moreover, the degree of K over F is $[K : F] = \frac{n}{k}$.

Example. We compute $[\mathbb{F}_{5^{21}} : \mathbb{F}_{5^3}] = 7$ since $21 = 3 \cdot 7$. ▲

In Figure 7.1 we draw the poset diagram for the subfields of $\mathbb{F}_{2^{24}}$. It is the same as the poset diagram for the divisors of 24 (see Figure 1.13).

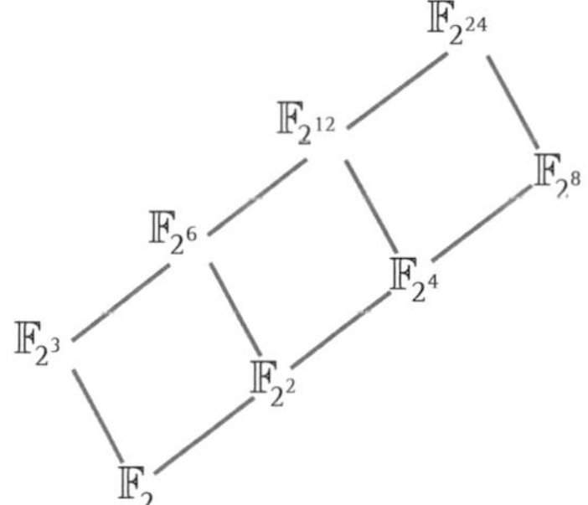

Figure 7.1 The poset of subfields of $\mathbb{F}_{2^{24}}$

Definition 7.5.1 *The* ***splitting field of a polynomial*** $f(x) \in F[x]$ *over the field* F *is the smallest extension field* E *of* F *such that* f *factors completely into linear factors from* $E[x]$, *that is,*

$$f(x) = c\prod_{i=1}^{n}(x - a_i), \quad \text{for } a_i, c \in E.$$

We say "the" splitting field since, as we will prove soon, it is unique up to field isomorphism fixing elements of F. Birkhoff and Maclane [9] say "root field" instead of "splitting field." Presumably that is the older terminology and it stands for the field where all the roots of the polynomial lie, whereas splitting field stands for the field over which the polynomial splits completely into linear factors.

Example 1. The splitting field of x^2+1 over $\mathbb{R}$ is $\mathbb{C}$, the complex numbers. ▲

Example 2. The splitting field of x^2-2 over $\mathbb{F}_5$ is

$$\mathbb{F}_{25}=\mathbb{F}_5\left[\sqrt{2}\right]=\left\{a+b\sqrt{2}\;\middle|\;a,b\in\mathbb{F}_5\right\}\cong\mathbb{F}_5[x]/\langle x^2-2\rangle.$$ ▲

Example 3. The splitting field of x^2+x+2 over $\mathbb{F}_3$. Since $f(x)=x^2+x+2$ has no roots in $\mathbb{F}_3$, it is irreducible by Proposition 5.5.1. It is not hard to see that it is also primitive. Let θ denote a root of $f(x)$. One computes the powers θ^j as we did in earlier examples using the feedback shift register idea and sees that the multiplicative group $\mathbb{F}_3(\theta)^*=\langle\theta\rangle$. It has order 8. Then $x^2+x+2=(x-\theta)(x-b)$, where $b=\theta^j$, for some j. It is not hard to see that $j=3$. The fact that $j=3$ will be no shock once we have understood the section on Galois theory. In any case we have thus shown that $\mathbb{F}_3(\theta)$ is the splitting field of x^2+x+2 over $\mathbb{F}_3$. ▲

Exercise 7.5.1 *Fill in the details in the last example. It helps to do the table of powers of θ. Since $2=b\theta$, it follows that $b=-1/\theta$.*

Exercise 7.5.2 *Find the splitting field E of the polynomial x^3+x+1 over $\mathbb{F}_2$. What is the degree $[E:\mathbb{F}_2]$?*

Theorem 7.5.1 *Any polynomial $f(x)\in F[x]$, where F is a field, has a splitting field E and this splitting field is unique up to field isomorphism fixing elements of F.*

Proof. Existence of E (induction on $\deg f$). If $\deg f=1$, f is already in the desired form $c(x-a)$. Now for the induction step, we know that we can construct a field E_1 containing F and a root θ of an irreducible factor $g(x)$ of f, namely, $E_1=F[x]/\langle g(x)\rangle$, with $\theta=[x]$. So we can factor $f(x)=(x-\theta)h(x)$ with $h\in E_1[x]$. By induction on $\deg f$ we may assume that $h(x)$ is completely factored into linear factors in an extension field E of E_1.

Uniqueness of E up to isomorphism. This follows from the next theorem in the special case that φ is the identity function. ▲

The proof of the following theorem is another proof by induction.

Theorem 7.5.2 *Suppose that E and E^φ are fields and $\varphi:E\to E^\varphi$ is a field isomorphism mapping E 1–1, onto E^φ. If $f(x)\in E[x]$ and $f^\varphi\in E^\varphi[x]$ is the polynomial obtained by applying φ to all the coefficients of f, let M be the splitting field of f over E, and M^φ be the splitting field of f^φ over E^φ. Then φ extends to an isomorphism between M and M^φ.*

Proof. (Induction on $\deg f$)
The result is clear if $\deg f=1$ or if $M=E$ which implies $M^\varphi=E^\varphi$.

Assume the result for f such that $\deg f\le n$ and prove it when $\deg f=n+1$.

We may assume f has an irreducible factor $g(x)\in E[x]$ such that $2\le\deg g\le n+1$. Define g^φ to be the polynomial obtained by applying φ to all the coefficients of g. Then $g^\varphi\in E^\varphi[x]$. Suppose that a is a root of $g(x)$ in M and b is a root of $g^\varphi(x)$ in M^φ. We have the ring isomorphism $\tau:E[x]\to E^\varphi[x]$ obtained by applying φ to the coefficients of the polynomials. Then $\tau(g(x))=g^\varphi(x)$ and τ induces an isomorphism $\tilde{\tau}:E[x]/\langle g(x)\rangle\to E^\varphi[x]/\langle g^\varphi(x)\rangle$.

Recall that $E[x]/\langle g(x)\rangle \cong E(a)$ and $E^{\varphi}[x]/\langle g^{\varphi}(x)\rangle \cong E^{\varphi}(b)$. So we compose all the maps and obtain a field isomorphism extending φ from $E(a)$ onto $E^{\varphi}(b)$. Now $f(x)=(x-a)h(x)$ and, by induction, φ extends to a field isomorphism between the splitting field F of h and the splitting field F^{φ} of h^{φ}. Since $M=F(a)$ and $M^{\varphi}=F^{\varphi}(b)$, we are done. ▲

Even though the splitting field of a polynomial $f(x)$ over F is only unique up to isomorphism, we will still say "the" splitting field, as we noted earlier. For $f(x)\in\mathbb{Q}[x]$, we can view the splitting field of f over $\mathbb{Q}$ as a subfield of the field $\mathbb{C}$ of complex numbers, by the fundamental theorem of algebra stated in Section 7.1.

For finite fields finding an analog of $\mathbb{C}$ in which one can do calculus – take limits, etc. – is not easy. It is not sufficient just to look at $\mathbb{F}_{p^\infty}$ from equation (7.12). One would also want to insure that Cauchy sequences of elements from $\mathbb{F}_{p^\infty}$ converge in the analog of $\mathbb{C}$. Such fields have been studied but we will not do that here. See the handbook [79].

Proposition 7.5.2 *If F is a field with p^n elements, then F must be the splitting field of $x^{p^n}-x$ over $\mathbb{F}_p=\mathbb{Z}_p$.*

Proof. By Lagrange's theorem from Section 3.3, any nonzero element of a field F with p^n elements is a root of the polynomial $x^{p^n-1}-1$, since the order of the multiplicative group F^* is p^n-1. So the elements of F are roots of $x\left(x^{p^n-1}-1\right)=x^{p^n}-x$. Moreover the polynomial $x^{p^n}-x$ has at most p^n distinct roots in F, since, by a corollary of the division algorithm, it has p^n roots counting multiplicity. Therefore this polynomial has exactly p^n roots in F, and F is the splitting field of $x^{p^n}-x$. ▲

Exercise 7.5.3 *In the proof of the preceding proposition, why is it that $x^{p^n}-x$ does not have roots of multiplicity larger than 1?*

Next we want to know whether a general polynomial in $F[x]$ has multiple roots. For this, one needs to take derivatives. We do not want to talk about limits since we usually think of F as a finite field, not the real numbers. So we define the formal derivative of a polynomial by the formula that was proved from the limit definition of the derivative in calculus.

Definition 7.5.2 *Suppose that F is any field. Then the **formal derivative** of*

$$f(x)=a_rx^r+a_{r-1}x^{r-1}+\cdots+a_1x+a_0,$$

with $a_j\in F$ is defined by

$$f'(x)=ra_rx^{r-1}+(r-1)a_{r-1}x^{r-2}+\cdots+a_1.$$

Exercise 7.5.4 *Show that the formal derivative has the following familiar properties of derivatives, for any $f,g\in F[x]$.*

(a) $(f+g)'=f'+g'$;
(b) $(fg)'=f'g+fg'$;
(c) $(f(x)^n)'=n\left(f(x)^{n-1}\right)f'(x)$.

Lemma 7.5.1 *A polynomial $f \in F[x]$ has a multiple root in an extension field E of F iff* $\deg\gcd(f,f') \geq 1$. *Here f' is the formal derivative of f.*

Proof. $\Longrightarrow$ Suppose that $f(x) = (x-a)^2 g(x)$, with $g(x) \in E[x]$ and $a \in E$. Then by the usual properties of derivatives from Exercise 7.5.4, we have $f'(x) = 2(x-a)g(x) + (x-a)^2 g'(x)$. It follows that $(x-a)$ divides $\gcd(f,f')$.
$\Longleftarrow$ Suppose $\deg\gcd(f,f') \geq 1$. Then $\gcd(f,f')$ is divisible by $x-a$ for some $a \in E$, where E is an extension field of F. Why does E exist? If $x-a$ divides $\gcd(f,f')$, then a is a root of both f and f'. So $f(x) = (x-a)h(x)$ for some $h \in E[x]$. But then $f'(x) = h(x) + (x-a)h'(x)$ and $0 = f'(a) = h(a)$. This means $h(x) = (x-a)k(x)$ for some $k \in E[x]$ and thus $(x-a)^2$ divides $f(x)$ and a is a multiple root of f. ▲

Exercise 7.5.5 *Answer the question in the proof of the preceding lemma.*

Theorem 7.5.3 *For every prime p and every $n = 1, 2, 3, \ldots$, there is a finite field with p^n elements which is isomorphic to the splitting field of $x^{p^n} - x$ over $\mathbb{F}_p$.*

Proof. The splitting field F of $x^{p^n} - x$ over $\mathbb{F}_p$ has p^n distinct roots of $f(x) = x^{p^n} - x$, since $x^{p^n} - x$ cannot have multiple roots using the preceding lemma and the fact that $\gcd(f,f') = \gcd(x^{p^n} - x, -1) = 1$. If we set $K = \{a \in F \mid a^{p^n} = a\}$, we can show that K is a subfield of F. We leave this as an exercise. We know that $x^{p^n} - x$ splits in K and thus $F = K$. Moreover, then F is a finite field that has p^n elements.

By Proposition 7.5.2, any other field E with p^n elements must be a splitting field of $x^{p^n} - x$ over $\mathbb{F}_p$. This makes E isomorphic to F by an isomorphism which fixes $\mathbb{F}_p$. ▲

Exercise 7.5.6 *Show that if F is the splitting field of $x^{p^n} - x$ over $\mathbb{F}_p$, then $K = \{a \in F \mid a^{p^n} = a\}$ is closed under addition and multiplication. Thus K is indeed a subfield of F.*

Hint. *Note that $(x+y)^p = x^p + y^p$, for all x, y in a field of characteristic p.*

Example. We have looked at $\mathbb{F}_8$, $\mathbb{F}_9$, and $\mathbb{F}_{25}$ in Sections 5.3, 5.6, and in this section. Next we want to consider

$$\mathbb{F}_{16} \cong \mathbb{F}_2[x]/\langle x^4 + x + 1\rangle \cong \{a\theta^3 + b\theta^2 + c\theta + d \mid a, b, c, d \in \mathbb{F}_2\},$$

where $\theta^4 + \theta + 1 = 0$. Here the degree of $\mathbb{F}_{16}$ over $\mathbb{F}_2$ is 4 and $\{\theta^3, \theta^2, \theta, 1\}$ is vector space basis of $\mathbb{F}_{16}$ as a vector space over $\mathbb{F}_2$. ▲

Exercise 7.5.7 *With the notation of the preceding example, show that:*

(a) the polynomial $x^4 + x + 1$ is irreducible in $\mathbb{F}_2[x]$;
(b) the polynomial $x^4 + x + 1$ in $\mathbb{F}_2[x]$ is primitive, that is, θ generates the multiplicative group $\mathbb{F}_{16}^*$.

Hint. *For part (b), you need to make a table of powers*

$$\theta^j = a_3\theta^3 + a_2\theta^2 + a_1\theta + a_0, \quad a_i \in \mathbb{F}_2,$$

using the feedback shift register idea from Section 5.6.

Now we can finally finish a proof.

Proof. **Completion of proof of Proposition 7.5.1** (of the fact that m divides n implies $\mathbb{F}_{p^m}$ is a subfield of $\mathbb{F}_{p^n}$). If $n = m \cdot r$, then $p^{m \cdot r} - 1 = (p^m - 1)(p^{m(r-1)} + p^{m(r-2)} + \cdots + p^m + 1)$. This says that $(p^m - 1)$ divides $(p^{m \cdot r} - 1) = (p^n - 1)$. It follows that $(x^{p^m - 1} - 1)$ divides $(x^{p^n - 1} - 1)$ in $\mathbb{F}_p[x]$. We leave the proof of this as an exercise. Since $\mathbb{F}_{p^r}$ is the splitting field of $x^{p^r} - x$ over $\mathbb{F}_p$, the proof is over. ▲

Exercise 7.5.8 *Prove that if m divides n, then the polynomial $\left(x^{p^m - 1} - 1\right)$ divides $\left(x^{p^n - 1} - 1\right)$ in $\mathbb{F}_p[x]$.*

Hint. *Use the formula for the sum of a geometric progression in the form:*

$$\frac{x^{s \cdot k} - 1}{x^s - 1} = (x^s)^{k-1} + (x^s)^{k-2} + \cdots + x^s + 1.$$

There are still several topics needed to complete our theory of finite fields. The first is to state a generalization of the result we gave as Exercise 6.3.10. The proof is essentially the same as that of the exercise.

Theorem 7.5.4 *The multiplicative group of a finite field is cyclic.*

Proof. We leave this as an exercise. ▲

Exercise 7.5.9 *Prove Theorem 7.5.4.*

Hint. *See Exercise 6.3.10.*

Exercise 7.5.10 *Find all the generators of the multiplicative group of $\mathbb{F}_9 \simeq \mathbb{F}_3[i]$, where $i^2 + 1 = 0$.*

There are lists of primitive polynomials in the books on finite fields such as Lidl and Niederreiter [69]. Here we give a short list containing one primitive polynomial over $\mathbb{F}_2$ for each small degree.

Some Primitive Polynomials over $\mathbb{F}_2$

$$\begin{aligned}
&x + 1, \quad x^2 + x + 1, \quad x^3 + x + 1, \quad x^4 + x + 1, \quad x^5 + x^2 + 1, \\
&x^6 + x + 1, \quad x^7 + x^3 + 1, \quad x^8 + x^4 + x^3 + x^2 + 1, \quad x^9 + x^4 + 1, \\
&x^{10} + x^4 + 1, \quad x^{11} + x^2 + 1
\end{aligned}$$

Exercise 7.5.11 *Show that the fields $\mathbb{F}_2[x]/\langle x^3 + x + 1\rangle$ and $\mathbb{F}_2[x]/\langle x^3 + x^2 + 1\rangle$ are isomorphic. Define the isomorphism.*

The following exercise is important for the next section.

Exercise 7.5.12 *Show that the mapping of $\mathbb{F}_{p^n}$ onto itself defined by $\sigma_p(x) = x^p$ is a field automorphism (called the **Frobenius automorphism**) fixing elements of $\mathbb{F}_p$ (viewing $\mathbb{F}_p$ as a subfield of $\mathbb{F}_{p^n}$). Such field automorphisms are also called **field conjugations**.*

Hint. *Use Lemma 5.3.4.*

It is a corollary of results of this section that the polynomial $x^{p^n} - x$ factors over $\mathbb{F}_p$ as a product of all the distinct irreducible polynomials having degrees that divide n.

Example. We seek to factor the polynomial $x^8 - x$ completely over $\mathbb{F}_2$. Since $8 = 2^3$, and there are only two positive divisors of 3, we expect to get irreducible polynomials of degrees 1 and 3. We already know two irreducible polynomials of degree 3 over $\mathbb{F}_2$. They are $x^3 + x + 1$ and $x^3 + x^2 + 1$. Both of them have as roots generators of $\mathbb{F}_8$. Thus they must divide $x^8 - x$. We know that 0 and 1 are also elements of $\mathbb{F}_8$. That gives the other two divisors of $x^8 - x$. So we find that

$$x^8 - x = x(x-1)\left(x^3 + x + 1\right)\left(x^3 + x^2 + 1\right).$$

If θ is a root of $x^3 + x + 1$, one can show that $x^3 + x + 1 = (x - \theta)(x - \theta^2)(x - \theta^4)$. The polynomial $x^3 + x^2 + 1$ has roots θ^3, θ^5, and θ^6. ▲

Exercise 7.5.13 *Check that $x^8 - x = x(x-1)(x^3 + x + 1)(x^3 + x^2 + 1)$ over $\mathbb{F}_2$ by multiplying out the polynomial on the right.*

Exercise 7.5.14 *Show that $\mathbb{F}_{p^n}$ is the splitting field of some irreducible polynomial of degree n over $\mathbb{F}_p$.*

Exercise 7.5.15 *Factor the polynomial $x^9 - x$ completely into irreducible factors over $\mathbb{F}_3$. Which factors are primitive?*

Exercise 7.5.16 *Show that for any finite extension E of a finite field there is an element $\theta \in E$ such that $E = F(\theta)$. We call such an extension* ***simple.***

Exercise 7.5.17 *Show that no finite field is algebraically closed. In fact, show that for every finite field F and every positive integer n, there is an irreducible polynomial over F of degree n.*

Exercise 7.5.18 *State whether each of the following statements is true or false and give a reason for your answer.*

(a) $\mathbb{F}_{p^n} \cong \mathbb{Z}_{p^n}$;
(b) $m|n \Longrightarrow \mathbb{F}_m \subset \mathbb{F}_n$;
(c) $\mathbb{Q}\left[\sqrt{5}\right] \cong \mathbb{Q}\left[\sqrt{7}\right]$.

7.6 Galois Theory for Finite Fields

Definition 7.6.1 *Suppose that $m|n$. The* ***Galois group*** *$G(\mathbb{F}_{p^n}/\mathbb{F}_{p^m})$ (which is read as the Galois group of $\mathbb{F}_{p^n}$ over $\mathbb{F}_{p^m}$) is defined to be the set of field automorphisms $\tau : \mathbb{F}_{p^n} \to \mathbb{F}_{p^n}$ such that $\tau(x) = x$, $\forall x \in \mathbb{F}_{p^m}$ (where $\mathbb{F}_{p^m}$ is viewed as a subfield of $\mathbb{F}_{p^n}$). Here by field automorphism we just mean that the map is a ring automorphism.*

It turns out that $G(\mathbb{F}_{p^n}/\mathbb{F}_{p^m})$ is a cyclic group, as we shall see. When $m = 1$, $G(\mathbb{F}_{p^n}/\mathbb{F}_p)$ is generated by the Frobenius automorphism σ_p of Exercise 7.5.12. Such field automorphisms are also called conjugations because they generalize complex conjugation, which is the generator of the Galois group $G(\mathbb{C}/\mathbb{R})$. Some references for Galois theory are Birkhoff and Maclane [9], Dornhoff and Hohn [25], Dummit and Foote [28], and Herstein [42]. The fundamental theorem of Galois theory basically says that there is a 1–1 correspondence between intermediate fields E such that $\mathbb{F}_{p^m} \subset E \subset \mathbb{F}_{p^n}$ and subgroups H of $G = G(\mathbb{F}_{p^n}/\mathbb{F}_{p^m})$. The subgroup H of G corresponding to E is $H = H(E) = \{\tau \in G | \tau x = x, \forall x \in E\}$. The intermediate field E corresponding to a subgroup H of G is $E = E(H) = \{x \in \mathbb{F}_{p^n} | \tau x = x, \forall \tau \in H\}$. The fundamental theorem of Galois theory says that the correspondence between intermediate fields and subgroups is 1–1, onto, and inclusion reversing. Moreover the degree of the extension $n = [\mathbb{F}_{p^n} : \mathbb{F}_{p^m}]$ is equal to the order of the Galois group G. So, for example, the poset diagram for fields E such that $\mathbb{F}_{p^m} \subset E \subset \mathbb{F}_{p^n}$ is the same as that for subgroups of $G = G(\mathbb{F}_{p^n}/\mathbb{F}_{p^m})$, except that all inclusion lines are reversed.

Example. What does Galois theory say about the extension $\mathbb{F}_{2^8}/\mathbb{F}_2$? What are the intermediate fields between $\mathbb{F}_2$ and $\mathbb{F}_{2^8}$? They are $\mathbb{F}_{2^1}, \mathbb{F}_{2^2}, \mathbb{F}_{2^4}, \mathbb{F}_{2^8}$. So the only non-trivial intermediate fields are $\mathbb{F}_4$ and $\mathbb{F}_{16}$. The Galois group $G = G(\mathbb{F}_{2^8}/\mathbb{F}_2) = \langle \sigma_2 \rangle$ is a cyclic group of order 8 generated by the Frobenius automorphism σ_2. The group G has exactly two non-trivial proper subgroups $\langle \sigma_2^2 \rangle$ and $\langle \sigma_2^4 \rangle$ which correspond to the two non-trivial intermediate fields: $G(\mathbb{F}_{2^8}/\mathbb{F}_{2^2}) = \langle \sigma_2^2 \rangle$ and $G(\mathbb{F}_{2^8}/\mathbb{F}_{2^4}) = \langle \sigma_2^4 \rangle$. ▲

Theorem 7.6.1 *Suppose that p is a prime and $n, m \in \mathbb{Z}^+$ such that $m | n$. The Galois group $G(\mathbb{F}_{p^n}/\mathbb{F}_p)$ is a cyclic group of order n generated by σ_p, where σ_p is the Frobenius automorphism defined by $\sigma_p(x) = x^p$, for all $x \in \mathbb{F}_{p^n}$.*

Proof. We know that σ_p is an automorphism of $\mathbb{F}_{p^n}$ by Exercise 7.5.12. Now suppose $\tau \in G(\mathbb{F}_{p^n}/\mathbb{F}_p)$ and suppose θ generates the multiplicative group $\mathbb{F}_{p^n}^*$. Then $\tau(\theta) = \theta^k$, for some integer k such that $0 < k < p^n - 1$. It follows that $\tau(x) = x^k$, for all $x \in \mathbb{F}_{p^n}$.

We need to show that $k = p^i$, for some i, for then $\tau = \sigma_p^i$ and we are done. We will prove $k = p^i$ by contradiction.

Assume, contradicting $k = p^i$, that $\tau(x) = x^k$ with $k = p^s k_0$, where p does not divide k_0 and $k_0 > 1$. Set $f(x) = \tau(1+x) - 1 - \tau(x)$. Then we can consider $f(x)$ as a polynomial in $\mathbb{F}_p[x]$ and we have the following equalities

$$\begin{aligned} f(x) &= (1+x)^{p^s k_0} - 1 - x^k \\ &= \left(1 + x^{p^s}\right)^{k_0} - 1 - x^{p^s k_0} \\ &= k_0 x^{p^s} + \binom{k_0}{2} \left(x^{p^s k_0}\right)^2 + \cdots + k_0 \left(x^{p^s}\right)^{k_0 - 1}. \end{aligned}$$

This polynomial cannot vanish identically on $\mathbb{F}_{p^n}$. Why? The problem is that $\mathbb{F}_{p^n}$ has p^n elements but the degree of the polynomial is smaller than p^n. But this gives a contradiction to the fact that – as a function on $\mathbb{F}_{p^n}$ – we have $f(x) = \tau(1+x) - 1 - \tau(x)$, which must be 0 for all $x \in \mathbb{F}_{p^n}$, since τ is a field automorphism. This contradiction completes our proof of the theorem. ▲

Exercise 7.6.1 *Suppose $\sigma \in G(\mathbb{F}_{p^n}/\mathbb{F}_p)$, $\alpha \in \mathbb{F}_{p^n}$, $f(x) \in \mathbb{F}_p[x]$ with $f(\alpha) = 0$. Show that $f(\sigma(\alpha)) = 0$. Thus elements of $G(\mathbb{F}_{p^n}/\mathbb{F}_p)$ permute the roots of polynomials $f(x) \in \mathbb{F}_p[x]$.*

Exercise 7.6.2 *Consider the smallest field containing* $\mathbb{F}_5$ *and roots of* $x^2-2=0$ *and* $x^2-3=0$. *What is the degree of E over* $\mathbb{F}_5$? *A primitive polynomial of degree 2 over* $\mathbb{F}_5$ *is* $f(x)=x^2+x+2$. *Let* θ *be a root of* $f(x)$. *What powers of* θ *represent* $\sqrt{2}$ *and* $\sqrt{3}$, *if any?*

Exercise 7.6.3 *Show that* x^4+x^2+2x+2 *is a primitive irreducible polynomial over* $\mathbb{F}_5$. *What is the degree of the extension of* $\mathbb{F}_5$ *generated by any root of this polynomial?*

Exercise 7.6.4 *Use the preceding exercise to find the intermediate fields between* $\mathbb{F}_{5^4}$ *and* $\mathbb{F}_5$. *Draw the poset diagram and then do the same for the corresponding Galois groups.*

Galois theory for extensions of characteristic 0 fields like the rationals is not so simple as that for finite fields. Non-commutative groups can be Galois groups. For example, one finds that the Galois group of $\mathbb{Q}\left[\sqrt[3]{2}, e^{2\pi i/3}\right]$ over $\mathbb{Q}$ is S_3. See Dummit and Foote [28] or Gallian [33]. As we said earlier, it is also possible to write down equations of degree 5 whose roots over $\mathbb{Q}$ have Galois group S_5. This is important because S_5 is not a solvable group (as defined below). Thus one has an example of a quintic (i.e., 5th degree) equation whose roots cannot be found by repeated radicals. We gave an example in Section 3.4. Of course, it is possible that there are other methods that lead to solutions of quintic equations: for example, using special functions such as modular functions.

A group G is said to be **solvable** iff it has a series of subgroups H_i such that

$$\{e\}=H_0\subset H_1\subset\cdots\subset H_r=G,$$

where H_i is normal in H_{i+1} and the quotient H_{i+1}/H_i is Abelian for each $i=0,1,\ldots,r-1$. Every commutative group is solvable. In 1963 Walter Feit and John G. Thompson proved that every finite group of odd order is solvable. This is easy to state but the proof required 254 pages. I remember hearing Feit give a colloquium on the subject when I was an undergraduate. It soon became clear to me that the proof of this famous theorem would not be easily understood.

R. Dedekind gave the first formal lectures on Galois theory in 1857. Many long-standing classical problems were solved using Galois theory. As we said, it was shown that one cannot solve all polynomial equations with rational coefficients by repeated radicals, although that works for polynomials of degrees 2, 3, and 4. The famous classical Greek problems of ruler and compass constructions (angle trisection, circle squaring, cube duplicating) were also proved impossible via Galois theory.

There are analogs of Galois theory for coverings of Riemann surfaces, topological manifolds and graphs. See Terras [117] for a graph theory version.

Exercise 7.6.5 *Suppose that F is a finite field and* $f(x)\in F[x]$ *with* $n=\deg f$. *Define the* ***reciprocal polynomial*** $f^*(x)=x^nf\left(\frac{1}{x}\right)$. *If*

$$f(x)=a_nx^n+a_{n-1}x^{n-1}+\cdots+a_1x+a_0,$$

then

$$f^*(x)=a_0x^n+a_1x^{n-1}+\cdots+a_{n-1}x+a_n.$$

Prove the following two facts about reciprocal polynomials, assuming that $f(x)$ is not a constant and $a_0a_n \neq 0$.

(a) The polynomial f is irreducible over F if and only if the reciprocal polynomial f^* is irreducible over F.
(b) If $F = \mathbb{F}_q$, a finite field, the polynomial f is primitive (i.e., a root θ generates the multiplicative group $\mathbb{F}_q(\theta)^*$) iff the reciprocal polynomial f^* is primitive.

For the next exercises we need the following definition.

Definition 7.6.2 *If F is a field and A is an $n \times n$ matrix with entries in F, define the* ***characteristic polynomial*** *of A to be $p_A(x) = \det(A - xI)$, where I denotes the identity matrix.*

Exercise 7.6.6 *Show that a 3×3 triangular matrix satisfies its own characteristic polynomial.*

Exercise 7.6.7 *Consider any matrix A over a field F. Show that there is an invertible $n \times n$ matrix U over some finite-degree extension field E of F such that $U^{-1}AU$ is upper triangular.*

Hint. *Use induction on n. You need to find a basis of the vector space E^n such that the matrix of the linear transformation $v \to Av$ is upper triangular. Start the basis with an eigenvector v_λ of A, meaning that $Av_\lambda = \lambda v_\lambda$. The eigenvalue λ is a root of the characteristic polynomial $p_A(x) = \det(A - xI)$ of A and thus is in some finite-degree extension $F[\lambda]$. With respect to any basis containing v_λ as its first element, the matrix of A looks like*

$$\begin{pmatrix} \lambda & B \\ 0 & C \end{pmatrix}, \quad \text{where C is } (n-1) \times (n-1)$$

Exercise 7.6.8 *Prove the* ***Cayley–Hamilton theorem****, which states that A satisfies its characteristic polynomial. That is, $p_A(A) = 0$.*

Hint. *Use Exercise 7.6.7.*

Exercise 7.6.9 *If F is a field and $f(x) = x^n - a_{n-1}x^{n-1} - \cdots - a_1x - a_0$ is a polynomial in $F[x]$, define the* **companion matrix** *by formula (7.2) in Section 7.1. Prove the following:*

(a) $\det(C_f - xI) = (-1)^{n-1} f(x)$;
(b) $\det(xC_f - I) = (-1)^n f^*(x)$, *where f^* denotes the reciprocal polynomial to f in Exercise 7.6.5.*

Exercise 7.6.10 *Suppose that $\mathbb{F}_q$ is a finite field and $f(x)$ is a monic irreducible polynomial with coefficients in $\mathbb{F}_q$. Let θ denote a root of $f(x)$ in an extension field of $\mathbb{F}_q$. Let $f(x) = x^k - a_{k-1}x^{k-1} - \cdots - a_1x - a_0$. The elements of the multiplicative group $\mathbb{F}^*_{q^k} = \mathbb{F}_q(\theta)^*$ have the form $u = s_0 + s_1\theta + \cdots + s_{k-1}\theta^{k-1}$. Thus $\theta u = s'_0 + s'_1\theta + \cdots + s'_{k-1}\theta^{k-1}$. Prove that if C_f denotes the companion matrix of f, then, writing the column vector $w = {}^{\mathrm{T}}(s_0, s_1, \ldots, s_{k-2}, s_{k-1})$ and $w' = {}^{\mathrm{T}}(s'_0, s'_1, \ldots, s'_{k-2}, s'_{k-1})$, we have $w' = C_f w$. Note that this says that repeated multiplication of the companion matrix of f produces the log table* *for the multiplicative group of* $\mathbb{F}_q(\theta)^*$ *which we considered in many examples such as the one in Section 5.6 where the field was* $\mathbb{F}_2$ *and the polynomial was* $x^3 - x - 1$. *In our log tables, if we start with* $w_0 = {}^{\mathrm{T}}(1, 0, 0, \ldots, 0)$, *which corresponds to the element* 1 *of the field extension, then the jth row of the table of powers* θ^j *is the transpose of the vector* $w_j = C_f^j w_0$, *for* $j = 0, 1, 2, \ldots$.

Exercise 7.6.11 *Given that* $x^3 + 2x + 1$ *is a primitive polynomial over* $\mathbb{F}_3$, *compute the table of powers of a root* θ *of* $x^3 + 2x + 1$, *using the result of the preceding exercise.*

Exercise 7.6.12 *Use the solution of the preceding problem to show that the 12 roots of*

$$x^{12} + x^{11} + \cdots + x + 1 = \frac{x^{13} - 1}{x - 1}$$

in $\mathbb{F}_3[\theta]$ *are* $\theta^2, \theta^4, \theta^6, \ldots, \theta^{22}, \theta^{24}$.

Exercise 7.6.13 *Show that the Galois group of* $\mathbb{C}$ *over* $\mathbb{R}$ *is generated by* $\tau(x + iy) = x - iy$, *for* $x, y \in \mathbb{R}$.

Perhaps we should say a bit more about Évariste Galois, as his story is a dramatic one. Eric Temple Bell [8] titles his chapter on Galois "Genius and Stupidity." Galois' life was short (1811–1832) – cut short by a duel. His works are also short (60 pages) but have led to much of modern algebra. Much of the work was written quickly (at age 20) the night before the duel. The famous quote from this night is: "I have no time." Tragically Cauchy lost Galois' memoir, and Fourier died before presenting another of Galois' memoirs to the French Academy. It was 14 years after the death of Galois that Liouville received his works and the Galois papers were published. See Edna Kramer [59] for more of this history.

We have not proved the fundamental theorem of Galois theory. You can find a proof in Birkhoff and Maclane [9].

8

Applications of Rings

8.1 Random Number Generators

A sequence of random numbers should be something like a sequence of 0s and 1s obtained by flipping a coin with 0 for tails and 1 for heads. How should one know if the coin is fair or the sequence is random? You would expect that the relative proportion of 0s in a large number of flips should approach 0.5. Similarly you would expect that the proportion of two successive 0s would be 0.25 in a large number of flips.

References for this section include: D. Austin [3], R. P. Brent [11], B. Cipra [16], P. Diaconis [24], M. Goresky and A. Klapper [36], D. E. Knuth [54], R. Lidl and H. Niederreiter [69], G. Marsaglia [74], H. Niederreiter [81], W. H. Press *et al.* [86], and A. Terras [116].

There are many uses for sequences of random numbers: for example, simulations of natural phenomena using "Monte Carlo" methods, systems analysis, software testing, and cryptography. Recently Monte Carlo methods have been used to test whether gerrymandering has occurred in designing congressional election districts in states like Wisconsin or North Carolina (see Gregory Herschlag, Robert Ravier, and Jonathan C. Mattingly [41] or Jonathan C. Mattingly and Christy Vaughn [75]). The idea is to use Markov chain Monte Carlo methods to create random redistrictings of a state and then compare election results. We discuss Markov chains in Section 8.4.

Monte Carlo methods also allow you to approximate $\frac{1}{V}\int_D f(x)\,dx$, where V is the volume of the domain D in $\mathbb{R}^n$, by the average value of f on a "random" finite set of points in D. Sadly, the first use of Monte Carlo methods seems to have been in work of Metropolis, Ulam and von Neumann that led to the atomic bomb. These methods appear in the list contained in Barry Cipra's article [16] discussing the top 10 algorithms of the twentieth century.

Where do random numbers come from? In the old days there were tables (e.g., that of the RAND corporation from 1955). One can also get random sequences of 0s and 1s from tossing a fair coin or from times between clicks of a geiger counter near some radioactive material. Algebra gives us random numbers (technically, **pseudo random numbers**) much more simply, as D. H. Lehmer found in 1949. We will consider a simple example. Take a random walk through the multiplicative group $\mathbb{Z}_{17}^* = \langle 3 \pmod{17}\rangle$. In Figure 8.1 we list the elements in order of the powers $3^j \pmod{17}$. The figure is the directed Cayley graph $X(\mathbb{Z}_{17}^*, \{3 \pmod{17}\})$. This is very non-random! If instead, we order the vertices according to the usual ordering of the elements of $\mathbb{Z}_{17}^*$, thinking of the group as a set of integers $\{1, 2, 3, \ldots, 16\}$, then we get the view of the same directed graph that is found in Figure 8.2. This second view of the same Cayley graph looks much more random. If you imagine doing the same thing with a truly large prime p instead of 17, you would certainly expect to get a random listing of integers by taking the random walk. Assuming you take a primitive root mod p as your multiplier, the walk would go through all elements of the multiplicative group mod p.

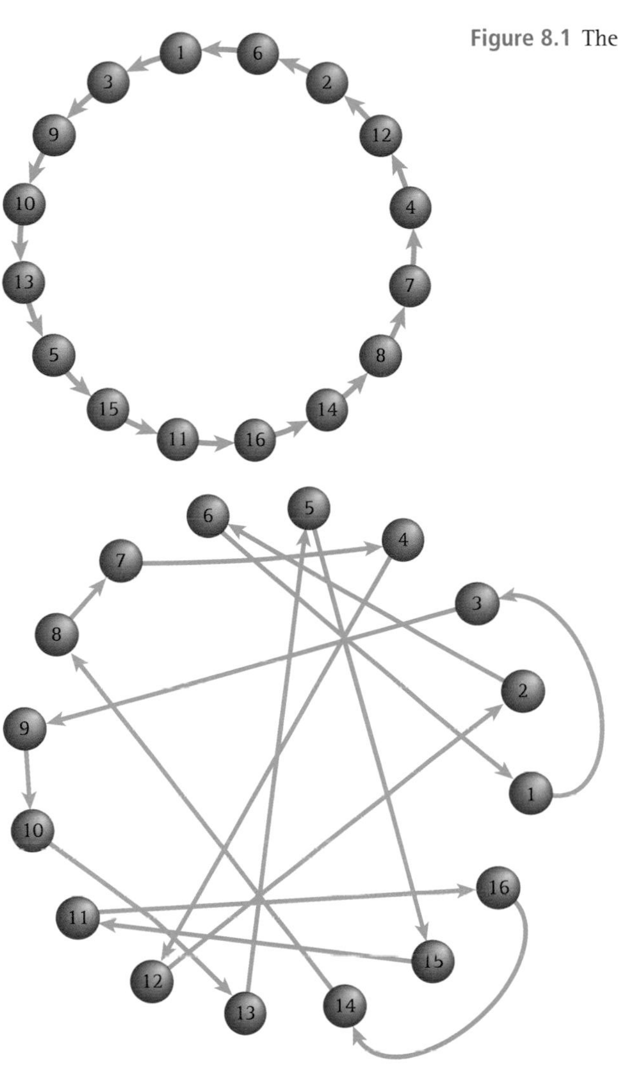

Figure 8.1 The Cayley graph $X(\mathbb{Z}_{17}^{*}, \{3 \pmod{17}\})$

Figure 8.2 The same graph as in Figure 8.1 except that now the vertices are given the usual ordering $1, 2, 3, 4, \ldots, 16$

The method of generating random numbers created by D. H. Lehmer in 1949 is called the **linear congruential method.** To get a sequence $\{X_n\}_{n \geq 1}$ of (pseudo-) random integers, you fix four numbers: m, the modulus; a, the multiplier; c, the increment; and X_0, the starting point. Then you generate the sequence recursively with the formula

$$X_{n+1} = aX_n + c \pmod{m}. \tag{8.1}$$

This is called a **linear congruential random number generator.** Let us refer to it as $L(X_0, a, c, m)$. If $m = 17$, $a = 3$, $c = 0$, we get the sequence represented in Figures 8.1 and 8.2. In the old days, popular choices were $m = 2^{31} - 1$ (a Mersenne prime) and $m = 2^{31}$. There was an infamous random number generator RANDU, which was built into the IBM mainframe computers of the 1960s. RANDU took $m = 2^{31}$, $a = 65\,539$, $c = 0$. For many years Matlab took $m = 2^{31} - 1$, $a = 7^5$, $c = 0$.

Clearly there are choices of a, c, m that lead to very non-random results. For example, look at $a = c = 1$. Then if $X_0 = 0$, you get $X_1 = 1$, $X_2 = 2, \ldots$ – marching through $1, 2, 3, \ldots, m$ in order. Certainly we want to hit all the numbers in $1, 2, 3, \ldots, m$ – or at least as many as possible. We must think about this problem a bit. The main theorem is a result of Hull and Dobell [46]. By the period t of the linear congruential random number generator $L(X_0, a, c, m)$, we mean the minimum positive integer t such that $X_k = X_{k+t}$, for large enough k. Note that the period cannot be m if $c = 0$, for example. The period could then be the Euler phi-function $\phi(m)$ if one could choose a to be a primitive root mod m, that is, a generator of $\mathbb{Z}_m^*$. Of course, $\mathbb{Z}_m^*$ is not cyclic for all – or even most – values of m. The following theorem has the object of producing a period of m.

Theorem 8.1.1 ***(Hull and Dobell).*** *The period of the linear congruential random number generator defined by (8.1) is m if and only if all of the following three conditions hold:*

(1) $\gcd(c, m) = 1$;
(2) if a prime p divides m, then p divides $(a - 1)$;
(3) if 4 divides m, then 4 divides $(a - 1)$.

We follow the discussion of the theorem in Knuth [54, pp. 15–21]. In order to understand this theorem, we should first prove a lemma.

Lemma 8.1.1 *Assume that* $k \in \mathbb{Z}^+$ *and* $a \in \mathbb{Z}$. *Define* $\frac{a^k - 1}{a-1} \pmod{m}$ *to mean*

$$1 + a + \cdots + a^{k-1} \pmod{m},$$

if $\gcd(a - 1, m) \neq 1$. *Then if a sequence of numbers* X_k *is given using the recursion in (8.1), we have the formula*

$$X_k \equiv a^k X_0 + c\frac{a^k - 1}{a - 1} \pmod{m}.$$

Exercise 8.1.1 *Prove the preceding lemma by induction on k.*

Now we consider a few examples of the random number generator

$$X_{n+1} = aX_n + c \pmod{m},$$

before attempting a proof of the theorem of Hull and Dobell.

Example 1. Consider $m = 8, c = 5, a = 3, X_0 = 0$ in (8.1). You get the sequence $0, 5, 4, 1, 0$. The period is 4 not 8. Condition (3) of the theorem is not valid. ▲

Example 2. Consider $m = 8, c = 3, a = 5, X_0 = 0$ in (8.1). You get the sequence $0, 3, 2, 5, 4, -1, -2, -7, 0$. The period is 8. The conditions of the theorem are valid. ▲

Example 3. Consider $m = 18, c = 5, a = 7, X_0 = 0$ in (8.1). You get the sequence

$$0, 5, 4, 15, 2, 1, 12, 17, 16, 9, 14, 13, 6, 11, 10, 3, 8, 7, 0.$$

The period is 18 and the conditions of the theorem are valid. ▲

Exercise 8.1.2 *Find the period of the linear congruential random number generator* $L(X_0, a, c, m)$ *defined by formula (8.1) if* $X_0 = 0, a = 3, c = 1, m = 16$.

Exercise 8.1.3 *Find the period for the linear congruential random number generator* $L(X_0, a, c, m)$ *defined by formula (8.1) if* $X_0 = 0, a = 13, c = 7, m = 36$.

Exercise 8.1.4 *Find the period for the linear congruential random number generator $L(X_0, a, c, m)$ defined by formula (8.1) if $X_0 = 0, a = 5, c = 2, m = 45$.*

Exercise 8.1.5 *Find the period for the linear congruential random number generator $L(X_0, a, c, m)$ defined by formula (8.1) if $X_0 = 0, a = 16, c = 2, m = 45$.*

Now we want to reduce our proof to the case of prime powers. To do this, we need the following exercise.

Exercise 8.1.6 *Suppose that $m = m_1 m_2$ with $\gcd(m_1, m_2) = 1$. Using the notation set up after formula (8.1) for linear congruential random number generators, show that the period of $L(X_0, a, c, m)$ is the least common multiple of the periods of $L(X_0, a, c, m_1)$ and $L(X_0, a, c, m_2)$.*

Now we want to prove the necessity of the three conditions in Theorem 8.1.1 for the case $m = p^e$, where p is an odd prime. It suffices to consider the case $X_0 = 0$. Why? Moreover, if $\gcd(c, m) \neq 1$, the period cannot be m. To prove the necessity of condition (2), we assume p does not divide $a - 1$. Then by Lemma 8.1.1, if the period were $p^e, e \geq 1$, we would have $a^{p^e} - 1 \equiv 0 \pmod{p^e}$. But this contradicts $a^{p^e} \equiv a \pmod{p}$.

To understand the necessity of condition (3), note that if $m = 2^e$ and $a \equiv 3 \pmod 4$, we have

$$\frac{a^{2^{e-1}} - 1}{a - 1} = 1 + a + \cdots + a^{2^{e-1}-1} \equiv 0 \pmod{2^e}.$$

To see this, note that $a^2 \equiv 1 \pmod 8$. So $a^4 \equiv 1 \pmod{16}, a^8 = 1 \pmod{32}, \ldots$ If $a \equiv 3 \pmod 4$, then $a - 1$ is twice an odd number. It follows that $\frac{1}{2}\left(a^{2^{e-1}} - 1\right) \equiv 0 \pmod{2^e}$. Thus we get a contradiction to $a \equiv 3 \pmod 4$.

For the sufficiency of the conditions, see Knuth [54].

Exercise 8.1.7 *Fill in all the details in the proof of the necessity of the three conditions for the case $m = p^e$, where p is an odd prime, in Theorem 8.1.1.*

What is the meaning of random? That is a question for statisticians who have devised tests to tell whether our lists are reasonably random. This is not a statistics course and thus we refer you to some of the references listed at the beginning of this section. Usually the applied mathematician wants a sequence of random real numbers in the interval $[0, 1]$ which approximates being uniformly distributed. To get this from X_n, you just divide by the modulus m which you used to generate them. Of course some properties of random sequences in $1, 2, 3, \ldots, m$ will be impossible to produce using linear congruential random number generators: for example, repetitions of numbers. For such properties you can consider sequences produced by linear recurrences over finite fields such as (8.2) below. We will consider some statistical properties of random number generators arising from finite fields at the end of this section.

In the late 1960s applied mathematicians using random numbers for Monte Carlo methods became angry when they discovered that if they used the linear congruential random integers X_n to produce vectors in $[0, 1]^n$, for $n > 1$, by writing $v = \frac{1}{m}(X_1, \ldots, X_n)$, they would have vectors lying in hyperplanes. Thus Marsaglia wrote a paper [74] proving that such vectors will fall into less than $(n!m)^{1/n}$ hyperplanes. For example if the modulus $m = 2^{32}$ and

$n = 10$, we get less than 41 hyperplanes. Of course no one had really tested the random vectors produced in this way for their uniform distribution in the hypercubes. So perhaps it should not have been a shock.

We wish to do a simple experiment along these lines. Again we take a fairly small prime, namely $p = 499$, and note that $\mathbb{Z}_{499}^* = \langle 7 \pmod{499} \rangle$. We compute a vector $v \in [0, 1]^{498}$ whose jth component is the real number $\frac{1}{499}$ times $7^j \pmod{499}$, identifying $7^j \pmod{499}$ as an integer between 1 and 498. Here we use Mathematica's `PowerMod` as described in Section 4.1 on public-key cryptography. If we do Mathematica's `ListPlot[v]` for this vector of points in $[0, 1]$, we will get a fairly random looking set of points in the plane. See Figure 8.3.

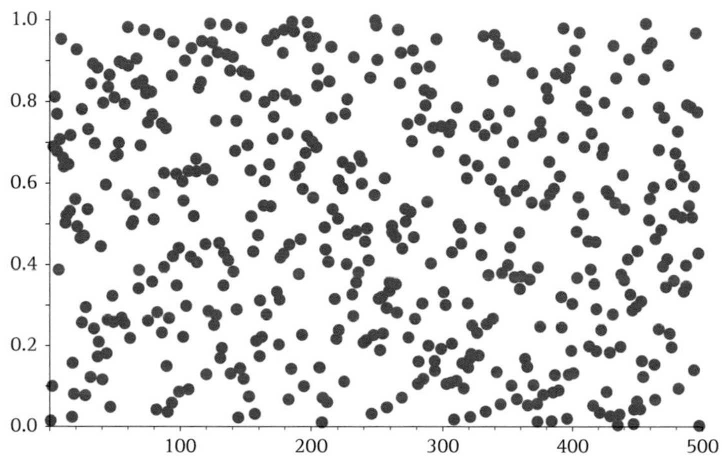

Figure 8.3 Plot of points $P_j = (j, v_j)$ whose second component is the real number $\frac{1}{499}$ times $7^j \pmod{499}$, identifying $7^j \pmod{499}$ as an integer between 1 and 498

However, there is a pitfall in the method. We are really only allowed to think of this as a one-dimensional thing. For if we try to plot points $(v_j, v_{j+1}) \in [0, 1]^2$, we get Figure 8.4.

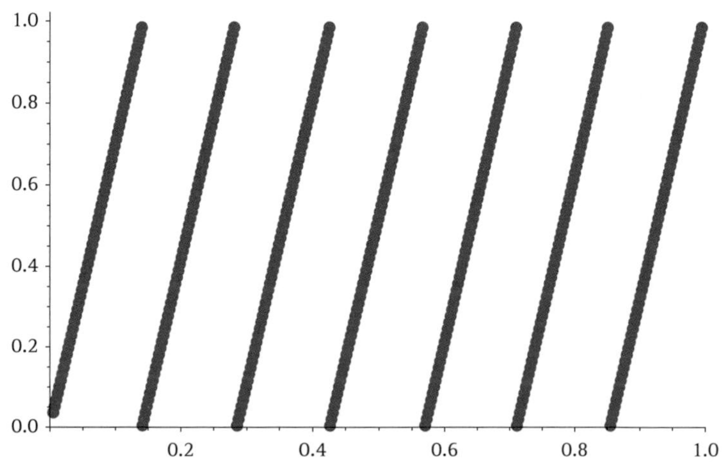

Figure 8.4 Plot of points $P_j = (v_j, v_{j+1})$ whose first component is the real number $\frac{1}{499}$ times $7^j \pmod{499}$, identifying $7^j \pmod{499}$ as an integer between 1 and 498

The same sort of thing happens in three dimensions, taking points $(v_j, v_{j+1}, v_{j+2}) \in [0, 1]^3$ to give Figure 8.5.

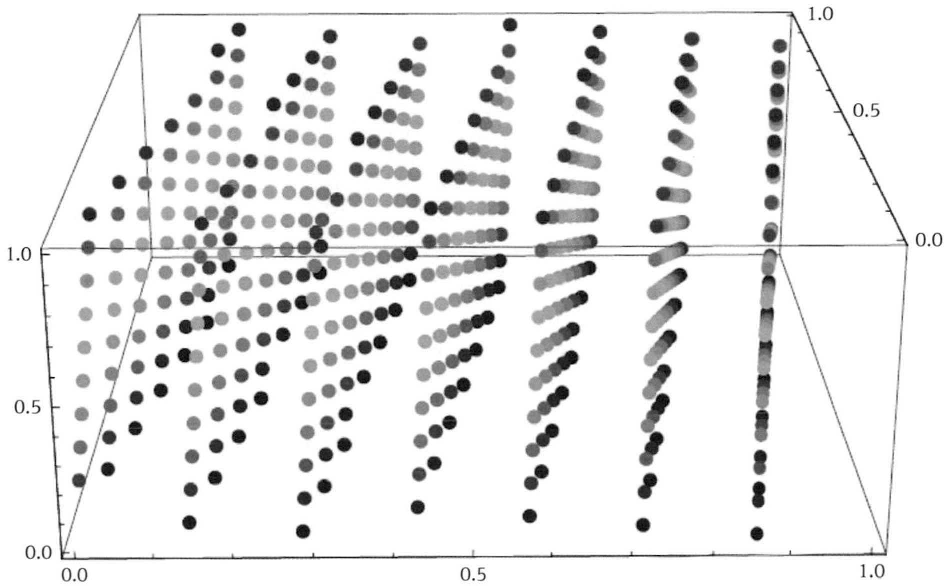

Figure 8.5 Plot of points $P_j = (v_j, v_{j+1}, v_{j+2})$ whose first component is the real number $\frac{1}{499}$ times $7^j \pmod{499}$, identifying $7^j \pmod{499}$ as an integer between 1 and 498

Suppose now we compute another random vector for a different prime modulus. We form $w \in [0, 1]$, with w_j being the real number $\frac{1}{503}$ times $5^j \pmod{503}$, identifying $5^j \pmod{503}$ as an integer between 1 and 502. Then we plot points $P_j = (v_i, w_i) \in [0, 1]$ in Figure 8.6.

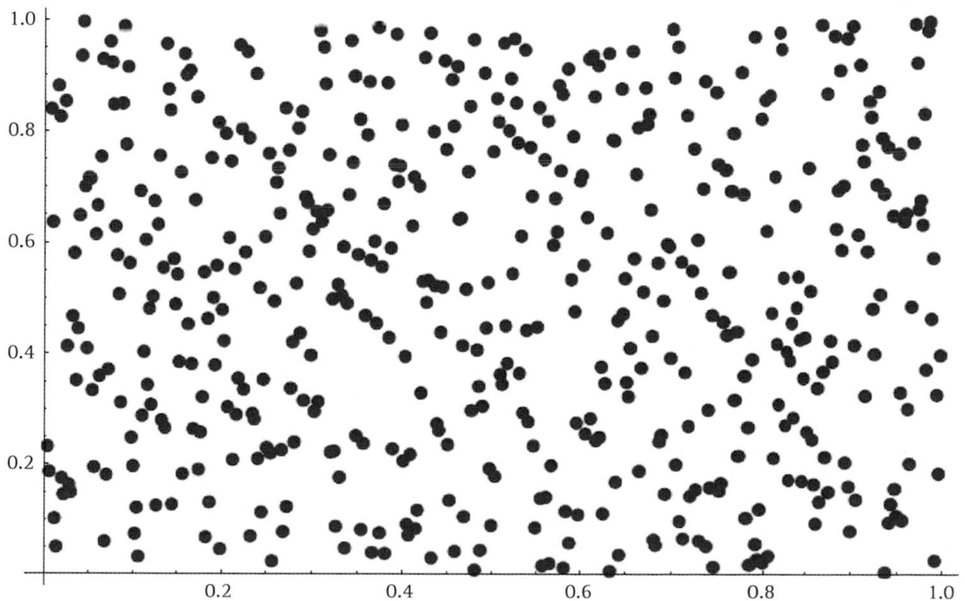

Figure 8.6 Plot of points $P_j = (v_j, w_j)$ whose first component is the real number $\frac{1}{499}$ times $7^j \pmod{499}$, identifying $7^j \pmod{499}$ as an integer between 1 and 498 and whose second component is the analog with 499 replaced with 503

So the points formed using two random number generators look more random although connecting some dots might create some creatures. We can do a 3D plot of points formed using a third random number generator. That is, we create another vector $z \in [0, 1]$, with z_j being the real number $\frac{1}{521}$ times $3^j \pmod{521}$, identifying $3^j \pmod{521}$ as an integer between 1 and 520. This gives us Figure 8.7 plotting points $P_j = (v_i, w_i, z_i) \in [0, 1]^3$.

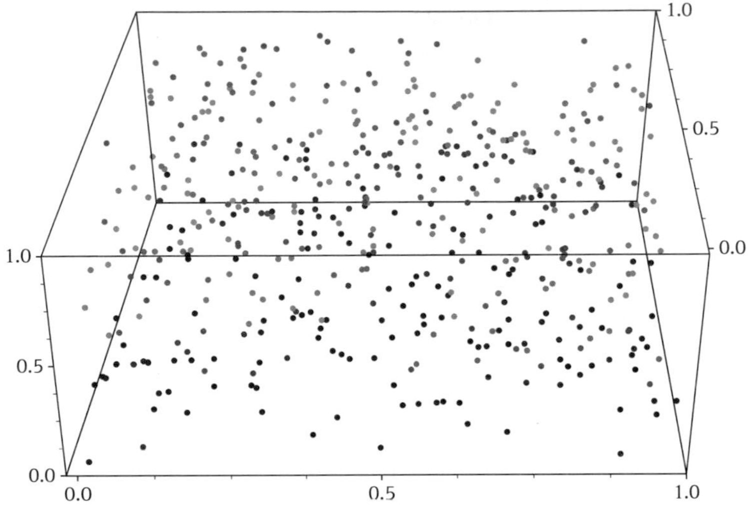

Figure 8.7 Points (v_i, w_i, z_i) from three vectors v, w, z formed from powers of generators of $\mathbb{F}_p^*$ for $p = 499, 503$, and 521, respectively

Now we have no obvious hyperplanes.

Exercise 8.1.8 *Find the equations of the lines in Figure 8.4.*

Exercise 8.1.9 *Find the equations of the hyperplanes in Figure 8.5.*

Applied mathematicians were in an uproar over the hyperplanes and did not want to use more than one generator, presumably worrying about slowing down the whole process. So new methods for random numbers arose. In 1995 Matlab switched to a Marsaglia generator. Brent [11] noticed that the Marsaglia Xor Shift generator can be viewed as a linear feedback shift register. Press *et al.* [86] spend many pages criticizing the linear congruential generators and then list them as 2/3 of their methods. Mathematica gives many basic methods. Not surprisingly Wolfram's favorites – cellular automata – appear. You are also allowed to create your own generator.

Derrick H. Lehmer (1905–1991) and his wife Emma were well-known number theorists of the last century. Derrick's father was also a number theorist who had built a prime generating machine. Like his father, D. H. was on the University of California, Berkeley, faculty – except for that short time in the early 1950s when he was fired for refusing to sign the university Regents' loyalty oath. That was the era of Joe McCarthy's un-American activity committee and the blacklists. Many readers may not remember this era when it was feared that communists were everywhere. I recommend reading the biography of D. H. Lehmer on Wikipedia. During the early 1950s D. H. was Director of the National Bureau of Standards' Institute for Numerical Analysis. In 1952, the loyalty oath was declared unconstitutional by the state Supreme Court and D. H. returned to the University of California, Berkeley, and chaired the mathematical department from 1954 to 1957. Emma Lehmer, being a woman, was never allowed to join the faculty. Yes, it was the bad old days.

I remember the Lehmers as two of the principal organizers of the West Coast Number Theory conference, a conference that I first attended as a young assistant professor in the 1970s. It is still meeting each year. Under their leadership this conference was democratically run. There was no elite bunch of organizers deciding who could speak and who could not. Instead, at the first night of the conference, anyone who wanted to give a talk would put their title on a piece of paper and the conference would be organized by putting talks on nearby subjects together.

Some random number generators generalize the Lehmer linear congruential generator by involving other finite fields or rings than $\mathbb{Z}_n$. They can be produced from feedback shift registers discussed earlier when we constructed log tables for $\mathbb{F}_{p^n}^*(\theta)$, where θ is a root of a primitive polynomial $f(x)$ of degree n in $\mathbb{F}_p[x]$. See Section 5.6 and Figure 5.4. These methods are also related to the construction of the Fibonacci numbers in Section 1.4. The sequences $\{s_n\}$ considered here involve elements of some finite field $\mathbb{F}_q$. The main reference that we use is Lidl and Niederreiter [69]. A reference covering more general theory than that over finite fields is Goresky and Klapper [36]. Sadly, going between references always involves changes of notation with minus signs disappearing, matrix transposes appearing and so forth. Reader beware! There are also differences in finite state machine diagrams involving up versus down as well as right versus left.

Define a **linear recurrent** (also linearly recurring) **sequence** to be $\{s_n\}$ obtained using the recursion formula below once you are given **coefficients** $b, a_0, a_1, \ldots, a_{k-1} \in \mathbb{F}_q$ and **initial values** $s_0, s_1, \ldots, s_{k-1} \in \mathbb{F}_q$,

$$s_{n+k} = a_{k-1}s_{n+k-1} + \cdots + a_0 s_n + b, \quad n = 0, 1, 2, \ldots. \tag{8.2}$$

Some would call formula (8.2) a **difference equation**. Recall that the Fibonacci numbers were defined by such a recursive equation $f_{k+2} = f_{k+1} + f_k$ over the field $\mathbb{Q}$ in Section 1.4. When $b = 0$, the recursion (8.2) is called **homogeneous**. We will restrict ourselves to that case. **Feedback shift registers** are electronic circuits that can be used to produce the s_n. An **impulse response sequence** has initial values (for $k \geq 2$) given by $s_0 = s_1 = \cdots = s_{k-2} = 0$ and $s_{k-1} = 1$.

Associated to the linear recurrent sequence from (8.2) we have the **characteristic polynomial**

$$f(x) = x^k - a_{k-1}x^{k-1} - \cdots - a_1 x - a_0$$

in $\mathbb{F}_q[x]$. The coefficients of this polynomial are -1 times the coefficients of the homogeneous sequence – ignoring b.

Example 1. Consider the field $\mathbb{F}_3$ and the polynomial $x^3 - x^2 - x + 1$. The homogeneous recurrence (8.2) is

$$s_{n+3} = s_{n+2} + s_{n+1} - s_n.$$

We will create a table of an impulse response sequence s_n. That means we start with $0, 0, 1$.

n	0	1	2	3	4	5	6	7	8	9	10	11	12	13	14	15	16	17
s_n	0	0	1	1	−1	−1	0	0	1	1	−1	−1	0	0	1	1	−1	−1

We find that the sequence has **period** 6. That is, $s_{n+6} = s_n$, for all $n \geq 0$. That might not be what you expected if you thought that the polynomial $x^3 - x^2 - x + 1$ would have roots generating a finite field of order 27. However $x^3 - x^2 - x + 1 = (x^2 - 1)(x - 1)$ over F_3. When the polynomial is reducible, the period need not divide the order of the multiplicative group of the field extension generated by a root of an irreducible polynomial, 26 in this case. ▲

Example 2. Consider the field $\mathbb{F}_3$ and the polynomial $x^2 - x - 1$. This is a primitive polynomial and so we expect period $3^2 - 1 = 8$. Again, we create a table for the impulse response sequence. The feedback shift register diagram for this example is shown in Figure 8.8. The table of values coming from the recurrence:

$$s_{n+2} = s_{n+1} + s_n.$$

is given below.

n	0	1	2	3	4	5	6	7	8	9	10	11	12	13	14	15	16	17
s_n	0	1	1	−1	0	−1	−1	1	0	1	1	−1	0	−1	−1	1	0	1

The period is indeed 8: that is, $s_{n+8} = s_n$, for all $n \geq 0$. ▲

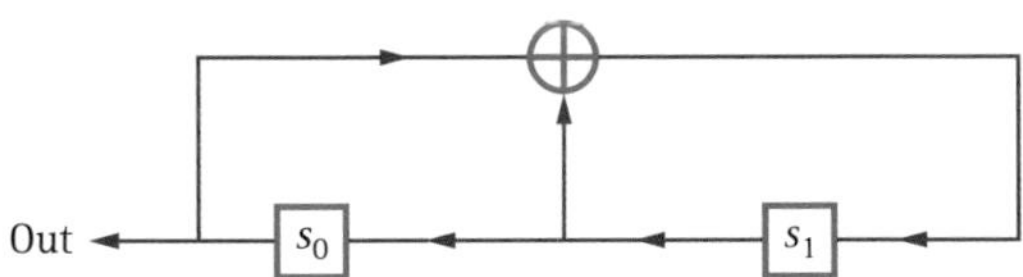

Figure 8.8 Feedback shift register corresponding to example 2

Exercise 8.1.10 *Consider the field $\mathbb{F}_2$ and the polynomial $x^5 + x + 1$. Produce a table for the homogeneous linear recurrent sequence associated to this characteristic polynomial by (8.2). Take the initial entries $0, 0, 0, 0, 1$ for the impulse response sequence. What is the period? Note that $x^5 + x + 1 = (x^3 + x^2 + 1)(x^2 + x + 1)$. Draw the feedback shift register corresponding to this example.*

Exercise 8.1.11 *Same as the preceding exercise – except use the primitive polynomial $x^5 + x^2 + 1$ over $\mathbb{F}_2$.*

As we said in Exercise 7.6.10, the successive states of the finite state machine corresponding to a polynomial can be found by multiplication by a matrix called the companion matrix of the polynomial. Now we want to see what happens for our linear recurrent sequences. To study the linear recurrent sequence $\{s_n\}$ defined by recursion (8.2), we associated the characteristic polynomial $f(x)=x^k - a_{k-1}x^{k-1} - \cdots - a_1x - a_0$ in $\mathbb{F}_q[x]$. Associated to this polynomial is the **companion matrix** whose definition we recall

$$C_f = \begin{pmatrix} 0 & 0 & 0 & \cdots & 0 & a_0 \\ 1 & 0 & 0 & \cdots & 0 & a_1 \\ 0 & 1 & 0 & \cdots & 0 & a_2 \\ 0 & 0 & 1 & \cdots & 0 & a_3 \\ \vdots & \vdots & \vdots & \ddots & \vdots & \vdots \\ 0 & 0 & 0 & \cdots & 1 & a_{k-1} \end{pmatrix}. \tag{8.3}$$

Now consider **state vectors** $v_n = (s_n, s_{n+1}, \ldots, s_{n+k-1})$. These are row vectors with entries in $\mathbb{F}_q$. The vector v_0 is the initial values vector. Suppose we multiply v_n and C_f. We obtain

$$v_nC_f = (s_n, s_{n+1}, \ldots, s_{n+k-1}) \begin{pmatrix} 0 & 0 & 0 & \cdots & 0 & a_0 \\ 1 & 0 & 0 & \cdots & 0 & a_1 \\ 0 & 1 & 0 & \cdots & 0 & a_2 \\ 0 & 0 & 1 & \cdots & 0 & a_3 \\ \vdots & \vdots & \vdots & \ddots & \vdots & \vdots \\ 0 & 0 & 0 & \cdots & 1 & a_{k-1} \end{pmatrix}$$

$$= \left(s_{n+1}, s_{n+2}, \ldots, s_{n+k-1}, \sum_{j=0}^{k-1} a_j s_{n+j} \right) = v_{n+1}.$$

Note that, in general, this is not the same as the result of Exercise 7.6.10 where we computed the log table for a finite field extension generated by a primitive polynomial f using the companion matrix of f. There we had the same matrix but we used column vectors w_n and thus looked at $w_{n+1} = C_f w_n$.

Suppose we consider a homogeneous linear recurrent sequence given by (8.2) with $b=0$ and having companion matrix (8.3). Suppose that $a_0 \neq 0$. Then, by the following exercise, $\det(C_f) \neq 0$, which means that C_f is an invertible $k \times k$ matrix and thus an element of the finite group $GL(k, \mathbb{F}_q)$ known as the general linear group over $\mathbb{F}_q$. What does the periodicity of the linear recurrent sequence have to do with the order of the companion matrix C_f in the general linear group $GL(k, \mathbb{F}_q)$?

Exercise 8.1.12 *Show that if the companion matrix is given by (8.3) and* $a_0 \neq 0$, *then* $\det(C_f) \neq 0$.

Hint. $\det(C_f) = (-1)^{k-1} a_0$.

Exercise 8.1.13 *What is the order of* $GL(k, \mathbb{F}_q)$?

Hint. *The columns must be nonzero and linearly independent. How many ways can you choose the first column in* $\mathbb{F}_q^k$? *Then once the first column is fixed, how many ways are there to choose the second column?*

It follows from our computations that if $\{s_n\}$ is defined by (8.2) with $b=0$, $a_0 \neq 0$, and the characteristic polynomial of the sequence is f with companion matrix C_f, then if the companion matrix has order r in $GL(k, \mathbb{F}_q)$, certainly the sequence repeats after r steps but it may have a smaller period. There is no reason that the period should divide r in general. If, however, f is a primitive polynomial over $\mathbb{F}_q$ then the order of C_f is $q^k - 1$. In general, if $b=0$, $a_0 \neq 0$, one might expect that the period of the sequence would divide the order of $GL(k, \mathbb{F}_q)$. But the state vectors v_n will not in general be the same as the vectors w_n in our log tables for the multiplicative group of the finite field $\mathbb{F}_q(\theta)$ for a root θ of f.

Next we find a formula for the s_n in our linear recurrent sequences in terms of powers of the roots of the characteristic polynomial of the sequence. This is exactly analogous to the connection between Fibonacci numbers and the golden ratio

$$\phi = \frac{1+\sqrt{5}}{2} \cong 1.618\,033\,988\,749\,894\,848\,2. \tag{8.4}$$

Artists and photographers believe that a **golden rectangle** with its width to height ratio of ϕ to 1 is particularly pleasing to the human eye. That $\phi = 1 + \frac{1}{\phi}$ means that you can remove a square from a golden rectangle and have another smaller golden rectangle. Moreover the process can be continued, leading to an inspiration to artists. See Young [128] for some pictures and more stories about the Fibonacci numbers. Another reference is [70].

Now we consider the Fibonacci numbers to suggest what happens for our linear recurrent sequences. Recall that the Fibonacci numbers f_n satisfy the recursion $f_{n+2} = f_{n+1} + f_n$, $n = 0, 1, 2, 3, \ldots$, starting with $f_0 = 0$ and $f_1 = 1$. Of course – unlike the linear recurrent sequences in finite fields – the sequence of Fibonacci numbers is definitely not periodic. The sequence $\{f_n\}$ is a strictly increasing sequence, once n is larger than 1. Recalling how we connect a recursion (8.2) with its characteristic polynomial, we see that the polynomial connected with the Fibonacci recursion is $x^2 - x - 1$ over the field of rational numbers $\mathbb{Q}$. The roots of this polynomial are the golden ratio defined by (8.4) and its conjugate $\phi' = (1-\sqrt{5})/2$. We claim that $f_n = \beta_1 \phi^n + \beta_2 (\phi')^n$. To find the β_i, just look at the cases $n=0$ and $n=1$. This gives two linear equations in two unknowns. The solution is $\beta_1 = 1/\sqrt{5} = -\beta_2$. Thus we find that, for all $n = 0, 1, 2, 3, \ldots$, the Fibonacci numbers are

$$f_n = \frac{1}{\sqrt{5}} \left\{ \phi^n - (\phi')^n \right\}. \tag{8.5}$$

Exercise 8.1.14 *Show that the right-hand side of formula (8.5) satisfies the same recursion as the Fibonacci numbers and gives the same initial values for $n=0$ and $n=1$, namely $f_0 = 0$ and $f_1 = 1$.*

With this example in mind, we will try to do the analog in finite fields. Assume that $\{s_n\}$ is a linear recurrent sequence in the finite field $\mathbb{F}_q$ defined by the recursion (8.2) with $b=0$. Suppose that the characteristic polynomial of this recursion is $f(x)$ and that $f(x)$ has k distinct roots $\theta_1, \ldots, \theta_k$ in a splitting field E over $\mathbb{F}_q$. Then there exist $\beta_i \in E$ for $i = 1, \ldots, k$, such that we have the following formula for s_n:

$$s_n = \sum_{i=1}^{k} \beta_i \theta_i^n. \tag{8.6}$$

The proof can be found in Lidl and Niederreiter [69, p. 196]. One proceeds as in the Fibonacci number example – making use of Cramer's rule to solve k linear equations in k unknowns.

Example. Consider the linear recurrent sequence associated to the polynomial $x^2 - x - 1$ over $\mathbb{F}_3$. The recurrence is $s_{n+2} = s_{n+1} + s_n$, $n = 0, 1, 2, 3, \ldots$ Assume that the initial values are $s_0 = 0$ and $s_1 = 1$. The polynomial $x^2 - x - 1$ is a primitive polynomial over $\mathbb{F}_3$. Its roots in a splitting field $\mathbb{F}_9$ are θ and θ^3, making use of what we know from Galois theory. Thus formula (8.6) says that

$$s_n = \beta_1 \theta^n + \beta_2 \theta^{3n} = \theta^{n+2} - \theta^{3n+2}. \quad \blacktriangle$$

Exercise 8.1.15 *Fill in the details of the computation of the coefficients β_1 and β_2 in the last formula for s_n associated to the polynomial $x^2 - x - 1$. Assume that the initial values are $s_0 = 0$ and $s_1 = 1$.*

Next let us say a bit about **statistical tests** of the randomness of the linear recurrent sequences. We take our discussion from Lidl and Niederreiter [69, pp. 283ff]. We assume that the sequence has a period r large enough for the application to pseudorandom numbers. Then there are three tests to consider:

1. distribution of the s_n,
2. distribution of blocks of s_n,
3. correlations of the s_n with s_{n+h}.

Goresky and Klapper [36] note that such tests were formulated by S. Golomb in the 1960s.

In order to consider the first two tests, define for a block $b \in \mathbb{F}_q^m$,

$$Z_m(b) = \#\{n | 0 \le n \le r - 1,\ s_{n+i-1} = b_i,\ \forall i, i = 1, \ldots, m\}.$$

Here r denotes the period of the sequence. Then one wants $m = 1$ for test 1 and one hopes that $Z_1(b)$ is more or less constant, that is, uniform. For test 2 if $m \ge 2$, one hopes that $Z_m(b)$ is more or less constant. For test 3 one considers a Fourier transform.

Assume that we are looking at a homogeneous linear recurrent sequence $\{s_n\}$ for which the characteristic polynomial is a monic primitive polynomial in $\mathbb{F}_q[x]$ of degree $k > m$. Then the period is $r = q^k - 1$. Assuming the initial states are not all 0, we run through all $q^k - 1$ nonzero vectors in a period. This means that $Z_m(b)$ is the number of nonzero vectors $v \in \mathbb{F}_q^k$ that have b as their first m coordinates. Therefore, if $k > m$,

$$Z_m(b) = \begin{cases} q^{k-m} - 1 & \text{if } b = 0, \\ \\ q^{k-m} & \text{if } b \neq 0. \end{cases} \tag{8.7}$$

This result says that our hopes for tests 1 and 2 are fulfilled – assuming the characteristic polynomial is primitive and we thus get the maximal period.

For test 3, one makes use of the Fourier transform on $\mathbb{F}_q$. It is a little easier to assume $q = p$ is a prime as we have discussed the finite Fourier transform in Section 4.2. We look at a homogeneous linear recurrent sequence $\{s_n\}$ for which the characteristic polynomial is a monic primitive polynomial in $\mathbb{F}_p[x]$ of degree k. Let $r = p^k - 1$ denote the period of the sequence. Then consider the following sum involving the fixed non-trivial character $\chi(b) = \chi_1(b) = e^{2\pi i b/p}$, for $b \in \mathbb{Z}$. The sum – called the **autocorrelation** – is defined to be

$$\mathrm{Cor}(h) = \sum_{j=0}^{r-1} \chi(s_j - s_{j+h}).$$

With the assumptions of the preceding paragraph, assuming that the initial states of our sequence are not all 0 as well, if h is not congruent to 0 mod r, we find that $\mathrm{Cor}(h) = -1$. To see this, use formula (8.7) on the sequence $u_n = s_n - s_{n+h}$, which satisfies the same recursion as s_n. Then if the period r does not divide h, we have

$$\mathrm{Cor}(h) = \left(q^{k-1} - 1\right)\chi(0) + q^{k-1}\sum_{b\in\mathbb{F}_p^*}\chi(b) = -1 + \sum_{b\in\mathbb{F}_p}\chi(b) = -1. \tag{8.8}$$

We used the orthogonality relation for additive characters of $\mathbb{F}_p$ from Lemma 4.2.1 in the last equality. The same argument will work for an arbitrary finite field, once one has considered the additive characters of the field. More details are in Lidl and Niederreiter [69, p. 284] – also Goresky and Klapper [36]. Golomb says a sequence like the one considered here has an **ideal correlation function.**

Example. Consider the recursion $s_{n+2} = s_{n+1} + s_n$ over $\mathbb{F}_3$ with initial values $s_0 = 0$, $s_1 = 1$. Then we saw above that the period is 8 and we computed the s_n row of the table below:

n	0	1	2	3	4	5	6	7
s_n	0	1	1	-1	0	-1	-1	1
s_{n+1}	1	1	-1	0	-1	-1	1	0

First we consider test 3. If we set $\zeta = e^{2\pi i/3}$ and $\chi(a) = \zeta^a$, then we find that in this case $\mathrm{Cor}(1) = 2\chi(0) + 3\chi(-1) + 3\chi(1) = -1$. This agrees with equation (8.8). As for test 1, we see that over a period, $s_n = 0$ twice; $s_n = 1$ three times; and $s_n = -1$ three times. This agrees with equation (8.7), when $k = 2$ and $m = 1$. The distribution is pretty uniform. Lastly consider test 2 for $m = 2$. The number of pairs (s_n, s_{n+1}) that are equal to $(0, 0)$ is 0. The number of pairs (s_n, s_{n+1}) that are equal to any (b_1, b_2) except $(0, 0)$ is 1. This agrees with equation (8.7), when $k = m = 2$. Again the distribution is pretty uniform. ▲

Exercise 8.1.16 *Imitate the preceding example for the linear recurrent sequence for which the characteristic polynomial is $x^5 + x^2 + 1$ over $\mathbb{F}_2$. In test 3, the character χ is defined by $\chi(a) = (-1)^a$. So in this case* $\mathrm{Cor}(1)$ *is the number of agreements of s_n with s_{n+1} minus the number of disagreements.*

8.2 Error-Correcting Codes

References for this section include Larry L. Dornhoff and Franz E. Hohn [25], William J. Gilbert and W. Keith Nicholson [35], Joseph A. Gallian [33], Vera Pless [85], Kenneth Rosen [92], Judy Walker [123], and Audrey Terras [116].

Suppose that we must send a message of 0s and 1s from our computer on earth to Professor Bolukxy's computer on Xotl. No doubt errors will be introduced by transmission over such a long distance and some random 1 will turn into a 0. In order for Professor Bolukxy to figure out my message, there must be some redundancy built in. Error-correcting codes are created for that purpose. The original signal $s \in \mathbb{F}_2^n$ will be encoded as $x \in \mathbb{F}_2^{n+r}$. If errors are added in transmission of the encoded signal, Professor Bolukxy will use a decoder to find the most likely original signal $s' \in \mathbb{F}_2^n$ – hoping that there is enough redundancy to do so. Such methods are used in compact discs as well as communications with spacecraft. The goal of error correction is really the opposite of the goal of cryptography. Here we

Figure 8.9 Sending a message of 0s and 1s to Professor Bolukxy on the planet Xotl

want our message to be understood. The simplest method might be to repeat the message a number of times. However, that would be inefficient.

R. W. Hamming (1915–1998) of Bell Telephone Labs published his codes in 1950. He had been working on a computer using punched cards. Whenever the machine detected an error, the computer would stop. Hamming got frustrated and began work on a way to correct errors. Hamming also introduced the Hamming distance defined below to get a measure of the error in a signal. And he worked on the Manhattan project – doing simulations to model whether the atomic bomb would ignite the atmosphere. I cannot resist including a quote from Hamming's book [10]: "we will avoid becoming too involved with mathematical rigor, which all too often tends to become rigor mortis."

Definition 8.2.1 *A **linear code** C is a vector subspace C of $\mathbb{F}_q^n$.*

Here $\mathbb{F}_q$ denotes the field with q elements. If the dimension of C as a vector space over $\mathbb{F}_q$ is k, we call C an **$[n, k]$-code**. Since all codes we consider are linear, we will drop the word "linear" and just call them "codes." Here q will be 2 mostly. Such codes are called "**binary**." If $q=3$, the code is "ternary."

Definition 8.2.2 *The **Hamming weight** of a codeword $x\in C$ is $|x|$, which is the number of components of x that are nonzero. The **distance** between $x, y\in C$ is defined to be $d(x,y)=|x-y|$.*

Exercise 8.2.1 *For the vector space $V=\mathbb{F}_q^n$, show that the Hamming distance $d(x,y)$ has the following properties for all $x, y, u\in V$:*

(a) $d(x,y)=d(y,x)$;
(b) $d(x,y)\geq 0$ *and* $d(x,y)=0 \Longleftrightarrow x=y$;

(c) ***(triangle inequality)*** $d(x, y) \leq d(x, u) + d(u, y)$;
(d) $d(x, y) = d(x + u, y + u)$;
(e) $d(x, y) \in \mathbb{Z}^+ \cup \{0\}$.

The first three properties of the Hamming distance in this exercise make it a **metric** on V. The fourth property makes it a **translation-invariant metric** on V.

Exercise 8.2.2

(a) For the prime p, show that the Hamming weight on $x \in \mathbb{F}_p^n$ satisfies

$$|x| \equiv x_1^{p-1} + \cdots + x_n^{p-1} \pmod{p}.$$

(b) Consider $\mathbb{F}_p^n$ as a group under addition and form the Cayley graph $X(\mathbb{F}_p^n, S)$, where

$$S = \{ s \in \mathbb{F}_p^n \mid |s| = 1 \}.$$

Draw the graph for $p = 5$ and $n = 2$.

Definition 8.2.3 *If C is an $[n, k]$-code such that the minimum distance of a nonzero codeword from 0 is d, we say that C is an* ***$[n, k, d]$-code***.

The following theorem assumes you decode a received vector as the nearest codeword using the Hamming distance.

Theorem 8.2.1 *If $d = 2e + 1$, an $[n, k, d]$-code C corrects e or fewer errors.*

Proof. Suppose distinct $x, y \in C$ are such that $d(x, y) \geq 2e + 1$. If the received word r has at most e errors, it cannot be in the Hamming ball of radius e about both x and y, since that would imply $0 < |x - y| = d(x, y) \leq d(x, r) + d(r, y) \leq e + e = 2e$. So the code can correct e errors – assuming that we can find the nearest codeword to r. ▲

We need to add a few more definitions to our coding vocabulary.

Since an $[n, k]$ code C is a k-dimensional vector space over $\mathbb{F}_q$, the code C has a k-element basis. Therefore we can form a matrix whose rows are the basis vectors. This is called a **generator matrix** G of the code C. A generator matrix of an $[n, k]$-code is a $k \times n$ matrix of rank k with elements in $\mathbb{F}_q$. The code C is the image of the map sending the row vector $v \in \mathbb{F}_q^k$ to vG. Since C has more than one basis, it also has many generator matrices. The **standard generator matrix** has the form

$$G = (I_k \; A), \tag{8.9}$$

where the first k columns form the $k \times k$ identity matrix I_k. If the generating matrix is in standard form, with no errors, decoding is easy; just take the first k entries of the codeword. We know that we can use elementary row operations over $\mathbb{F}_q$ to put any generator matrix into row echelon form and that this must be the standard form $(I_k \; A)$ since this matrix must have rank k.

To describe the encoding envisioned here, we take the original message viewed as a row vector $s \in \mathbb{F}_q^k$ and then we encode the message as sG, adding redundancy to be able to correct errors.

A **parity check matrix** H of a $[n, k]$-code C is a matrix with n columns and rank $n-k$ such that $x \in C$ if and only if $xH = 0$. (Many texts take transpose H instead.) If $G = (I_k \; A)$, then

$$H = \begin{pmatrix} -A \\ I_{n-k} \end{pmatrix}. \tag{8.10}$$

Exercise 8.2.3 *For code C with generator matrix $G = (I_k \; A)$ as in (8.9), show that $x \in C$ iff $xH = 0$, with H as in (8.10).*

Hint. *It is easy to see that C lies in the kernel of the linear transformation L sending $x \in \mathbb{F}_q^n$ to xH. Since*

$$\dim \ker L + \dim L\left(\mathbb{F}_q^n\right) = n,$$

we find that the dimension of $\ker L$ *is k and thus obtain the equality of the kernel of L and the code C.*

The parity check matrix is quite useful for decoding. See Dornhoff and Hohn [25] or Gallian [33] for more information.

Our next question is: Where does all our theory of finite fields come in?

Definition 8.2.4 *A linear* **cyclic code** *is a linear code C with the property that if $c = (c_0, c_1, \ldots, c_{n-2}, c_{n-1})$ is a codeword then so is the cyclic permutation of c given by $c' = (c_{n-1}, c_0, \ldots, c_{n-3}, c_{n-2})$.*

Let R denote the factor ring $R = \mathbb{F}_q[x]/\langle x^n - 1\rangle$. Represent elements of R by polynomials with coefficients in $\mathbb{F}_q$ of degree $<n$. Identify codeword $c = (c_0, c_1, \ldots, c_{n-2}, c_{n-1})$ with (the coset of) the polynomial $c_0 + c_1 x + \cdots + c_{n-1}x^{n-1}$.

Theorem 8.2.2 *A linear code C in R is cyclic if and only if it is an ideal in the ring $R = \mathbb{F}_q[x]/\langle x^n - 1\rangle$, using the identification of the preceding paragraph.*

Proof. First note that a subspace W of R is an ideal if $xW \subset W$, because this implies $x^j W \subset W$, for all $j = 1, 2, 3, \ldots$ Thus $RW \subset W$.

Now suppose that C is an ideal and $c_0 + c_1 x + \cdots + c_{n-1}x^{n-1} \in C$. Then C contains

$$\begin{aligned} x\left(c_0 + c_1 x + \cdots + c_{n-1}x^{n-1}\right) &= c_0 x + c_1 x^2 + \cdots + c_{n-1}x^n \\ &= c_{n-1} + c_0 x + c_1 x^2 + \cdots + c_{n-2}x^{n-1} \left(\operatorname{mod}(x^n - 1)\right). \end{aligned}$$

The last equality happens because x^n is congruent to 1 modulo $\langle x^n - 1\rangle$. So C is cyclic. We leave the converse as an exercise. ▲

Exercise 8.2.4 *Complete the proof of the preceding theorem.*

Question. What are the ideals A in the ring $R=\mathbb{F}_q[x]/\langle x^n-1\rangle$?

Answer. Just as we found in Section 5.4 for ideals in $\mathbb{Z}_{12}$, they are principal ideals $\langle g(x)\rangle$, where $g(x)$ divides x^n-1. We call $g(x)$ the generator of A. If $g(x)=c_0+c_1x+\cdots+c_rx^r$ has degree r, then the corresponding code is an $[n, n-r]$-code and a generator matrix for the code (as defined above) is the $(n-r)\times n$ matrix:

$$\begin{pmatrix} c_0 & c_1 & c_2 & \cdots & c_r & 0 & 0 & \cdots & 0 \\ 0 & c_0 & c_1 & \cdots & c_{r-1} & c_r & 0 & \cdots & 0 \\ 0 & 0 & c_0 & \cdots & c_{r-2} & c_{r-1} & c_r & \cdots & 0 \\ \vdots & \vdots & \vdots & \ddots & \vdots & \vdots & \vdots & \ddots & \vdots \\ 0 & 0 & 0 & \cdots & c_0 & c_1 & c_2 & \cdots & c_r \end{pmatrix}. \tag{8.11}$$

Exercise 8.2.5 *Show that the code described above has dimension $n-r$.*

Hint. *$g(x)=c_0+c_1x+\cdots+c_rx^r$ has degree r, the cosets of the vectors $g(x)x^j$, $j=0,\ldots,$ $n-r-1$, are linearly independent in the ring R. These vectors span the ideal $A=\langle g(x)\rangle$ since elements of A have the form $f(x)g(x)$, for some polynomial $f(x)$ of degree less than or equal to $n-r-1$.*

Example: The Hamming [7, 4, 3]-code. Note that the polynomial x^7-1 can be completely factored into irreducibles over $\mathbb{F}_2$ as follows:

$$x^7-1=(x-1)(x^3+x+1)(x^3+x^2+1)\in\mathbb{F}_2[x].$$

Take $g(x)=x^3+x+1\in\mathbb{F}_2[x]$ to generate our ideal I in $R=R=\mathbb{F}_2[x]/\langle x^7\quad 1\rangle$ corresponding to the code. The codewords in C in are listed below.

0	0	0	0	0	0	0	1	1	1	1	1	1	1
1	1	0	1	0	0	0	0	0	1	0	1	1	1
0	1	1	0	1	0	0	1	0	0	1	0	1	1
0	0	1	1	0	1	0	1	1	0	0	1	0	1
0	0	0	1	1	0	1	1	1	1	0	0	1	0
1	0	0	0	1	1	0	0	1	1	1	0	0	1
0	1	0	0	0	1	1	1	0	1	1	1	0	0
1	0	1	0	0	0	1	0	1	0	1	1	1	0

A generator matrix corresponding to $g(x)=x^3+x+1\in\mathbb{F}_2[x]$ as in (8.11) is

$$G=\begin{pmatrix} 1 & 1 & 0 & 1 & 0 & 0 & 0 \\ 0 & 1 & 1 & 0 & 1 & 0 & 0 \\ 0 & 0 & 1 & 1 & 0 & 1 & 0 \\ 0 & 0 & 0 & 1 & 1 & 0 & 1 \end{pmatrix}.$$

▲

Exercise 8.2.6

(a) *Use elementary row operations to put the generator matrix G above into row-reduced echelon (standard) form.*
(b) *Explain why the codewords listed above are correct by making a table listing the* 16 *elements of $x \in \mathbb{F}_2^4$ (the possible messages) in the first column and then listing the corresponding xG (the encodings of the messages) in the second column.*
(c) *Explain why the minimum weight of the Hamming $[7, 4, 3]$-code is 3.*

Exercise 8.2.7 *Imitate the preceding example except use the polynomial $x^3 + x^2 + 1$ to build the code instead of $x^3 + x + 1$.*

Suppose $g(x)h(x) = x^n - 1$, in $\mathbb{F}_2[x]$, with $g(x)$ of degree r, the generator polynomial of a code C and $h(x)$ of degree $k = n - r$. Then we get a parity check matrix for the code from the following matrix associated to the polynomial $h(x) = h_0 + h_1 x + \cdots + h_k x^k$:

$$\begin{pmatrix} h_k & 0 & \cdots & 0 \\ h_{k-1} & h_k & \cdots & 0 \\ \vdots & \vdots & \ddots & \vdots \\ h_0 & h_1 & \cdots & h_k \\ 0 & h_0 & \cdots & h_{k-1} \\ \vdots & \vdots & \ddots & \vdots \\ 0 & 0 & \cdots & h_0 \end{pmatrix}. \tag{8.12}$$

Exercise 8.2.8

(a) *Show that the matrix given in (8.12) is indeed a parity check matrix for our code with generator polynomial $g(x)$ as described above and generator matrix given in (8.11).*
(b) *Use (8.12) to find a parity check matrix for the Hamming $[7, 4, 3]$-code with generator polynomial $g(x) = x^3 + x + 1 \in \mathbb{F}_2[x]$ in the example above.*

There is a method for constructing codes that correct lots of errors called BCH codes. See Dornhoff and Hohn [25, p. 442] for the mathematical details. Here we only sketch a bit of the theory.

Suppose that $\gamma \in E$ and $E \supset F$ are finite fields. Recall that the minimal polynomial of γ over F is the polynomial $f \in F[x]$ of least degree such that $f(\gamma) = 0$.

Recall that we obtained the Hamming $[7, 4, 3]$-code by looking at the generator polynomial $g(x) = x^3 + x + 1$. This is the minimal polynomial of an element γ of $\mathbb{F}_8$ whose other roots are γ^2 and γ^4. So we could say that any polynomial $f(x)$ is in our code C iff $f(\gamma^j) = 0$, $j = 1, 2, 4$. For any polynomial whose roots include the roots of $g(x)$ must be divisible by $g(x)$.

Definition 8.2.5 *A* ***primitive nth root of 1*** *in a field K is a solution γ to $\gamma^n = 1$ such that $\gamma^m \neq 1$, for $1 \leq m < n$.*

Theorem 8.2.3 ***(Bose-Chaudhuri and Hoquenghem, 1960)*** *Suppose* $\gcd(n,q)=1$. *Let* γ *be a primitive nth root of* 1 *in an extension field of* $\mathbb{F}_q$. *Suppose the generator polynomial* $g(x)$ *of a cyclic code of length n over* $\mathbb{F}_q$ *has* $\gamma,\gamma^2,\ldots,\gamma^{d-1}$ *among its roots. Then the minimum distance of a nonzero code element from* 0 *is at least d.*

For a proof see Dornhoff and Hohn [25, pp. 442–443].

A **Reed–Solomon code** is a BCH code with $n=q-1$. It is also assumed that $\gamma\in\mathbb{F}_q$. These codes are used by the makers of CD players, NASA, and others. They can be used to correct amazing numbers of errors. If you suppose $q=2^8$ so that $n=255$, a 5-error-correcting code has $g(x)=(x-\gamma)(x-\gamma^2)\cdots(x-\gamma^{10})$ of degree 10, where γ is a primitive nth root of unity. Elements of $\mathbb{F}_{2^8}$ are eight-dimensional vectors over $\mathbb{F}_2$. This code can be used as a code of length $8*255=2040$ over $\mathbb{F}_2$, which can correct any "burst" of 33 consecutive errors. For any 33 consecutive errors over $\mathbb{F}_2$ will affect at most five of the elements of $\mathbb{F}_8$. See Dornhoff and Hohn [25, p. 444] for more of an explanation of this error-correction ability.

Feedback shift registers are of use in encoding and decoding cyclic codes. See Dornhoff and Hohn [25, pp. 449ff] and Pless [85].

Exercise 8.2.9 *Suppose that E is the splitting field of* $f(x)=x^n-1$ *over* $\mathbb{F}_q$. *Here n is a positive integer and q is a power of a prime. Suppose in addition that* $\gcd(n,q)=1$. *Show that* $f(x)$ *has n distinct roots in E.*

Exercise 8.2.10 *Show that* $\gcd(n,q)=1$, *iff there is a primitive nth root of* 1 *in the splitting field of* x^n-1 *over* $\mathbb{F}_q$.

Example: Codes from the Hadamard Matrix. The code used in the 1969 NASA Mariner 9 spacecraft which orbited Mars comes from the **Hadamard matrix** $H_{2^5}=\left((-1)^{u\cdot v}\right)_{u,v\in\mathbb{F}_2^5}$, with u,v ordered as for the corresponding numbers in binary and $u\cdot v=\sum_{i=1}^5 u_iv_i$. This matrix is pictured in Figure 8.10. ▲

The code is found by forming the new matrix $G=\Phi\begin{pmatrix}H_{2^5}\\-H_{2^5}\end{pmatrix}$, where Φ replaces 1s with 0s and -1s with 1s. The rows of G are the codewords of the $[32,6,16]$ **Reed–Muller code** used in the Mariner Mars probe.

Exercise 8.2.11 *How many errors can the* $[32,6,16]$ *Reed–Muller code correct?*

Exercise 8.2.12 *Consider the* $[4,3,2]$ *Hadamard matrix code with generator matrix* $G=\Phi\begin{pmatrix}H_4\\-H_4\end{pmatrix}$, *using the notation of the last example. Show that the dimension of the code is indeed* 3. *Then show that the minimum weight of vectors in the code is indeed* 2.

The general **Hadamard matrix** $H_{2^n}=\left((-1)^{u\cdot v}\right)_{u,v\in\mathbb{F}_2^n}$ has the inductive (or recursive) definition

$$H_{2^{n+1}}=\begin{pmatrix}H_{2^n} & H_{2^n}\\ H_{2^n} & -H_{2^n}\end{pmatrix},\quad\text{with } H_2=\begin{pmatrix}1 & 1\\ 1 & -1\end{pmatrix}. \tag{8.13}$$

Figure 8.10 The matrix H_{32} where the 1s and -1s have become red and purple

The matrix H_m is defined to be a matrix of 1s and -1s such that $H_m{}^{\mathrm{T}} H_m = mI$, where I is the identity matrix. For H_m to exist with $m > 2$, it is necessary that 4 divide m. The smallest values of m without a construction of H_m are $m = 428$ and 668, according to the chapter on combinatorial designs in the handbook edited by K. H. Rosen [92].

Why did Hadamard study these matrices? He wanted a matrix H_m with entries h_{ij} such that $|h_{ij}| \leq 1$ and $|\det(H_m)|$ is maximal (i.e., $m^{m/2}$). In Terras [116, p. 172], we note that the Hadamard matrix H_{2^n} is a matrix for the linear transformation given by the Fourier transform (or DFT) on the group $\mathbb{F}_2^n$.

H. B. Mann [73] gives more information on the code used in the Mariner Mars probe, as well as on the history of error-correcting codes. In this book one finds a limerick inspired by the coding theorist Jessie MacWilliams:

Delight in your algebra dressy
But take heed from a lady named Jessie
Who spoke to us here of her primitive fear
That good codes just might be messy.

W. W. Rouse Ball and H. S. M. Coxeter [95] give more recreational aspects of Hadamard matrices. See also F. Jessie MacWilliams and Neil J. A. Sloane [72].

Exercise 8.2.13 *Check that the definition (8.13) implies that $H_{2^n}{}^{\mathrm{T}} H_{2^n} = 2^n I$, where I is the identity matrix.*

Exercise 8.2.14 *Consider the* $[12, 6]$ *extended ternary (i.e., over* $\mathbb{F}_3$*) Golay code with generator matrix* $(I_6\ A)$*, where*

$$A = \begin{pmatrix} 0 & 1 & 1 & 1 & 1 & 1 \\ 1 & 0 & 1 & 2 & 2 & 1 \\ 1 & 1 & 0 & 1 & 2 & 2 \\ 1 & 2 & 1 & 0 & 1 & 2 \\ 1 & 2 & 2 & 1 & 0 & 1 \\ 1 & 1 & 2 & 2 & 1 & 0 \end{pmatrix}.$$

Show that the minimum Hamming weight of a codeword is 6.

Hint. *You may want to look at some of the references such as Pless [85] for results on weights of vectors in such self-dual codes.*

The code in the preceding exercise was found by M. J. Golay in 1949. The ternary $[11, 6]$ cyclic code can be found by factoring (over $\mathbb{F}_3[x]$)

$$x^{11} - 1 = (x-1)\left(x^5 - x^3 + x^2 - x - 1\right)\left(-x^5 - x^4 + x^3 - x^2 + 1\right).$$

This is also a **quadratic residue code**. To obtain a ternary quadratic residue code we proceed as follows. Let p be a prime such that 3 is a square mod p. Suppose ζ is a primitive pth root of unity in some field containing $\mathbb{F}_3$ as in Definition 8.2.5. Then let $\square$ denote the set of squares in $\mathbb{F}_p^*$ and let $\square'$ be the set of non-squares in $\mathbb{F}_p^*$. Define the polynomials

$$q(x) = \prod_{j \in \square} \left(x - \zeta^j\right) \quad \text{and} \quad n(x) = \prod_{j \in \square'} \left(x - \zeta^j\right).$$

One can show that the polynomials $q(x)$ and $n(x)$ have coefficients in $\mathbb{F}_3$ and that $x^p - 1 = (x-1)q(x)n(x)$.

We mentioned the Fourier transform on the group $\mathbb{F}_2^n$ in our discussion of Hadamard matrices. Let us say a bit more about this subject. Functions $f\colon \mathbb{F}_2^n \to \{0, 1\}$ are called Boolean functions or switching functions. The Fourier analysis of such functions has many applications in computer science as well as the theory of voting. A recent reference is the book of O'Donnell [83]. There one finds, for example, a use of Fourier analysis on the group $\mathbb{F}_2^n$ to prove Arrow's theorem. This theorem says that in an election with three candidates for office any voting rule – such as majority rule – can produce a paradoxical result unless one of the n voters is a dictator who decides the election. Fourier analysis in the case of the additive group $\mathbb{F}_2^n$ works much the same as it did for the group $\mathbb{Z}_n$ in Section 4.2. For those of us made miserable by the last US election, we recommend considering the mathematical theory of voting as a way to work through the grief. If nothing else, your mind will be diverted into the construction of a dictionary relating the notation for Fourier analysis on $\mathbb{F}_2^n$ in [83] and the notation in [116]. There is also a theory of influence that puts California as the most influential state which makes me wonder how realistic the mathematical theory of voting really is. Complicated voting rules such as that of the electoral college are certainly worth studying, in retrospect, as is the mathematics of gerrymandering.

Let $G = \mathbb{F}_2^n$ under addition. The dual group consists of characters $\chi_a(x) = (-1)^{{}^{\mathrm{T}}ax}$, where ${}^{\mathrm{T}}ax = \sum_{1 \le i \le n} a_i x_i$, if the column vectors $x = (x_i)$ and $a = (a_i)$ both come from $\mathbb{F}_2^n$. The characters are real valued in this case – unlike that of Section 4.2. The group operation for characters is pointwise multiplication. Analogously to equation (4.2), we define the **Fourier transform** of a function $f\colon \mathbb{F}_2^n \to \mathbb{R}$ to be $\widehat{f}\colon \mathbb{F}_2^n \to \mathbb{R}$, with

$$\widehat{f}(a) = \frac{1}{2^n} \sum_{x \in \mathbb{F}_2^n} f(x) \chi_a(x).$$

In the next exercise we ask you to prove the analog of Theorem 4.2.1.

Exercise 8.2.15

(a) *Define convolution of functions on $\mathbb{F}_2^n$ by an analogous equation to (4.1). Then show that $\widehat{f * g} = \widehat{f} \cdot \widehat{g}$.*

(b) *State and prove the inversion formula for the Fourier transform on $\mathbb{F}_2^n$.*

Define the Hamming shell to be

$$S_r = \{x \in \mathbb{F}_2^n \mid \|x\| = r\}.$$

Define

$$\delta_{S_r}(x) = \begin{cases} 1, & x \in S_r, \\ 0, & x \notin S_r. \end{cases}$$

The Fourier transform for $\mathbb{F}_2^n$ has a special name, the Krawtchouk polynomial, when applied to the function δ_{S_r}. It is given by

$$K_r(\|a\|; n) = 2^n \widehat{\delta_{S_r}}(b) = \sum_{\substack{x \in \mathbb{F}_2^n \\ \|x\| = r}} (-1)^{^T ax}.$$

I discuss its properties in Terras [116, p. 178].

Exercise 8.2.16 *Consider an election involving voting for two candidates. If there are n voters, you can view the voting rule as a function $f\colon \mathbb{F}_2^n \to \{0, 1\}$. The usual such rule is **majority rule** which works to produce a well-defined winner if n is odd. When n is even, there could be a tie. Define $\mathrm{maj}(x) = 1$ iff the Hamming weight of x satisfies $\|x\| \geq \lceil \frac{n}{2} \rceil$, and $f(x) = 0$, otherwise. Here $\lceil x \rceil$ is the **ceiling** of x – the smallest integer $\geq x$. The rule is well defined when n is even, but is perhaps not really a good voting rule in that case. Compute the Fourier transform $\widehat{\mathrm{maj}}$ if $n = 3$ or 5.*

Hint. $\mathrm{maj}_n = \sum_{r = \lceil \frac{n}{2} \rceil}^{n} \delta_{S_r}.$

8.3 Finite Upper Half Planes and Ramanujan Graphs

In this section we construct a finite analog of the real non-Euclidean Poincaré upper half plane H consisting of points $z = x + iy$, with $x, y \in \mathbb{R}$ and $y > 0$. The Poincaré upper half plane has a distance element ds defined by:

$$ds^2 = \frac{dx^2 + dy^2}{y^2}.$$

One can show that ds is invariant under the group of **fractional linear transformations**

$$z \to \frac{az+b}{cz+d}, \quad \text{with } a, b, c, d \in \mathbb{R} \text{ and } ad-bc>0.$$

Moreover, methods from Section 4.3 can be used to show that the geodesics or distance-minimizing curves for this Poincaré distance are half lines and half circles perpendicular to the real axis. Viewing these as the straight lines of our geometry makes Euclid's fifth postulate false. Thus we get Henri Poincaré's model of non-Euclidean geometry. See Terras [118] for more information. Number theorists are enamoured of functions on H which have invariance properties under the action of the modular group $SL(2,\mathbb{Z})$ of fractional linear transformations with integer a, b, c, d and $ad-bc=1$. We gave the density plot of the absolute value of such a function in Figure 2.7 in Section 2.1. Such functions are called **modular forms** and they play a key role in modern number theory.

Here we want to consider a finite analog of the Poincaré upper half plane. Suppose that $\mathbb{F}_q$ is a finite field of *odd* characteristic p. This implies that $q=p^r$. Suppose δ is a fixed non-square in $\mathbb{F}_q$. The **finite upper half plane** over $\mathbb{F}_q$ is defined to be

$$H_q=\left\{z=x+y\sqrt{\delta}\ \middle|\ x,y\in\mathbb{F}_q, y\neq 0\right\}.$$

We will write for $z=x+y\sqrt{\delta}\in x,y\in\mathbb{F}_q\left[\sqrt{\delta}\right]$, with $x,y\in\mathbb{F}_q$, the **real part** of $z=\mathrm{Re}(z)=x$, the **imaginary part** of $z=\mathrm{Im}(z)=y$. Our finite analog of complex conjugate is given by defining the **conjugate** of z to be $\overline{z}=x-y\sqrt{\delta}=z^q$ and the **norm** of z to be $Nz=z\overline{z}$.

Perhaps you will object to the use of the word "upper." Since we have no good notion of $>$ for finite fields, we use the word "upper" thinking, for example, if $q=p$, the y-coordinate of a point is in the set $\{1,2,\ldots,p-1\}$ of "positive" numbers. That is perhaps a cheat and we should really view H_q as a union of an upper and a lower half plane, with the y-coordinate of a point in the set $\{-\frac{p-1}{2},\ldots,-1,1,\ldots,\frac{p-1}{2}\}$. You be the judge.

The **general linear group** $GL(2,\mathbb{F}_q)$ of matrices $g=\begin{pmatrix} a & b \\ c & d\end{pmatrix}$ with $ad-bc\neq 0$ acts on $z\in H_q$ by the **fractional linear transformation**:

$$gz=\frac{az+b}{cz+d}. \tag{8.14}$$

Exercise 8.3.1 *Show that if we consider the action of $g\in GL(2,\mathbb{F}_q)$ on $z\in H_q$ defined by equation (8.14) then* $\mathrm{Im}(z)\neq 0$ *implies* $\mathrm{Im}(gz)\neq 0$. *Then show that this does indeed give a group action of $GL(2,\mathbb{F}_q)$ on H_q.*

Exercise 8.3.2

(a) Show that, with the group action in equation (8.14),

$$K=\left\{g\in GL(2,\mathbb{F}_q)\mid g\sqrt{\delta}=\sqrt{\delta}\right\}=\left\{\begin{pmatrix} a & b\delta \\ b & a\end{pmatrix}\ \middle|\ a,b\in\mathbb{F}_q \text{ with } a^2-\delta b^2\neq 0\right\}.$$

(b) Show that K is a subgroup of $G=GL(2,\mathbb{F}_q)$ which is isomorphic to the multiplicative group $\mathbb{F}_q\left(\sqrt{\delta}\right)^*$.

(c) Show that we can identify G/K with H_q.

Hint on (b). *The isomorphism is given by* $\begin{pmatrix} a & b\delta \\ b & a \end{pmatrix} \longmapsto a + b\sqrt{\delta}$.

The subgroup K of $GL(2, \mathbb{F}_q)$ considered in the preceding exercise is analogous to the **orthogonal subgroup** of the general linear group $GL(2, \mathbb{R})$, namely, $O(2, \mathbb{R}) = \{g \in GL(2, \mathbb{R}) \mid {}^T gg = I\}$ consisting of rotation matrices.

The **finite Poincaré distance** on H_q is defined to be

$$d(z, w) = \frac{N(z - w)}{\operatorname{Im} z \operatorname{Im} w}.$$

The distance has values in $\mathbb{F}_q$. Thus we are not talking about a metric here. There is no possibility of a triangle inequality such as we had with the Hamming metric in Section 8.2. Perhaps we should call this distance a pseudo-distance instead.

Exercise 8.3.3 *Let* $z = x + y\sqrt{\delta}$ *and* $w = u + v\sqrt{\delta}$, *with* $x, y, u, v \in \mathbb{F}_q$ *and* $yv \neq 0$. *Show that*

$$d(z, w) = \frac{(x - u)^2 - \delta\,(y - v)^2}{yv}.$$

Exercise 8.3.4 *Show that* $d(gz, gw) = d(z, w)$ *for all* $g \in GL(2, \mathbb{F}_q)$ *and all* $z, w \in H_q$.

We can draw a contour map of the distance function by making a grid representing the finite upper half plane and coloring the point $x + y\sqrt{\delta}$ according to the value of

$$d(z, \sqrt{\delta}) = \frac{x^2 - \delta(y - 1)^2}{y}.$$

When $q - 163$ we get Figure 8.11. This figure should be compared with the analogous figure obtained using an analog of the Euclidean distance on a finite plane given in Figure 5.1. I see monsters in Figure 8.11. My website has a movie of such figures for various values of p.

Exercise 8.3.5 *Make a figure analogous to Figure 8.11 using the "Euclidean" distance* $d((x, y), (0, 0)) = x^2 + y^2$, *for* $(x, y) \in \mathbb{F}_{163} \times \mathbb{F}_{163}$.

Next we want to define some graphs attached to this stuff.

Definition 8.3.1 *Let* $a \in \mathbb{F}_q$ *and define the* ***finite upper half plane graph*** $X_q(\delta, a)$ *to have vertices the elements of* H_q *and then draw an edge between two vertices* z, w *iff* $d(z, w) = a$.

Example: The Octahedron. Let $q = 3$, $\delta = -1 \equiv 2 \pmod 3$, and $a = 1$. We will write $i = \sqrt{-1}$. To draw the graph $X_3(-1, 1)$ we need to find the points adjacent to i for example. These are the points $z = x + iy$ such that

$$d(z, i) = \frac{N(z - i)}{y} = 1.$$

Figure 8.11 Color at point $z = x + y\sqrt{\delta}$ in H_{163} is found by computing the Poincaré distance $d(z, \sqrt{\delta})$

This is equivalent to solving $x^2 + (y-1)^2 = y$, for $x, y \in \mathbb{Z}_3$ with $y \neq 0$. Solutions are the four points $1+i, 1-i, -1+i, -1-i$. To find the points adjacent to any point $a + bi \in H_3$, just apply the matrix $\begin{pmatrix} b & a \\ 0 & 1 \end{pmatrix}$ to the points $\pm 1 \pm i$ that we just found. The graph $X_3(-1, 1)$ is drawn on the left in Figure 8.12. It is an octahedron. ▲

The adjacency matrix A of the octahedron graph is the 6×6 matrix below of 0s and 1s where the i, j entry is 1 iff vertex i is adjacent (i.e., joined by an edge) to vertex j.

$$A = \begin{pmatrix} 0 & 1 & 1 & 0 & 1 & 1 \\ 1 & 0 & 1 & 1 & 0 & 1 \\ 1 & 1 & 0 & 1 & 1 & 0 \\ 0 & 1 & 1 & 0 & 1 & 1 \\ 1 & 0 & 1 & 1 & 0 & 1 \\ 1 & 1 & 0 & 1 & 1 & 0 \end{pmatrix}.$$

The eigenvalues $\lambda \in \mathbb{C}$ of A are the solutions of $\det(A - \lambda I) = 0$. The set of eigenvalues is $\operatorname{spec}(A) = \{4, -2, -2, 0, 0, 0\}$. Note that the second largest eigenvalue in absolute value,

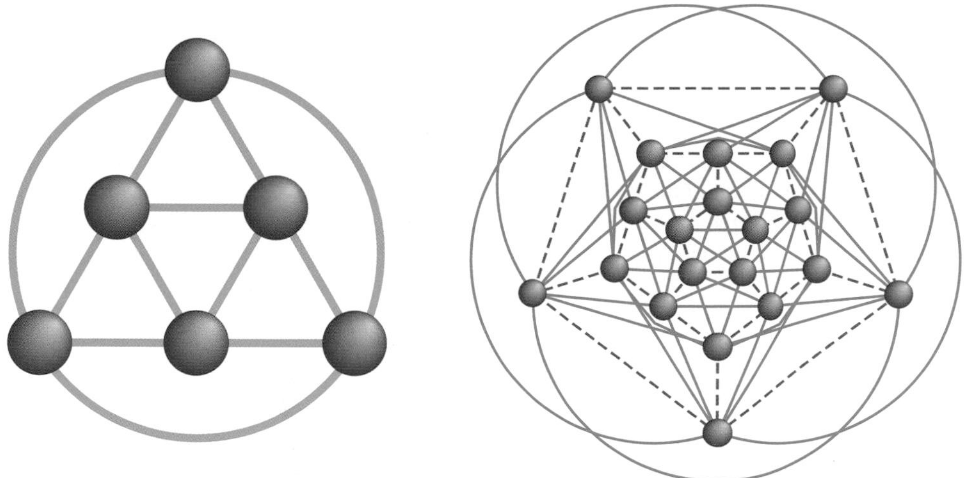

Figure 8.12 The graph on the left is $X_3(-1, 1)$, an octahedron, and that on the right is $X_5(2, 1)$ with the edges in green. The pink dashed lines on the right are the dodecahedron

which is $|\lambda| = 2$, satisfies $|\lambda| \leq 2\sqrt{3} \cong 3.46$. This means that the graph $X_3(-1, 1)$ is what is called a **Ramanujan graph.**

Figure 8.12 also shows $X_5(2, 1)$ on the right. The solid lines are the edges of the graph. The dotted lines are the edges of a dodecahedron. We can view the graph $X_5(2, 1)$ as that which you get by putting a five-pointed star on each face of a dodecahedron.

A graph X is called **k-regular** if there are k edges coming out of every vertex. We say that a k-regular graph is a **Ramanujan graph** if for all eigenvalues λ of the adjacency matrix such that $|\lambda| \neq k$, we have $|\lambda| \leq 2\sqrt{k-1}$. This definition was made by Lubotzky, Phillips and Sarnak in a paper from 1988. It turns out that such graphs provide good communication networks as the random walk on them converges rapidly to uniform. We will say more about that sort of thing in the next section. In the 1980s Margulis and independently Lubotzky, Phillips and Sarnak found infinite families of Ramanujan graphs of fixed degree. We will say a bit more about these examples at the end of this section See also Guiliana Davidoff, Peter Sarnak, and Alain Valette [23] or Terras [116]. Denis Charles, Eyal Goren, and Kristin Lauter [13] give applications of Ramanujan graphs to cryptography.

Of course, one really wants infinite families of Ramanujan graphs of fixed small degree. The finite upper half plane graphs $X_q(\delta, a)$ have degree $q + 1$ provided that $a \neq 0$ or 4δ. These finite upper half plane graphs were proved to be Ramanujan by N. Katz using work of Soto-Andrade and estimates of exponential sums. See Terras [116] for more of the history. Ramanujan graphs are also good expander graphs, meaning that if they form a gossip network, the gossip gets out fast. Sarnak [99] states: "it is in applications in theoretical computer science where expanders have had their major impact. Among their applications are the design of explicit superefficient communication networks, constructions of error-correcting codes with very efficient encoding and decoding algorithms, derandomization of random algorithms, and analysis of algorithms in computational group theory ..." The subject of expander graphs of this sort has an accessible introduction in the book by Mike Krebs and Tony Shaheen [61].

Now we can explain Figure 5.2 in Section 5.1. The picture is that of points (x, y), with $x, y \in \mathbb{F}_{121}$ and $y \neq 0$. Take $\delta \in \mathbb{F}_{121}$ to be a non-square. View a point (x, y) as $z = x + y\sqrt{\delta} \in$

$H_{121} \subset \mathbb{F}_{121}[\sqrt{\delta}]$. Let 2×2 matrices $g = \begin{pmatrix} a & b \\ c & d \end{pmatrix} \in GL(2, \mathbb{F}_{11})$ act on $z = x + y\sqrt{\delta}$ by fractional linear transformation $gz = \frac{az+b}{cz+d}$. Color two points z and w the same color if there is a matrix $g \in GL(2, \mathbb{F}_{11})$ such that $w = gz$. This gives the picture in Figure 5.2. Figure 8.13 is another version of that figure. This figure is reminiscent of tessellations of the real Poincaré upper half plane H obtained by translating a fundamental domain $D \cong \Gamma \backslash H$ around using elements of the discrete group Γ which is usually some group like a subgroup of the modular group $SL(2, \mathbb{Z})$. There are some beautiful tessellations on Helena Verrill's website: www.math.lsu/~verrill/. Such tessellations of the unit disc inspired many pictures by the artist M. C. Escher. The tessellation of the Poincaré upper half plane by $SL(2, \mathbb{Z})$ is visible in Figure 2.7.

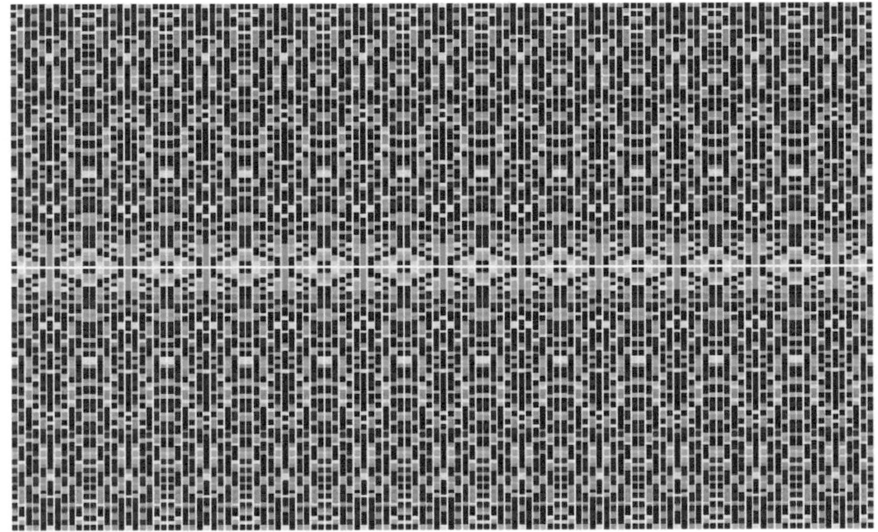

Figure 8.13 Another version of Figure 5.2

Exercise 8.3.6 *Apply Burnside's lemma from Section 3.7 to $GL(2, \mathbb{F}_p)$ acting on H_{p^2} to find out how many colors need to be used in creating the analog of Figure 5.2 for an arbitrary odd prime p.*

As we have said, Lubotzky, Phillips, and Sarnak along with Margulis constructed the first infinite families of Ramanujan graphs. We consider these examples briefly. The construction involves the **projective general linear group** over $\mathbb{Z}_q$, for prime q, defined by $PGL(2, \mathbb{Z}_q) = GL(2, \mathbb{Z}_q)/Z$, where Z denotes the center of $GL(2, \mathbb{Z}_q)$. Here, as usual, $GL(2, \mathbb{Z}_q)$ consists of all 2×2 matrices with entries in $\mathbb{Z}_q$ and nonzero determinant.

Exercise 8.3.7 *Describe the elements of the center of $GL(2, \mathbb{Z}_q)$.*

The Lubotzky, Phillips, and Sarnak construction requires two distinct odd primes p and q – both congruent to 1 modulo 4. Then take i to be any integer such that $i^2 \equiv -1 \pmod{q}$.

It can be shown (using an old formula of Jacobi proved using theta functions) that this implies that there are $p+1$ vectors

$$(a,b,c,d)\in\mathbb{Z}^4 \text{ such that } p=a^2+b^2+c^2+d^2 \tag{8.15}$$

where a is odd and positive, while b,c,d are all even.

If, in addition, p is not a square mod q, then define the Cayley graph to be $X^{p,q}=X(PGL(2,\mathbb{Z}_q),S)$, where S is the subgroup consisting of the matrices

$$\begin{pmatrix} a+ib & c+id \\ -c+id & a-ib \end{pmatrix},$$

with (a,b,c,d) as in (8.15). When $q>2\sqrt{p}$, Davidoff, Sarnak, and Valette [23] show that S does have degree $p+1$. When p is a square mod q, then, in the definition of $X^{p,q}$, the group $PGL(2,\mathbb{Z}_q)$ is replaced with $PSL(2,\mathbb{Z}_q)=SL(2,\mathbb{Z}_q)/Z$, where Z is the center of $SL(2,\mathbb{Z}_q)$. Here $SL(2,\mathbb{Z}_q)$ is the determinant 1 matrices in $GL(2,\mathbb{Z}_q)$. Davidoff, Sarnak and Valette [23] show that if $p\geq 5$, $q>p^8$, then the graph $X^{p,q}$ is connected. They also give proofs of lower bounds on some of the interesting graph-theoretic constants of these graphs but do not manage to give a proof of Ramanujanicity. A proof that these graphs are indeed Ramanujan can be found in Sarnak [98]. Quaternion algebras are involved in the creation of these graphs as well as the proofs. One also needs the Ramanujan conjecture on the size of Fourier coefficients of modular forms – at least a special case that was known before Deligne's proof of the conjecture. See Sarnak [98] for more information.

Even the smallest examples of the Lubotzky, Phillips, Sarnak graphs $X^{p,q}$ are quite large. Thus it is of interest that Sarnak's student Patrick Chiu managed to do the case of 3-regular graphs (see [14]). In the case that $q=3$, he gets the smallest such Ramanujan graph $X^{2,3}$ which can be constructed in a similar way to the Lubotzky, Phillips, and Sarnak graphs above. The generating set S is

$$S=\left\{\begin{pmatrix} 1 & 0 \\ 0 & 2 \end{pmatrix}, \begin{pmatrix} 1 & 2 \\ 2 & 0 \end{pmatrix}, \begin{pmatrix} 0 & 1 \\ 1 & 1 \end{pmatrix}\right\}.$$

The Cayley graph is then $X^{2,3}=X(G,S)$ with $G=PGL(2,\mathbb{Z}_3)$.

Exercise 8.3.8 *Draw the Cayley graph $X^{2,3}$ and compute the spectrum of its adjacency matrix. Prove that it is a Ramanujan graph.*

Exercise 8.3.9 *Show that in the Lubotzky, Phillips, and Sarnak construction above, when p is a square* mod q, *the matrices*

$$\begin{pmatrix} a+ib & c+id \\ -c+id & a-ib \end{pmatrix}$$

do not generate $PGL(2,\mathbb{Z}_q)$, since they are in a subgroup of index 2. This means that the Cayley graph $X(PGL(2,\mathbb{Z}_q),S)$ is not connected in this case.

Exercise 8.3.10 *Show that in the Lubotzky, Phillips, and Sarnak construction above, when p is not a square* mod q*, the Cayley graph $X^{p,q} = X(PGL(2,\mathbb{Z}_q),S)$ has $q(q^2-1)$ vertices, while the graph $X^{p,q}$ has half that many vertices when p is a square* mod q.

Exercise 8.3.11 *Consider the case $p=5$, $q=13$ of the Lubotzky, Phillips, and Sarnak construction above. What is the degree of this graph $X^{5,13}$? How many vertices does $X^{5,13}$ have? Is the graph connected? If you have a computer available, compute the spectrum of the adjacency matrix of $X^{5,13}$ and check whether the graph is indeed a Ramanujan graph. If possible, draw the graph.*

Before leaving the subject of Ramanujan graphs, it is perhaps time to say a bit about Srinivasa Ramanujan. He lived from 1887 to 1920 and is one of the few mathematicians who has had a movie devoted to his life. That movie was based on the book by Robert Kanigel [52]. During his short life Ramanujan produced several large notebooks of mathematical formulas, without proofs – perhaps because he could not afford the extra paper – or maybe because it was enough that the formulas had been revealed to him by his goddess. Most of his life was spent in India but for five years starting in 1914 he was in England, working with the analytic number theorist G. H. Hardy. The sort of number theory done by Ramanujan was rather different from that of Hardy and the Ramanujan conjecture on the size of Fourier coefficients of modular forms seems to have been considered a "backwater" by Hardy, though (once generalized beyond all bounds) it has been a central topic in modern number theory. See Terras [118] for more information on this subject.

8.4 Eigenvalues, Random Walks on Graphs, and Google

Notation. All the vectors in this section will be **column vectors** in $\mathbb{C}^n$. Thus our matrices $A \in \mathbb{C}^{n\times n}$ will act on the left – taking vectors $v \in \mathbb{C}^n$ to the vector Av, as in formula (1.3). For a matrix $M \in \mathbb{C}^{n\times n}$, the transpose of M is denoted ${}^{\mathrm{T}}M$.

We have considered the spectrum of a matrix in Section 4.1. First we review the definitions. Given an $n \times n$ matrix A whose entries are complex numbers, we say that $\lambda \in \mathbb{C}$ is an **eigenvalue** of A iff $\det(A-\lambda I)=0$, where I is the $n \times n$ identity matrix. This is the same thing as saying that the matrix $A-\lambda I$ is singular; or that $Ax = \lambda x$ for some nonzero column vector $x \in \mathbb{C}^n$. Then we say x is an **eigenvector** of A corresponding to the eigenvalue λ. The set of all the eigenvalues of the matrix A is called the **spectrum** of A. We denote it spec (A). The name eigenvalue comes from D. Hilbert in 1904. Many other words have been used. P. Halmos (in [39, p. 102]) said: "Almost every combination of the adjectives proper, latent, characteristic, eigen, and secular, with the nouns root, number, and value has been used in the literature ..." Modern computer software such as Matlab, Mathematica, Scientific Workplace will find approximations to eigenvalues of large matrices.

Exercise 8.4.1

(a) Find the eigenvalues of the following matrices:

$$\begin{pmatrix} 1 & 1 \\ 0 & 1 \end{pmatrix}, \begin{pmatrix} 2 & 0 \\ 0 & 1 \end{pmatrix}, \begin{pmatrix} 0 & 1 \\ 1 & 0 \end{pmatrix}.$$

Do this exercise by hand – no computers allowed.

(b) Show that, for any square complex matrix A, $\mathrm{spec}(A) = \mathrm{spec}({}^{\mathrm{T}}A)$.

If the following exercises seem too terrible, you can find them in most linear algebra books.

Exercise 8.4.2

(a) Show that for any matrix $A \in \mathbb{C}^{n\times n}$ there is a unitary matrix U (meaning that ${}^{\mathrm{T}}\overline{U}\,U = I$) and an upper triangular matrix T, with $U, T \in \mathbb{C}^{n\times n}$, such that $A = {}^{\mathrm{T}}\overline{U}\,TU$. This is called the Schur decomposition of A. Since ${}^{\mathrm{T}}\overline{U} = U^{-1}$, this says that the matrix A is similar to T, that is, conjugate to T in the general linear group $GL(n, \mathbb{C})$.

(b) Then show that if $A = {}^{\mathrm{T}}\overline{U}\,TU$ as in part (a), the diagonal entries of T are the eigenvalues of A.

Hint on (a). *We know that* $\det(A - \lambda I) = 0$ *has a root λ_1 by the fundamental theorem of algebra. Therefore there is a corresponding eigenvector $v_1 \neq 0$ such that $Av_1 = \lambda_1 v_1$. Upon multiplying v_1 by a scalar, we may assume that $\|v_1\| = 1$. Complete v_1 to an orthonormal basis $\{v_1, v_2, \ldots, v_n\}$ of $\mathbb{C}^n$ using the Grams–Schmidt process that is to be found in Strang [115], for example. Then $U_1 = (v_1 v_2 \cdots v_n)$ is a unitary matrix. And $U_1^{-1} A U_1 = \begin{pmatrix} \lambda_1 & * \\ 0 & A_2 \end{pmatrix}$, where $A_2 \in \mathbb{C}^{(n-1)\times(n-1)}$. Use induction on n to complete the proof.*

Exercise 8.4.3

(a) Suppose that the matrix $A \in \mathbb{C}^{n\times n}$ is **Hermitian** *meaning that ${}^{\mathrm{T}}\overline{A} = A$. Show that then the upper triangular matrix T in the Schur decomposition of A from the preceding exercise can be taken to be diagonal. This is the* **spectral theorem**.

(b) Show that the eigenvalues of a Hermitian matrix are real numbers.

Exercise 8.4.4 *Suppose that $A \in \mathbb{C}^{n\times n}$ has n pairwise distinct eigenvalues $\lambda_1, \ldots, \lambda_n \in \mathbb{C}$. Then the corresponding eigenvectors are linearly independent over $\mathbb{C}$. This implies that there is an invertible matrix $V \in \mathbb{C}^{n\times n}$ such that $A = V^{-1}DV$, where D is diagonal with ith diagonal entry λ_i.*

There are many applications of these concepts to engineering, physics, chemistry, statistics, economics, music – even the internet. Eigenvalues associated to structures can be used to analyze their stability under some kind of vibration such as that caused by an earthquake. The word "spectroscopy" means the use of spectral lines to analyze chemicals – as discussed earlier in Section 4.2. We will investigate one such application in this section – that of the Google search engine. References for this section include: Google's website, G. Strang [115], C. D. Meyer [77], R. A. Horn and C. R. Johnson [45], Amy N. Langville and Carl D. Meyer [66], D. Cvetković, M. Doob, and H. Sachs [22], and A. Terras [116].

This section concerns real and complex linear algebra, the sort you learn as a beginning undergrad, for the most part, except for the Perron–Frobenius theorem. We will not be thinking about matrices with elements in finite fields in this section. Usually our matrices will have elements that are non-negative real numbers. That happens because our matrices will be Markov matrices from elementary probability theory. Markov invented this concept in 1907. Markov chains are random processes that retain no memory of their past states. An example is a random walk on the pentagon graph in Figure 8.14 below. References for the subject are J. C. Kemeny and J. L. Snell [53] and J. R. Norris [82].

A **Markov matrix** $M \in \mathbb{R}^{n \times n}$ means that the entries are in the interval $[0, 1]$ and the columns sum to 1. From such a matrix you get a Markov chain of probability vectors v, where **probability vector** means that the entries are in $[0, 1]$ and sum to 1. If we are given a **probability vector** $v_0 \in [0, 1]^n$ then we get a **Markov chain** corresponding to a Markov matrix $M \in \mathbb{R}^{n \times n}$ defined inductively by:

$$v_0, v_1 = Mv_0, v_2 = Mv_1, \ldots, v_{k+1} = Mv_k, \ldots. \tag{8.16}$$

All the vectors v_n are probability vectors. It follows that $v_j = M^j v_0$. At time n, the vector v_n has jth component which should be interpreted as the probability that the system is in its jth state at time n. In the example which follows – a random walk on a pentagon graph – the jth component is the probability that the random walker is at vertex j of the graph. In the case of the Google Markov matrix, the jth component is the probability that a web-surfer is at the jth website.

Exercise 8.4.5 *Show that if M is a Markov matrix and v is a probability vector, then Mv is also a probability vector.*

Since, in general, a Markov matrix need not be symmetric, its eigenvalues need not be real numbers. That makes the analysis of the behavior of the Markov chain a little more delicate. The eigenvalues will be random complex numbers. Random matrix theory makes predictions about their distribution in the complex plane. See [119, Section 1.3.5] for a brief discussion of random matrix theory for Hermitian matrices and [117, Chapter 26] for a discussion of some non-Hermitian examples coming from graph theory. The spectral theorem of the exercise above is not sufficient for our understanding of non-symmetric Markov matrices. One needs the Perron theorem (see Theorem 8.4.1 below) or more generally the Perron–Frobenius theorem.

Example: A Random Walk on a Pentagon. Consider the pentagon graph. This is the Cayley graph $X(\mathbb{Z}_5, \{\pm 1 \pmod 5\})$. It is undirected. The associated Markov matrix for the random walk in which a creature moves from vertex x to vertex $x+1 \pmod 5$ or $x-1 \pmod 5$ with equal probability is

$$M = \begin{pmatrix} 0 & 0.5 & 0 & 0 & 0.5 \\ 0.5 & 0 & 0.5 & 0 & 0 \\ 0 & 0.5 & 0 & 0.5 & 0 \\ 0 & 0 & 0.5 & 0 & 0.5 \\ 0.5 & 0 & 0 & 0.5 & 0 \end{pmatrix}. \tag{8.17}$$

See Figure 8.14. Note that $M = {}^{T}M$: that is, M is real symmetric. Thus it has real eigenvalues which must include 1 since the vector ${}^{T}w = (1, 1, 1, 1, 1)$ satisfies $Mv = v$. Scientific Workplace tells me that the other eigenvalues are approximately: -0.8090, 0.3090, each with multiplicity 2. ▲

Exercise 8.4.6 *Use the theory of the Fourier transform on $\mathbb{Z}_5$ from Section 4.2 to compute all the eigenvalues of the matrix M in (8.17) in the same way that we computed the spectrum of benzene.*

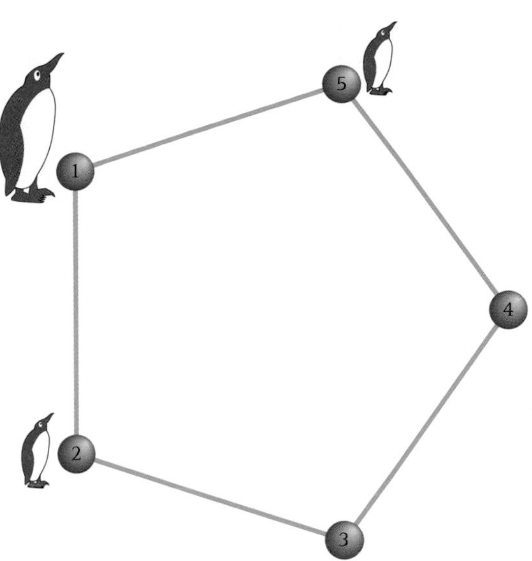

Figure 8.14 A random walk on a pentagon. At time $t = 0$, the big penguin is at vertex 1. At time $t = 1$ the penguin has probability $\frac{1}{2}$ of being at vertex 2 and probability $\frac{1}{2}$ of being at vertex 5. So the penguins at these vertices are half size

If we start our random walker at vertex 1, that corresponds to the probability vector ${}^{\mathrm{T}}v_0 = (1, 0, 0, 0, 0)$. Then at time $t = 1$ the creature is either at vertex 2 or 5 with equal probability. That corresponds to the probability vector $v_1 = Mv_0$, ${}^{\mathrm{T}}v_1 = (0, 0.5, 0, 0, 0.5)$. Then at time $t = 2$ we have the probability vector $v_2 = M^2 v_0$, ${}^{\mathrm{T}}v_2 = (0.5000, 0, 0.2500, 0.2500, 0)$. At time $t = 3$ we have $v_3 = M^3 v_0$, ${}^{\mathrm{T}}v_3 = (0, 0.3750,\ 0.1250,\ 0.1250,\ 0.3750)$. Continue in this manner up to time $t = 10$ and you find that $v_{10} = M^{10} v_0$, ${}^{\mathrm{T}}v_{10} \cong (0.2480, 0.1611, 0.2148, 0.2148, 0.1611)$. Already we see that we are approaching the eigenvector $\frac{1}{5}w = (0.2, 0.2, 0.2, 0.2, 0.2)$, which is the probability that the poor creature is totally lost, also known as the uniform probability distribution. The speed of convergence to the uniform probability vector is governed by the second largest eigenvalue which is 0.8090 in this case. You need a time t such that 0.8090^t is negligible (depending on what metric you use on the space of vectors in $\mathbb{R}^5$). Anyway, for our example, at time $t = 30$, the probability vector is $v_{30} = M^{30} v_0$, ${}^{\mathrm{T}}v_{30} \cong (0.2007, 0.1994, 0.2002, 0.2002, 0.1994)$ which is close enough to $u = (0.2, 0.2, 0.2, 0.2, 0.2)$ not to be able to notice the difference in a picture. Note that $0.8090^{30} \cong 0.00173$. The actual Euclidean distance between the two vectors is

$$\|v_{30} - u\|_2 = \sqrt{(0.0007) + 2(0.0006)^2 + 2(0.0002)^2} \cong 0.03.$$

If our graph were the web, we would be saying all websites have the same rank since all the coefficients of the steady-state vector u are equal.

Exercise 8.4.7

(a) Prove that 1 is an eigenvalue of any symmetric or non-symmetric Markov matrix M.
(b) Show that if λ is any eigenvalue of a symmetric or non-symmetric Markov matrix M, then $|\lambda| \leq 1$.

(c) *Suppose, in addition that if* $\lambda \neq 1$ *is an eigenvalue of the symmetric Markov matrix M, then* $|\lambda| < 1$. *Then there is is an orthonormal basis* $B = \{v_1, \ldots, v_n\}$ *of* $\mathbb{R}^n$ *consisting of eigenvectors of M such that* $Mv_1 = v_1$. *Prove that then*

$$\lim_{k \to \infty} M^k v = u, \quad \text{where } u = {}^{\mathrm{T}}\left(\frac{1}{n}, \ldots, \frac{1}{n}\right), \tag{8.18}$$

for any probability vector v. Note that the speed of convergence in this limit is measured by the second largest eigenvalue of M in absolute value. Here we measure the distance between two vectors v and u in $\mathbb{R}^n$ *by the usual distance* $\|v-u\|_2 = \sqrt{{}^{\mathrm{T}}(v-u)(v-u)}$.

Hint. *Using the orthonormal basis of eigenvectors* v_j *of M, write* $v = \sum_{j=1}^{n} \alpha_j v_j$, *for* $\alpha_j \in \mathbb{R}$. *Note that* $\alpha_j = \langle v, v_j \rangle$, *the inner product of v and* v_j. *Apply* M^k *to both sides and take the limit as* $t \to \infty$.

The following exercise gives examples of symmetric Markov matrices for which the hypothesis for the eigenvalues $\lambda \neq 1$ in part (c) of the preceding exercise is false. The conclusion is false as well. The Google matrix will have to be constructed to avoid such behavior.

Exercise 8.4.8

(a) *What happens if you replace the pentagon in the preceding example with a square?*

(b) *More generally consider the random walk on the Cayley graph* $X(\mathbb{Z}_n, \{\pm 1 \pmod{n}\})$ *in which the random walker at vertex x has equal probability of moving to vertex* $x+1 \pmod{n}$ *or to vertex* $x-1 \pmod{n}$. *Show that if you want the random walk to converge to the uniform probability vector* $u = {}^{\mathrm{T}}\left(\frac{1}{n}, \ldots, \frac{1}{n}\right)$, *you will need to take n odd or change the random walk to allow the walker to have three choices at each step, one being to stay at vertex x.*

Hint. *Recall from Section 4.2 on the finite Fourier transform that we can use the additive characters* $\chi_a(x) = e^{2\pi i a x/n}$ *to find the eigenvalues of the adjacency matrices of such Cayley graphs to be* $2\cos(2\pi a/n)$, *for* $a = 0, 1, \ldots, n-1$, *in the case that n is odd, when* $2\cos(2\pi a/n) \neq -1$.

Now we want to apply similar reasoning to a random walk on an extremely large directed graph. If you websurf to www.google.com and type in some words such as "eigenvalues," you will get a long list of websites, ordered according to importance. How does Google produce the ordering? Google had to take over 8.1 billion webpages and rank them – as of 2006, according to Amy N. Langville and Carl D. Meyer [66]. This was up from 2.7 billion in 2002. If you Google the question "how big is the Google matrix?" you get many answers. Most recently (September, 2016) I found that Google was indexing ≥ 30 trillion webpages and that the number increases by a factor of 30 every five years. Presumably this number does not include the "Deep Web." Do not expect this text to discuss that subject. There is a website – www.internetlinestats.com – which claims to give the current numbers. It was around 1 billion websites when I looked in May of 2016. Note that a website may contain many webpages. We should also note that the algorithm we discuss mostly comes from the ten year old book by Langville and Meyer [66] and thus may bear only slight resemblance to the algorithm used by Google at the moment. Much is top secret of course.

Google seemingly does its page rank computation once a month. Google is a name close to googol which means 10^{100}. Google was invented by two computer science doctoral students (Sergey Brin and Larry Page) at Stanford – in the mid-1990s. They only use ideas from a standard undergraduate linear algebra course, plus a bit of elementary probability. They view the web as a directed graph with a web surfer randomly hopping around. The main idea is that the more links a website has to it, the more important it must be (these links are called "inlinks"). Figure 8.15 shows a tiny web with only five websites. The webpages are the vertices of a directed graph. An arrow from vertex x to vertex y means that vertex x contains a link to vertex y. So in the example of Figure 8.15 you might think vertex 5 is the most important, since it has the most arrows going to it. In short, if x_k is the number of links to site k, then $x=(1,2,1,2,3)$.

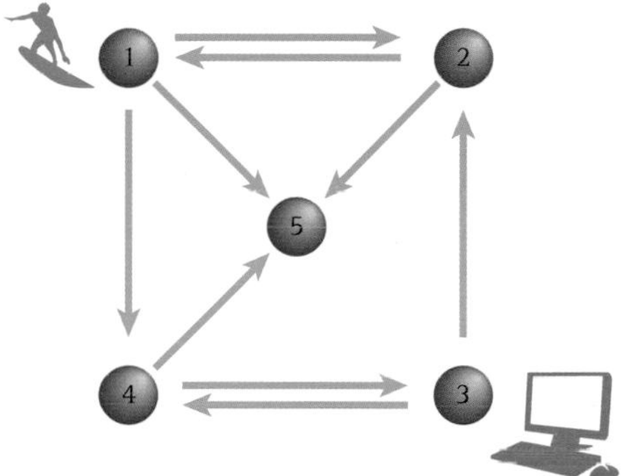

Figure 8.15 Surfing a very small web

On the other hand, node 5 is what is called a "dead end." It has no links to any other site. If we imagine a web surfer bouncing around from webpage to webpage, that surfer will land at node 5 and have nowhere to go. Many webpages are like this: for example, pdfs, gifs, jpgs.

We want to create a Markov matrix to give the transition matrix for a random web surfer. Let us first ignore the problem of node 5 and just look at the matrix H whose i,j entry is

$$h_{ij}=\begin{cases} \dfrac{1}{\#(\text{arrows going out from site } j)} & \text{if there is an arrow from site } j \text{ to site } i \\ 0 & \text{otherwise.} \end{cases} \tag{8.19}$$

For the webgraph of Figure 8.15, we get

$$H=\begin{pmatrix} 0 & \frac{1}{2} & 0 & 0 & 0 \\ \frac{1}{3} & 0 & \frac{1}{2} & 0 & 0 \\ 0 & 0 & 0 & \frac{1}{2} & 0 \\ \frac{1}{3} & 0 & \frac{1}{2} & 0 & 0 \\ \frac{1}{3} & \frac{1}{2} & 0 & \frac{1}{2} & 0 \end{pmatrix}. \tag{8.20}$$

This is almost a Markov matrix except that the entries of the last column do not sum to 1. For the Google method to work, it would be nice to have an actual positive Markov matrix M – meaning that every entry of M is positive. For we want it to satisfy the hypotheses of

Theorem 8.4.1 below which would imply that the largest eigenvalue is $\lambda_1 = 1$ and that λ_1 is an eigenvalue with a non-negative probability eigenvector corresponding to the steady state (i.e., limiting) behavior of the associated Markov chain. This theorem also importantly says that the other eigenvalues of M satisfy $|\lambda_j| < 1$.

You may want to use Matlab or Mathematica or Scientific Workplace (or whatever) to do the matrix computations in the following exercises.

Exercise 8.4.9

(a) Find the largest eigenvalue and a corresponding positive eigenvector for the matrix H in (8.20) above. Since H is not a Markov matrix, do not expect $\lambda_1 = 1$.
(b) What is the interpretation of this as far as ranking the websites? Which site is most important?
(c) Follow a web surfer who starts at site 1 through 20 iterations. That is, compute $H^k v$, $v = {}^{\mathrm{T}}(1,0,0,0,0)$, $k = 1,2,3,\ldots,10$.
(d) What seems to be happening to the vector $H^k v$ in the limit as $k \to \infty$?
(e) Can you use the eigenvalues of H to explain what is happening in part (d)?

Exercise 8.4.10 *To produce a Markov matrix, one Google idea is to replace the last column in the matrix H of (8.20) above by a column with $1/5$ in each row. Call the new matrix S. Now it comes closer to satisfying the hypothesis of Theorem 8.4.1 below.*

(a) Write down the matrix S. Does some S^k have all positive entries?
(b) Compute a probability eigenvector of S corresponding to the eigenvalue 1. Which site does this eigenvector say is the most important?
(c) Follow a web surfer who starts at site 1 through 20 iterations. What is the limit of $S^k v$, for $v = {}^{\mathrm{T}}(1,0,0,0,0)$, as $k \to \infty$? Compare your answers with those in part (b).

Google has one more trick. The matrix S obtained in Exercise 8.4.10 need not be such that all its entries are positive, which is the hypothesis of Perron's theorem (Theorem 8.4.1) below (although the weaker hypothesis that S^k has all positive entries for some k will also work – but is harder to check on a matrix which is 30 trillion $\times$ 30 trillion or even 8 billion $\times$ 8 billion). The new Google trick will also affect the second largest eigenvalue in absolute value. Suppose we have n nodes in our internet. In the formulas that follow the vectors are column vectors. Let b be the an n-vector whose jth component ($j = 1,\ldots,n$) is

$$b_j = \begin{cases} 1 & \text{if site } j \text{ has no arrows going out (i.e., it is a dead end),} \\ 0 & \text{otherwise.} \end{cases} \tag{8.21}$$

Define

$$S = H + \frac{1}{n} e\, {}^{\mathrm{T}}b, \tag{8.22}$$

where e is a n-vector of 1s and H is defined by (8.19). The **Google matrix** is given, for $0 < \alpha < 1$, by setting

$$G = \alpha S + (1-\alpha)\frac{1}{n} J, \tag{8.23}$$

where J is an $n \times n$ matrix all of whose entries are 1.

Exercise 8.4.11 *Prove that the Google matrix G is a Markov matrix.*

Exercise 8.4.12

(a) Write down G in formula (8.23) for the example of the small web in Figure 8.15 for $\alpha=0.9$ and then compute a probability eigenvector for G corresponding to the eigenvalue 1. Note that the entries of G are all positive. Which site does this eigenvector say is the most important?

(b) Follow a web surfer who starts at site 1 through 20 iterations. What is the limit of $G^k v$, for $v={}^{\mathrm{T}}(1,0,0,0,0)$, as $k\to\infty$? Compare your answers in Exercises 8.4.9 and 8.4.10.

Notes. In the formula for G, Google chooses $\alpha=0.85$ – or it did 10 years ago or so. It could be any number between 0 and 1. If $\alpha=0.85$, it means that 85% of the time the web surfer follows the hyperlink structure of the web and the other 15% of the time the web surfer jumps (teleports) to a random webpage. Since 1/(8.1 billion) is small, the alteration in the entries of the matrix H is not enormous. Of course the Google version of this matrix will be 8.1 billion $\times$ 8.1 billion – at least that was the approximate size in 2006. Now it may be 1000 times that large, or more. How does Google find the **dominant eigenvector** or **page rank** of G? It uses a very old method called the **power method** which works well for sparse matrices (meaning matrices most of whose entries are 0). That is, Google uses the fact that we noticed in the preceding problems. If you just keep multiplying some fixed probability vector v by G – in effect, computing $G^k v$ – this should converge to the probability eigenvector for the eigenvalue 1.

In the case that the Markov matrix M is symmetric, the power method basically takes advantage of equation (8.18) which says that for arbitrary probability vectors v, the vectors $M^n v$ approach $u={}^{\mathrm{T}}(\frac{1}{n},\cdots,\frac{1}{n})$, the steady-state of the Markov chain, as $n\to\infty$. Why? For a non-symmetric positive Markov matrix M, an analogous result comes from Theorem 8.4.1 below. But in the case of a non-symmetric Markov matrix the stationary state vector will not have all entries equal and that will of course give the website rankings. The power method was published by R. von Mises in 1929.

Exercise 8.4.13 *Suppose that the matrix $A\in\mathbb{R}^{n\times n}$ has n linearly independent eigenvectors $v_j\in\mathbb{R}^n$ with $Av_j=\lambda_j v_j$. Suppose that $|\lambda_1|>|\lambda_2|\geq\cdots\geq|\lambda_n|$. Show that, for any vector $w\in\mathbb{R}^n$,*

$$\lim_{k\to\infty}\frac{1}{|\lambda_1|^k}A^k w=\beta v_1,$$

for some scalar $\beta\in\mathbb{R}$.

Hint. *Write $w=\sum_{j=1}^{n}\gamma_j v_j$, for $\gamma_j\in\mathbb{R}$. Apply A^k to both sides and take the limit.*

However, the power method is "notoriously slow" for non-sparse matrices like G. So why does it work for Google? The first part of the answer has to do with the fact that is proved in the next exercise. Google only needs to compute the iterates of sparse matrices and not G itself.

The second part of the answer says that this method requires only about 50 iterations for the huge matrix Google is dealing with. Why should this be? This has to do with the size of the second largest eigenvalue λ_2 of G in absolute value. It turns out that for the Google matrix, $|\lambda_2| \cong \alpha$ with α from formula (8.23).

Choosing $\alpha = 0.85$, one finds that $0.85^{50} \cong 0.000296$. This means that Google can expect about 2–3 places of accuracy in the page rank vector after about 50 iterations of replacing v by Gv.

Exercise 8.4.14 *Here we are trying to understand part of the second-to-last paragraph. Consider a web with n sites. Let H be the matrix whose i,j entry is defined by formula (8.19). The Google matrix is as in formula (8.23) with b as in formula (8.21) and S as in formula (8.22). Again suppose that e is an n-column vector of 1s. Show that if v is any probability (column) vector, meaning its entries are ≥ 0 and sum to 1 (which implies ${}^{\mathrm{T}}e\, v = 1$), we have*

$$Gv = \alpha Hv + \frac{1}{n}(\alpha\, {}^{\mathrm{T}}b\, v + 1 - \alpha)e. \tag{8.24}$$

Note that H is sparse (with on average only about 10 nonzero elements in a column) and the scalar ${}^{\mathrm{T}}b\, v$ is easy to compute. It follows that iterating $v \to Gv$ will be quickly computed.

The next exercise is an attempt to explain "Google bombing." To do this, people are paid to set up link farms to fool Google into thinking a webpage is more important than it otherwise would appear to be. Google attempts to find such occurrences and then give such pages lower ranks. It was sued for doing so in 2002. The lawsuit was dismissed in 2003. See the book of Langville and Meyer [66] for more information. Now Google claims to be using many (200) factors to rank sites – not just the page rank.

Exercise 8.4.15 *In the example of the small web in Figure 8.15, suppose the site-1 people are angry to be rated below site 5. To increase the rating of site 1, they create a new site 6 with a link to site 1. Site 1 will also link to site 6. Does this help site 1's ranking?*

(a) Find the new H matrix from formula (8.19). Then form the S matrix in formula (8.22). Finally form the G matrix as in formula (8.24) with $\alpha = 0.85$.
(b) Find the probability eigenvector of the G matrix corresponding to the eigenvalue 1.
(c) Would it help if site 1 created another new site with a link to site 1? Would it help more if we added a new site with 10 links to site 1?
(d) What can Google do to minimize the effect of this sort of thing?

The Perron theorem was proved by Perron in 1907 and later generalized by Frobenius in 1912. The general version is called the Perron–Frobenius theorem. We give only a special case. To see the general version, look at Horn and Johnson [45].

Theorem 8.4.1 ***(Perron).*** *Suppose that the $n \times n$ Markov matrix M has all positive entries. Then the following hold.*

(1) 1 is an eigenvalue of M and the corresponding vector space of eigenvectors is one-dimensional.
(2) If the eigenvalues of M are listed as $\lambda_1 = 1, \lambda_2, \ldots, \lambda_n$, then $1 = |\lambda_1| > |\lambda_j|$, for all $j = 2, \ldots, n$.

(3) *There is an eigenvector v_1 corresponding to the eigenvalue* 1 *which is such that all its entries are* >0 *and they sum to* 1.
(4) *The vector v_1 is the steady state of the Markov chain with transition matrix M: that is,*

$$\lim_{r\to\infty} M^r x = v_1, \quad \textit{for any probability vector } x.$$

It is easy to see that if M is a Markov matrix, 1 is an eigenvalue of the transpose, ${}^{\mathrm{T}}M$, with eigenvector ${}^{\mathrm{T}}(1,1,\ldots,1)$. That is because the columns of M sum to 1. The eigenvalues of ${}^{\mathrm{T}}M$ are the same as the eigenvalues of M, since the determinant of a matrix is the same as the determinant of its transpose. For the rest of the proof, see Meyer [77, Chapter 8], or the last chapter of Horn and Johnson [45].

Exercise 8.4.16

(a) *Prove the Perron theorem in the case that the positive Markov matrix M is 2×2.*
(b) *Prove that if the spectrum of the Markov matrix S from Exercise 8.4.14 is $\{1,\lambda_2,\ldots,\lambda_n\}$, then the spectrum of the Google matrix $G=\alpha S+(1-\alpha)\frac{1}{n}J$ is $\{1,\alpha\lambda_2,\ldots,\alpha\lambda_n\}$.*

Hint on (b). *The proof can be found in Langville and Meyer [66, p. 46]. The trick is to make clever use of block matrix multiplication. We know $e={}^{\mathrm{T}}(1,\ldots,1)$ is an eigenvector of $R={}^{\mathrm{T}}S$ corresponding to the eigenvalue 1. Replace the Google matrix by its transpose ${}^{\mathrm{T}}G=\alpha R+(1-\alpha)\,u\cdot{}^{\mathrm{T}}v$, where $v=\frac{1}{n}e$. Now $\mathrm{spec}(R)=\mathrm{spec}(S)=\{1,\lambda_2,\ldots,\lambda_n\}$. If $Q=(e\ X)$ is a non-singular matrix with e as its first column and $Q^{-1}=\begin{pmatrix}{}^{\mathrm{T}}y\\ {}^{\mathrm{T}}Y\end{pmatrix}$, we have*

$$Q^{-1}RQ=\begin{pmatrix}1 & {}^{\mathrm{T}}yRX\\ 0 & {}^{\mathrm{T}}YRX\end{pmatrix} \quad \text{and} \quad Q^{-1}\,{}^{\mathrm{T}}GS=\begin{pmatrix}1 & *\\ 0 & \alpha\,{}^{\mathrm{T}}YRX\end{pmatrix}.$$

It follows from this that $\mathrm{spec}({}^{\mathrm{T}}YRX)=\{\lambda_2,\ldots,\lambda_n\}$.

Exercise 8.4.17

(a) *If a real matrix A has non-negative entries, write $A\geq 0$. If A has positive entries write $A>0$. Prove that if $A>0$ and $x\geq 0, x\neq 0$, then $Ax>0$.*
(b) *If $A-B\geq 0$, write $A\geq B$. Show that if $N\geq 0$ and $u\geq v\geq 0$, then $Nu\geq Nv\geq 0$.*
(c) *Is $\leq$ an equivalence relation on $n\times n$ real matrices? Is $\leq$ a partial order on $n\times n$ real matrices?*

You might still ask how Google finds the webpages with the words you typed. Google answers on its website that it has a large number of computers to "crawl" the web and "fetch" the pages and then form a humongous index of all the words it sees. So when we type in "eigenvalue" Google's computers search their index for that word and the page rank of the websites containing that word among "200 factors." That is what I found in 2011. Maybe things have changed a bit when you are reading this.

We have one last question to ask concerning Google. Are they really living up to their motto – Don't be evil? Is making money from ads that may contain lies possibly evil?

8.5 Elliptic Curve Cryptography

Before beginning our discussion of elliptic curves and the cryptography they enable, we list a few references: Ramanujachary Kumanduri and Cristina Romero [62, Chapter 19], Neal Koblitz [55], Joseph H. Silverman [105], [106], and Joseph H. Silverman and John Tate [107]. Google.com gave us many hits when we typed in "elliptic curve cryptography."

What is an elliptic curve? First it is *not* an ellipse. The connection with ellipses responsible for the name is in the computation of the arclength of an ellipse.

Let K be any field, for example $K = \mathbb{R}$, $\mathbb{C}$, or $\mathbb{Q}$ or $\mathbb{Z}/p\mathbb{Z}$, p prime. We will mostly be interested in finite fields here.

Definition 8.5.1 *Assume $a, b, c \in K$. An* **elliptic curve** *$E = E(K)$ is the set of points (x, y) with x, y in K such that*

$$y^2 = x^3 + ax^2 + bx + c.$$

We omit some technical conditions, which we will soon be forced to consider. You can also replace the y^2 on the left with some other quadratic polynomial in y.

The real points on $E(\mathbb{R})$ are of interest. They will help us to visualize what we are doing over finite fields. So we draw some pictures of elliptic curves over $\mathbb{R}$ in the plane. Figure 8.16 shows the real points (x, y) on the elliptic curve $y^2 = x^3 + x^2$. Figure 8.17 shows the real points (x, y) on the elliptic curve $y^2 = x^3 - x + 1$. Figure 8.18 plots the real points (x, y) on the elliptic curve $y^2 = x^3 - x$.

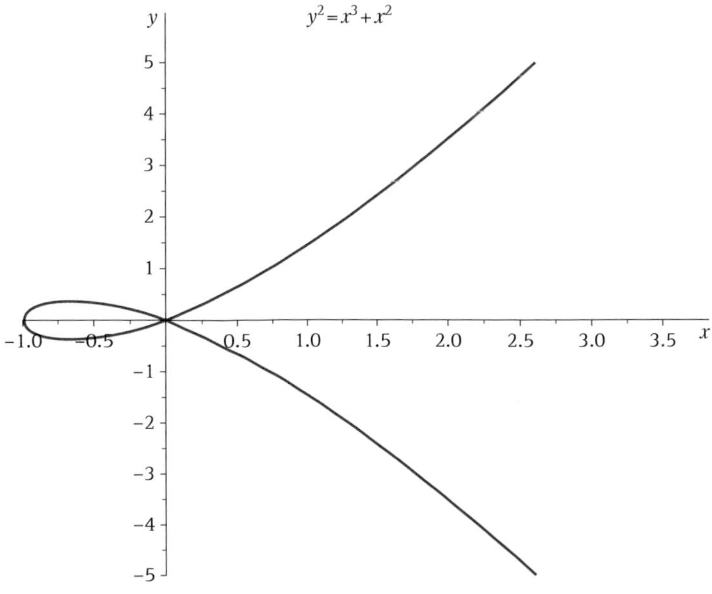

Figure 8.16 Real points (x, y) on the elliptic curve $y^2 = x^3 + x^2$

Exercise 8.5.1 *Plot the real points (x, y) on the elliptic curve $y^2 = x^3 + x$ and any other curves you find interesting.*

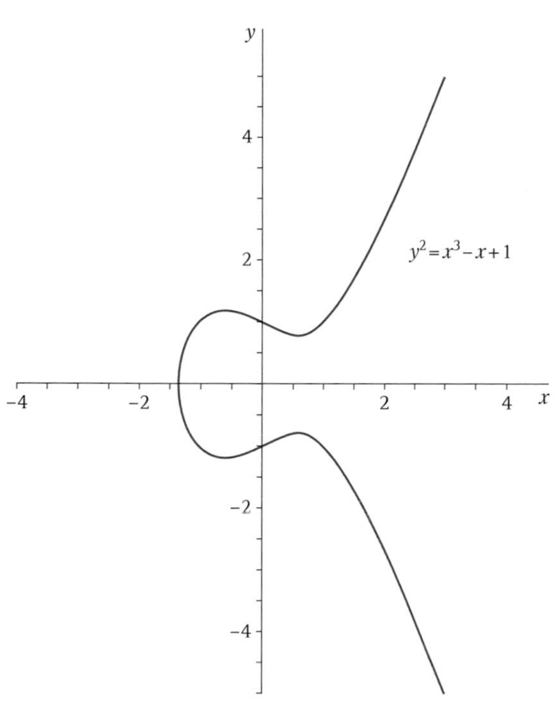

Figure 8.17 Real points (x, y) on the elliptic curve $y^2 = x^3 - x + 1$

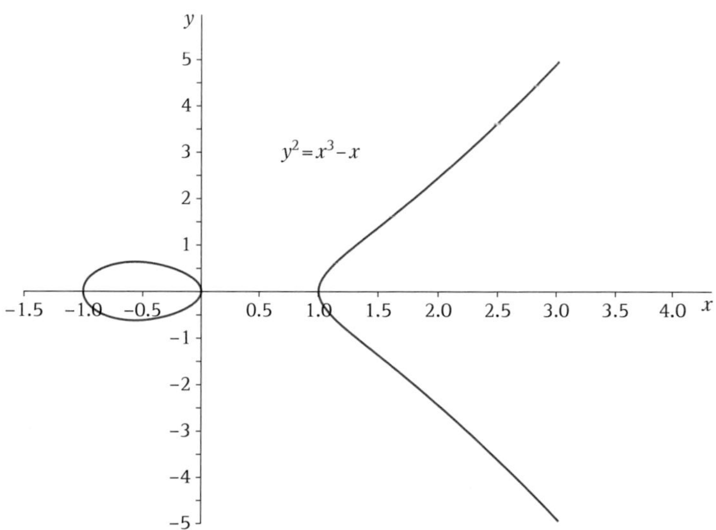

Figure 8.18 Real points (x, y) on the elliptic curve $y^2 = x^3 - x$

It is useful to replace the plane with the projective plane. In general if K is some field, projective n-space is obtained by looking at points $x = (x_0, x_1, x_2, \ldots, x_n) \in K^{n+1}$ with $x \neq 0$, and setting up an equivalence relation $x \sim t$ iff $x = \alpha t$, for some $\alpha \in K$. Here αt means the usual multiplication of a vector t by a scalar α.

Projective n-space over K is the set of equivalence classes of $K^{n+1}-0$ under the equivalence relation of the preceding paragraph: that is, $\mathbb{P}_n(K)=(K^{n+1}-0)/\sim$. This allows us to replace our elliptic curve $E(K)$ with a curve in the projective plane $\mathbb{P}_2(K)$:

$$y^2=x^3+ax^2+bx+c$$

becomes

$$(y/z)^2=(x/z)^3+a(x/z)^2+bx/z+c \tag{8.25}$$

or

$$y^2z=x^3+ax^2z+bxz^2+cz^3.$$

We will identify $(x,y,1)$ in $\mathbb{P}_2(K)$ with (x,y) in K^2. So we view K^2 as a subspace of the projective plane called **affine space.** The **line at infinity** in $\mathbb{P}_2(K)$ consists of the equivalence classes of points $(x,y,0)$ in $\mathbb{P}_2(K)$. The intersection of this line with the elliptic curve of equation (8.25) has $x=0$. Then the equivalence class in $\mathbb{P}_2(K)$ containing the point $(0,1,0)$ is called the **point at infinity**. View it as a point on the intersection of the y-axis and the line at infinity in $\mathbb{P}_2(K)$. See Silverman and Tate [107].

Definition 8.5.2 *Suppose $a,b,c\in\mathbb{C}$. If*

$$f(x)=x^3+ax^2+bx+c=(x-r_1)(x-r_2)(x-r_3),$$

then the ***discriminant*** *of f is $\Delta f=(r_1-r_2)^2(r_2-r_3)^2(r_1-r_3)^2$. One can show that the discriminant is*

$$\Delta f=a^2b^2+18abc-27c^2-4a^3c-4b^3.$$

This formula is proved for the case that $f(x)=x^3+bx+c$ in Birkhoff and Maclane [9, pp. 112–113] using the formula for the roots of the cubic. Mathematica will compute such things. An extra factor of 16 may appear in the discriminant of an elliptic curve. See, for example N. Koblitz [56, p. 26] or the extensive tables of J. E. Cremona, *Algorithms for Modular Elliptic Curves* (www.warwick.ac.uk/staff/J.E.Cremona/book/fulltext/index.html or the website http://l-functions.org).

In order to create a group associated with our elliptic curve, we need to be able to draw tangents to our elliptic curves. The tangent lines to the curve $y=f(x)$ will be undefined at points r where both $f(r)$ and $f'(x)$ vanish. That is, the tangent is undefined when r is a double root of f and thus the discriminant vanishes. An elliptic curve for which the discriminant is nonzero is called **non-singular**. For example, the curve $y^2=x^3+x^2$ in Figure 8.16 does not have a well-defined tangent at the origin and its discriminant is 0. We will want to avoid curves with such points and, when we consider curves over finite fields, we will avoid primes like 2 and 3 and those fields of characteristic p dividing the discriminant. We really want to avoid double and triple roots over the finite field in order to be able to add points on the curve. See also Silverman [105, p. 233] who explains in a footnote that there are ways of turning bad primes into good primes. See Silverman and Tate [107].

Dummit and Foote [28, pp. 534–536] give a bunch of exercises involving the resultant of two polynomials over any field explaining why the discriminant detects multiple roots. Let us consider the special case of interest here. Suppose $f(x)=x^3+ax^2+bx+c$ and

$g(x) = rx^2 + sx + t$, with a, b, c, r, s, t in a field F. We assume that $r \neq 0$. Then the **resultant** $R(f, g)$ is

$$\det \begin{pmatrix} 1 & a & b & c & 0 \\ 0 & 1 & a & b & c \\ r & s & t & 0 & 0 \\ 0 & r & s & t & 0 \\ 0 & 0 & r & s & t \end{pmatrix}.$$

If $f(x) = (x - \alpha)(x - \beta)(x - \gamma)$ and $g(x) = r(x - \rho)(x - \sigma)$ for roots $\alpha, \beta, \gamma, \rho, \sigma$ in some extension field of F, then one can show that we have

$$\begin{aligned} R(f, g) &= r^3(\alpha - \rho)(\alpha - \sigma)(\beta - \rho)(\beta - \sigma)(\gamma - \rho)(\gamma - \sigma) \\ &= r^3 f(\rho) f(\sigma) = g(\alpha) g(\beta) g(\gamma). \end{aligned} \tag{8.26}$$

It follows that $R(f, g) = 0$ iff $\{\alpha, \beta, \gamma\} \cap \{\rho, \sigma\} \neq \emptyset$. Thus $R(f, f') = 0$ iff f has a root of multiplicity greater than 1. In fact, the discriminant of the cubic polynomial $f(x)$ is $-R(f, f')$. You can easily compute these things with Mathematica or Scientific Workplace.

Exercise 8.5.2

(a) *Evaluate the resultant* $R(f, f')$ *for the polynomial* $f(x) = x^2 + ax + b$. *It will be a* 3×3 *determinant.*

(b) *Explain how to get the analog of formula (8.26) for* $R(f, g)$ *if* $g(x) = rx + s$, *with* a, b, r, s *in some field* F.

Exercise 8.5.3 *Evaluate the resultant* $R(f, f')$ *for the polynomial* $f(x) = x^3 + ax^2 + bx + c$. *It will be a* 5×5 *determinant.*

What is the group of an elliptic curve? To associate an Abelian group G to an elliptic curve $E(K)$ where K is any field – for us $K = \mathbb{Q}$ or $\mathbb{F}_q$ – the simplest way is to say that three points p, q, r on $E(K)$ add to 0 iff they lie on a straight line. See Figure 8.19. We define the identity 0 to be the point at infinity on the curve. Then if $p = (x, y)$, we see that $-p = (x, -y)$. Think of 0 as a point infinitely far up any vertical line. If you need to compute $2p = p + p$, then define the intersection of the curve and its tangent at p to be $-2p$. Of course, this makes sense over $\mathbb{R}$. To figure out what is happening over a finite field, we just use the formulas derived from those over the real field. We will make this more precise in the examples below. The curve must have a well-defined tangent at every point for our construction to work. Thus the curve in Figure 8.16 is a bad one, since there is no well-defined tangent at the origin.

Theorem 8.5.1 *The preceding definition makes the non-singular elliptic curve into an Abelian group.*

Proof sketches are given in Kumanduri and Romero [62, p. 496] and Silverman and Tate [107]. The big problem is the proof of the associative law. The simplest proof of this law uses results from algebraic geometry on numbers of points on intersections of curves. One could also make use of equation (8.32) for adding two distinct points A, B and then throw

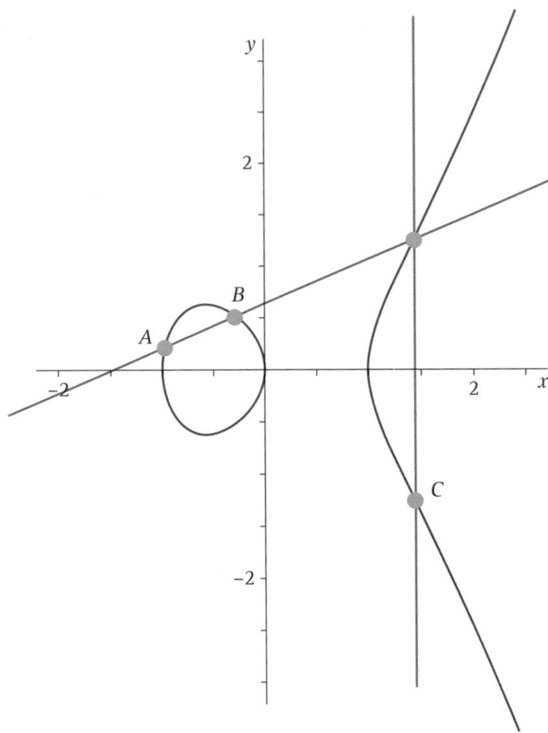

Figure 8.19 Addition $A+B=C$ on the elliptic curve $y^2=x^3-x$ over $\mathbb{R}$

in a third distinct point C. You might want to use the symbolic computational power of Mathematica to see the symmetry of the final result in A, B, C. Then there would be other cases to consider: for example, $A=B$.

Example. Look at $y^2+y=x^3-x^2$ over the field $\mathbb{Q}$. This curve has five rational points:

$$a=(0,0),\ \ b=(1,-1),\ \ c=(1,0),\ \ d=(0,-1) \text{ and the point at infinity denoted } 0.$$

Can you prove it? The finite real points are shown in Figure 8.20. ▲

The group G of this curve over $\mathbb{Q}$ turns out to be the cyclic group of order 5 generated by a. First, you can see that $2a=b$. In this case, you just need to see that the tangent to the curve at a (which is the x-axis) intersects the curve at c. So that means $a+a+c=0$ and thus $a+a=-c=b$. Next note that $b+c=0=$ the point at ∞, since the line through b and c is vertical and thus goes through the point at ∞.

Exercise 8.5.4 *Compute the addition table for the group* $G=\{0,a,b,c,d\}$ *of the curve* $y^2+y=x^3-x^2$. *The group G is a cyclic group generated by a. Assume that we have found all the rational points on the curve.*

There is a proof of the following theorem in Silverman and Tate [107].

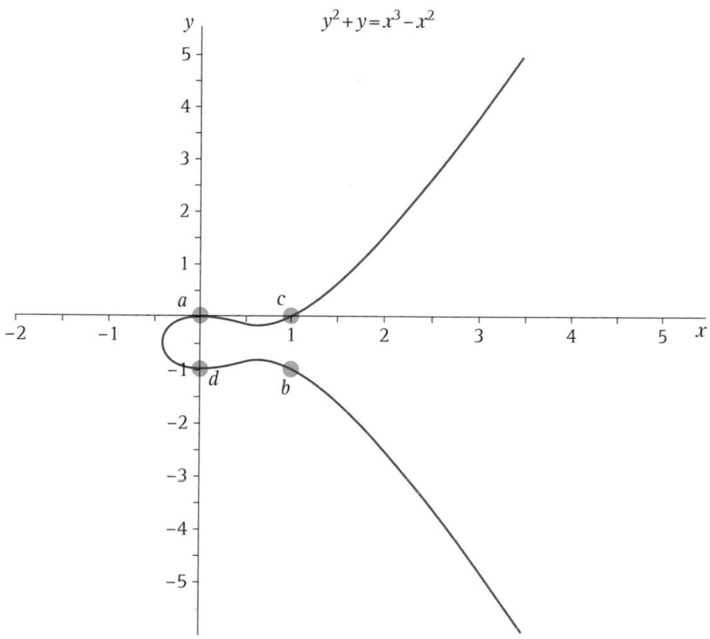

Figure 8.20 The rational points on the curve $y^2 + y = x^3 - x^2$ are a, b, c, d and the point at ∞

Theorem 8.5.2 *(Mordell).* *The group G associated to an elliptic curve $E(\mathbb{Q})$ is finitely generated (not necessarily finite) Abelian.*

Thus, by the fundamental theorem of finitely generated Abelian groups, the group G of $E(\mathbb{Q})$ is isomorphic to a direct sum $\mathbb{Z}_{a_1} \oplus \cdots \oplus \mathbb{Z}_{a_n} \oplus \mathbb{Z}^r$. Here r is called the **rank** of G. The finite part of G is called the **torsion subgroup.** Kumanduri and Romero [62, Section 19.5] find the torsion subgroup. The rank is harder.

The rank of an elliptic curve over $\mathbb{Q}$ is connected to the congruent number problem which is still open. The congruent number problem asks which positive integers n are such that there is a right triangle with rational sides whose area equals n. More precisely, the following two questions are equivalent:

Question A. For every $n \in \mathbb{Z}^+$ does there exist a right triangle with rational sides whose area equals n?

Question B. Is the rank of $y^2 = x^3 - n^2x$ over $\mathbb{Q}$ positive?

Remarks on Elliptic Curves over the Field $\mathbb{C}$

Historically it has been important to study elliptic curves over the complex numbers. We will not really want to do this here as it involves complex analysis. However, we cannot resist giving a sketch of the basics. For the theory of elliptic curves over $\mathbb{C}$, one needs the **Weierstrass $\wp$-function.** The function $\wp(z)$ is a holomorphic function of z in the complex plane except for a double pole at each point of a lattice $L = w_1\mathbb{Z} + w_2\mathbb{Z}$ in the plane. Moreover one has the differential equation:

$$\wp'(z)^2 = 4\wp(z)^3 - g_2\wp(z) - g_3.$$

This implies that the point $(\wp(z), \wp'(z))$ lies on an elliptic curve $E(C)$. The "curve" is a subset of $\mathbb{C}^2 \cong \mathbb{R}^4$ which has one complex parameter and thus two real parameters. The graph would have to be drawn in four real dimensions. If you need to know, the numbers g_2, g_3 are given by **Eisenstein series:**

$$g_2 = 60 \sum_{\substack{m,n\in\mathbb{Z} \\ (m,n)\neq(0,0)}} \frac{1}{(mw_1 + nw_2)^4} \quad \text{and } g_3 = 140 \sum_{\substack{m,n\in\mathbb{Z} \\ (m,n)\neq(0,0)}} \frac{1}{(mw_1 + nw_2)^6}.$$

If you set $w_2 = 1$, you get a function of w_1 in the upper half plane, which is a modular form. Mathematica knows $\wp(z)$ as `WeierstrassP[z,{g1,g2}]`. This gives a one-to-one correspondence from z to $(\wp(z), \wp'(z))$ which takes the torus $\mathbb{C}/L$ (where we identify points in $\mathbb{C}$ which differ by a lattice point in $L = w_1\mathbb{Z} + w_2\mathbb{Z}$) to the elliptic curve $E(\mathbb{C})$. The mapping is an isomorphism of Abelian groups. The Weierstrass $\wp$-function is not usually covered in undergraduate analysis. However, you can find a discussion in Koblitz [56] and many complex analysis books.

Elliptic Curves over the Field $\mathbb{F}_q$

For our application to cryptography, we need to discuss elliptic curves over finite fields $\mathbb{F}_q$. Mostly we consider the special case that $q = p$, where p is prime. How do we add points on a curve $E = E(\mathbb{F}_p)$ for $a, b, c \in \mathbb{Z}$,

$$y^2 = x^3 + ax^2 + bx + c \pmod{p}? \tag{8.27}$$

As we said earlier, we must avoid the prime 2 as well as the "bad primes" dividing the discriminant of the polynomial, which is:

$$a^2b^2 + 18abc - 27c^2 - 4a^3c - 4b^3.$$

We will also assume that the prime p is larger than 3. Now let us imitate the construction over $\mathbb{R}$. Over $\mathbb{F}_p$ the "curve" is just a finite set of points.

See Figure 8.21 for an example mod 59. The red points on the 59×59 grid correspond to (x, y) such that $y^2 = x^3 - x + 1 \pmod{59}$ and they do bear some resemblance to the graph of the same elliptic curve over the reals in Figure 8.17. Perhaps they would bear more resemblance if I had graphed y from -29 to 29 rather than from 1 to 59. Of course, the figure is periodic mod 59 in both x and y. You could fill up the plane with copies, or – better yet – put the whole thing on a torus.

In general, consider points on an elliptic curve $E(\mathbb{F}_p)$ such as the one in Figure 8.21. We want to write down equations for adding points. Let $P = (x_1, y_1)$ and $Q = (x_2, y_2)$, $P + Q = (x_3, y_3)$, with $x_1 \neq x_2$. Let $y - y_1 = \mu(x - x_1)$ be the line L through P and Q. Points y on the "line" L must satisfy

$$y = \mu x + \beta \pmod{p}, \quad \text{where } \mu = \frac{y_2 - y_1}{x_2 - x_1} \pmod{p} \text{ and } \beta = y_1 - \mu x_1 \pmod{p}. \tag{8.28}$$

In equation (8.28), you have to find the inverse of $x_2 - x_1 \pmod{p}$ to find the "slope" μ of the line L.

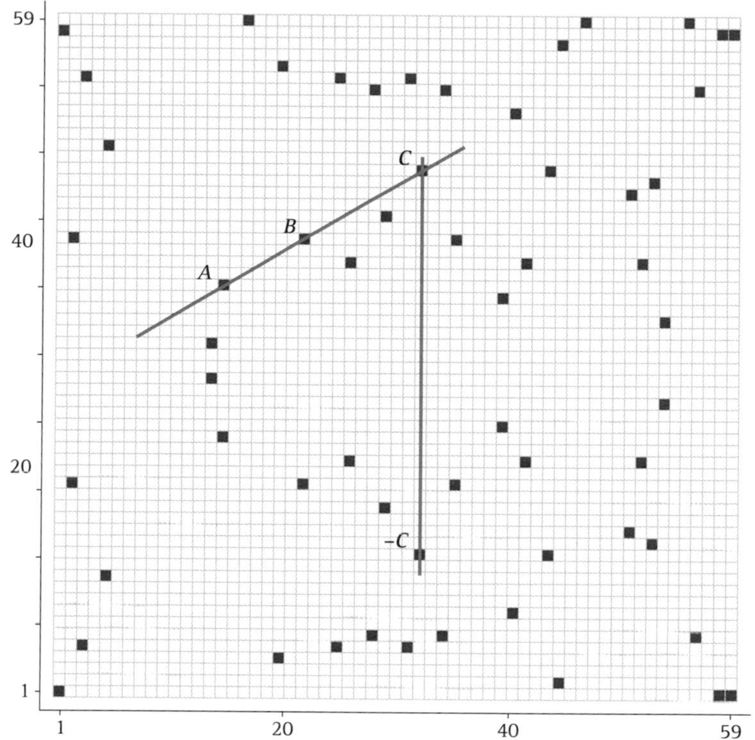

Figure 8.21 The pink squares indicate the points (x, y) on the elliptic curve $y^2 = x^3 - x + 1 \bmod 59$. Points marked are: $A = (15, 36), B = (22, 40), C = (32, 46)$, with $A + B = -C = (32, 13)$

To find the third point on L and E, which is $-(P + Q)$, we plug equations (8.28) into equation (8.27) for the elliptic curve $E(\mathbb{F}_p)$. You get a cubic equation for x:

$$(\mu x + \beta)^2 = x^3 + ax^2 + bx + c \pmod{p}.$$

So we can find our point $-(P + Q)$ by solving the cubic:

$$f(x) = x^3 + (a - \mu^2)x^2 + (b - 2\mu\beta)\,x + c - \beta^2 \pmod{p}. \tag{8.29}$$

We already have two roots of $f(x)$, namely x_1 and x_2. This means that

$$\begin{aligned} f(x) &= (x - x_1)(x - x_2)(x - x_3) \\ &= x^3 - (x_1 + x_2 + x_3)x^2 + (x_1x_2 + x_1x_3 + x_2x_3)x - x_1x_2x_3 \pmod{p}. \end{aligned} \tag{8.30}$$

Therefore if $x_1 \neq x_2$, $P = (x_1, y_1)$ and $Q = (x_2, y_2)$, we have $P + Q = (x_3, -y_3)$, with

$$\begin{aligned} x_3 &\equiv \mu^2 - a - x_1 - x_2 \pmod{p}, \\ y_3 &\equiv \mu(x_3 - x_1) + y_1 \pmod{p}, \quad \text{where } \mu = \frac{y_2 - y_1}{x_2 - x_1} \pmod{p}. \end{aligned} \tag{8.31}$$

Again in the formula for μ, you must divide mod p. When $x_1 = x_2$ but $P \neq Q$, the sum is 0, the point at infinity.

To figure out the rule when $P = Q$, look at the tangent to E at P. This is found by recalling that the "derivative" is the "slope" to the "tangent" and thus formally we have

$$2y\frac{dy}{dx} = 3x^2 + 2ax + b \pmod{p}.$$

It follows that the slope $\frac{dy}{dx}$ of the "tangent" to E at the point (x_1, y_1) is

$$\mu = \frac{3x_1^2 + 2ax_1 + b}{2y_1} \pmod{p}.$$

If $y_1 = 0$, the "tangent" is vertical and the third point on the curve is 0, the point at infinity. Once more, to find the sum $P + Q$, substitute the equation $y = \mu x + \beta$ into equation (8.27) for the elliptic curve. This time we have a double root at x_1. So instead of (8.30) we see that

$$f(x) = (x - x_1)^2(x - x_3) = x^3 - (2x_1 + x_3)x^2 + (x_1^2 + 2x_1x_3)x - x_1^2x_3 \pmod{p}. \quad (8.32)$$

This implies $2P = P + P = (x_3, -y_3)$ where

$$x_3 \equiv \mu^2 - a - 2x_1 \pmod{p},$$
$$y_3 \equiv \mu(x_3 - x_1) + y_1 \pmod{p}, \quad \mu = \frac{3x_1^2 + 2ax_1 + b}{2y_1} \pmod{p}. \quad (8.33)$$

Note that equation (8.33) is just (8.32) with $x_1 = x_2$, except for finding μ with formal derivatives.

As with elliptic curves over $\mathbb{Q}$, we should really give a proof that equations (8.32) and (8.32) define an Abelian group operation. One could use Mathematica to help check the associative law or just be content that, for our examples, we really have groups.

Example. The group of the elliptic curve E given by $y^2 = x^3 + 1 \pmod{5}$.

It is now easy to find the points on the curve. Substitute $x = 0, 1, 2, 3, 4$ and solve for $y \pmod{5}$. You find

$$A = (0,1),\ B = (0,-1),\ C = (2,3),\ D = (2,-3),$$
$$F = (4,0) \quad \text{and 0 is the point at } \infty.$$

We can use equations (8.32) and (8.33) to compute the group table for the group G of points on E. First note that $A = (0,1)$ and $B = (0,-1) \Longrightarrow A + B = 0$, the point at infinity.

To find $A + C$, note that the slope of the line L through A and C is $\mu = \frac{3-1}{2-0} \pmod{5}$. So $\mu \equiv 1 \pmod{5}$. Then, using (8.32), we see that

$$x_3 \equiv \mu^2 - a - x_1 - x_2 \equiv 1 - 0 - 0 - 2 \equiv -1 \equiv 4 \pmod{5}.$$
$$y_3 \equiv \mu(x_3 - x_1) + y_1 \equiv 1(4 - 0) + 1 \equiv 5 \equiv 0 \pmod{5}.$$

Thus $A + C = (4,0) = F$.

To find $C + C$, use equation (8.33) and note that in this case $\mu = \frac{3*4}{2*3} \equiv 3 * 4 \equiv 2 \pmod{5}$.

$$x_3 \equiv \mu^2 - a - 2x_1 \equiv 4 - 0 - 4 \equiv 0 \pmod{5}.$$
$$y_3 \equiv \mu(x_3 - x_1) + y_1 \equiv 2(0 - 2) + 3 \equiv -1 \pmod{5}.$$

It follows that $2C = A$. ▲

Exercise 8.5.5 *Compute the rest of the group table for the group G of the preceding example. Is G cyclic?*

Exercise 8.5.6 *Compute the group table for* $y^2 = x^3 - x + 1 \pmod{7}$. *You can draw a graph of the curve and count the points on it (not forgetting to add in the point at infinity) or use the formula in Exercise 8.5.10 below telling you how many points are on the curve.*

Example. Consider the elliptic curve $y^2 = x^3 - x + 1 \pmod{59}$ in Figure 8.21. This prime was chosen because $x^3 - x + 1 = (4 + x)(13 + x)(42 + x) \pmod{59}$. The prime 59 is the first prime for which the polynomial factors completely. We plot the points as red squares in a grid of $(x, y) \in \mathbb{F}_{59}^2$. For any prime, the point $(1, 1)$ – at the bottom left in the figure – is on the elliptic curve. Some points can be added visually. For example, the point $(17, 59)$ – at the top of the figure – is of order 2 because the y-coordinate is 0 (mod 59). There are two other points of order 0 because they have the same y-coordinate. By the main theorem on cyclic groups and their subgroups in Section 2.5, this implies that the group of this curve is not cyclic.

One can use formula (8.35) in Exercise 8.5.10 below to see that the curve $y^2 = x^3 - x + 1 \pmod{59}$ has 60 points. I used Mathematica which can compute the Legendre symbol defined in the same exercise. Or you can count the points in Figure 8.21. There are 59 points visible in Figure 8.21 plus the point at infinity, which can be thought of at infinitely far up the y-axis.

For some of the points in Figure 8.21, it is possible to draw a line to do the addition. For example the points $A = (15, 36)$, $B = (22, 40)$, $C = (32, 46)$ clearly form a line. Thus $A + B = -C = (32, 13)$, since $59 - 46 = 13$. Of course the "lines" are not always so easily seen – thanks to the periodicity mod 59 in each coordinate. ▲

Exercise 8.5.7 *Compute the discriminant of the elliptic curve in Figure 8.21. Primes p dividing this discriminant must be avoided when considering the curve* mod p.

Exercise 8.5.8 *Use Mathematica or the software of your choice to add points on the curve*

$$y^2 = x^3 - x + 1 \pmod{59}$$

in Figure 8.21. Show that the point $(1, 1)$ has order 30. With a little more effort you can create the group of this curve – a non-cyclic Abelian group of order 60. This can be done by first creating the cyclic subgroup H_2 generated by an element left out of $H_1 = \langle(1, 1)\rangle$, and then continuing with the cyclic subgroup H_3 generated by an element left out of H_2, and so on. The cycle diagrams generated in this way produce interesting figures. Recall that we considered such diagrams for the multiplicative groups $\mathbb{Z}_n^*$ in Section 2.5. As a culmination of this exercise, try to draw the cycle diagram. Which of the following groups is (isomorphic to) the group of this curve: $\mathbb{Z}_{30} \oplus \mathbb{Z}_2$, $\mathbb{Z}_6 \oplus \mathbb{Z}_{10}$, $\mathbb{Z}_{15} \oplus \mathbb{Z}_2 \oplus \mathbb{Z}_2$, or $\mathbb{Z}_{15} \oplus \mathbb{Z}_4$, or something else? Are these groups all non-isomorphic anyway?

Exercise 8.5.9 *Show that the elliptic curve $y^2 = x^3 + ax^2 + bx + c \pmod{p}$ has at most $2p + 1$ points.*

Remarks on the Number of Points on an Elliptic Curve Mod *p*

Define the **Legendre symbol** by for odd prime p by:

$$\left(\frac{n}{p}\right) = \begin{cases} 0 & \text{if } p \text{ divides } n, \\ 1 & \text{if } p \text{ does not divide } n \text{ and } n \equiv x^2 \pmod{p} \text{ has a solution } x, \\ -1 & \text{otherwise.} \end{cases} \tag{8.34}$$

This symbol is beloved by number theorists because of its appearance in the quadratic reciprocity law relating $\left(\frac{q}{p}\right)$ and $\left(\frac{p}{q}\right)$ for two distinct primes p and q. See Rosen [91] or

Terras [116, Chapter 8]. Gauss was so fond of this subject that he found eight proofs of the quadratic reciprocity law.

Exercise 8.5.10 ***(Formula for the Number of Points on an Elliptic Curve Modulo a Prime).***

(a) Let $f(x)=x^3+ax^2+bx+c$. Show that

$$1+\left(\frac{f(x)}{p}\right)=\text{the number of solutions } y \text{ to the congruence } y^2\equiv f(x) \pmod{p}.$$

(b) Then prove that the number of points on the elliptic curve $y^2\equiv x^3+ax^2+bx+c \pmod{p}$ is

$$N_p=p+1+\sum_{x=0}^{p-1}\left(\frac{f(x)}{p}\right). \tag{8.35}$$

(c) Use the formula from part (b) in order to see that there are six points on the curve $y^2=x^3+1 \pmod{5}$.

Theorem 8.5.3 ***(H. Hasse, 1933).*** *If N_p is the number of points on an elliptic curve* mod *p, set $a_p=p+1-N_p$. Then $|a_p|\le 2\sqrt{p}$.*

To prove this theorem, using Exercise 8.5.10, one must bound the sum $\sum_{x=0}^{p-1}(f(x)/p)$. One expects the Legendre symbols to be randomly $+1$ or -1. That leads to the heuristic reason for the bound. If you want a real proof, see Lang [64] or Silverman [106].

Much is known about elliptic curves. For example, it has also been proved that the group G of points on an elliptic curve (mod p) is a product of at most two cyclic groups. More references on the subject are: Jeff Hoffstein, Jill Pipher, and Joseph H. Silverman [44], Kristin Lauter [67], Karl Rubin and Alice Silverberg [96], Alice Silverberg [104], and Joseph H. Silverman [106].

Exercise 8.5.11 *Find the number of points on the curve $y^2=x^3-x+1 \pmod{p}$ for all primes p such that $5\le p\le 30$ and such that p does not divide the discriminant of x^3-x+1.*

Hint. *Mathematica knows how to compute the Legendre symbol $\left(\frac{n}{p}\right)$ via the command* `JacobiSymbol[n,p]`.

Exercise 8.5.12 *State whether the groups of the curves $y^2=x^3-x+1 \pmod{p}$ considered in the preceding exercise are cyclic.*

Exercise 8.5.13 *State whether the group of the curve $y^2=x^3-x+1 \pmod{163}$ is cyclic.*

Hint. *Mathematica can automate the addition of points. Start with $a=(1,1)$; $pr=163$. Compute $b=2a$ as follows.*

```
mu=Mod[(3* PowerMod[a[[1]],2,pr]-1)* PowerMod[2* a[[2]],-1,pr],pr];
q1=Mod[PowerMod[mu,2,pr]-2* a[[1]],pr];q2=Mod[-mu* (q1-a[[1]])
    -a[[2]],pr];b={q1,q2}.
```

From then on, one can keep adding a to b. For example if we want $b = 12a$, we use the command below to add a to b ten times.

```
Do[(mu=Mod[(b[[2]]-a[[2]])* PowerMod[b[[1]]-a[[1]],-1,pr],pr];
r1=Mod[PowerMod[mu,2,pr]-a[[1]]-b[[1]],pr];
r2=Mod[-mu*(r1-a[[1]])-a[[2]],pr];b={r1,r2}),{10}];b
```

Elliptic Curves over Finite Fields and Cryptography

In Section 4.1 we saw how to get public-key secret codes from the multiplicative group $\mathbb{Z}_{pq}^*$, when p and q are large primes. The security of such codes derives from the difficulty of factoring pq when p and q are two large primes. Elliptic curve cryptography makes use of the group G of an elliptic curve mod p. It seems that one can use smaller public-keys and still have secure messages using elliptic curve cryptography.

Note that our analog of raising an element $a \in \mathbb{Z}_{pq}^*$ to the kth power in G is multiplying an element a in G by k, or adding a to itself k times to get

$$\underbrace{a + \cdots + a}_{k \text{ times}} = ka.$$

To do this fast, one can proceed in an analogous way to that with powers. For example,

$$100 = 2^6 + 2^5 + 2^2$$

The result is that, in order to compute $100a$, we need six doublings and two additions.

We will want to encode our plaintext m as a point P_m on an elliptic curve E so that it will be easy to get m from P_m.

Remarks

(1) There does not exist an algorithm for writing down lots of points on $E(\mathbb{F}_p)$ in $\log p$ time.

(2) It is not sufficient to generate random points on $E(\mathbb{F}_p)$ anyway.

A Probabilistic Method to Encode Plaintext m as P_m on an elliptic curve $E(\mathbb{F}_p)$. We will illustrate the method with an example from Koblitz [55, p. 168]. The curve is

$$y^2 + y \equiv x^3 - x \pmod{751}.$$

This curve has 727 points, and 727 is a prime. Thus the group of the curve is cyclic.

Exercise 8.5.14 *Check the last statement. You will want to make use of formula (8.36) below completing the square on the left-hand side of the equation.*

Take a number $\kappa = 20$ (or larger). The number κ is chosen so that a failure rate of $1/2^\kappa$ is OK when seeking our point. We will need to represent numbers m between 0 and 35 (meaning the usual alphabet plus the digits from 0 to 9):

$$0, 1, 2, \ldots, 9, A, B, C, D, \ldots, X, Y, Z.$$

So we want $p > 35 * \kappa = 700$. Our $p = 751$ so that is OK.

Write x between 0 and 700 in the form $x = m * 20 + j$, where $1 \le j \le 20$. Then compute $y' = y + 376$ so that

$$2 * 376 \equiv 1 \pmod{751} \quad \text{and} \quad 376^2 \equiv 188 \pmod{751}.$$

$$y'^2 \equiv (y + 376)^2 \equiv y^2 + y + 188 \equiv x^3 - x + 188 \pmod{751}. \tag{8.36}$$

Thus we need a fast way to do square roots mod p. Luckily Mathematica does square roots mod p. So we do not need to program this algorithm (unless we hate Mathematica). Of course programs like SAGE actually know about elliptic curves.

If we can solve for y then set $P_m = (x, y)$. Otherwise replace j by $j + 1$ in the formula for x and try again. Since our curve has 727 points, probability says we should not have to increment more than 20 times. Set $f(x) \equiv x^3 - x + 188 \pmod{751}$. There is approximately a $\left(\frac{1}{2}\right)^{20}$ chance that $f(m * 20 + j)$ will not be a square for any $j = 1, 2, \ldots, 20$; assuming that the events $f(m * 20 + j) = \text{square}$ and $f(m * 20 + j + 1) = \text{square}$ are independent – an unproved but reasonable assumption.

Here is our alphabet table.

0	1	2	3	4	5	6	7	8	9	A	B	C	D	E	F	G	H
0	1	2	3	4	5	6	7	8	9	10	11	12	13	14	15	16	17

I	J	K	L	M	N	O	P	Q	R	S	T	U	V	W	X	Y	Z
18	19	20	21	22	23	24	25	26	27	28	29	30	31	32	33	34	35

We find some points on the curve corresponding to our alphabet entries.

(1) First $m = 0$ gives P_0. We look at $x = 0 * 20 + 1$ and plug that into

$$y'^2 \equiv (y + 376)^2 \equiv x^3 - x + 188 \equiv 1 - 1 + 188 \equiv 188 \equiv (376)^2 \pmod{751}.$$

Clearly the solution is $y = 0$. So $P_0 = (1, 0)$.

(2) Next we look at $m = 1$, and form $x = 1 * 20 + j$, with $j = 1, \ldots, 20$. For $j = 1$, we have $x = 21$ and solve

$$y'^2 \equiv (y + 376)^2 \equiv x^3 - x + 188 \equiv 21^3 - 21 + 188 \equiv 416 \equiv (618)^2 \pmod{751}.$$

Then $y'^2 \equiv (618)^2 \pmod{751}$ has two solutions. We take $y' \equiv 618 \pmod{751}$. Then $y \equiv y' - 376 \equiv 242 \pmod{751}$. The other solution is $y \equiv 508 \pmod{751}$. We will ignore it. So we get the point $P_1 = (21, 242)$. We could have equally well said $(21, 508)$.

(3) Similarly we find $P_2 = (41, 101)$.

(4) The next case is more interesting. If we set $m = 3$ and form $x = 3 * 20 + j$, with $j = 1$, we see that when $x = 61$, we cannot solve the congruence

$$y'^2 \equiv (y + 376)^2 \equiv x^3 - x + 188 \equiv 61^3 - 61x + 188 \equiv 306 \pmod{751}.$$

So we must increment j to 2 and look at $x = 62$. Luckily this guy is a square mod 751 and we find that $P_3 = (62, 214)$ or $(62, 536)$.

(5) Corresponding to the letter S is the number $m = 28$ (from our alphabet table). Then we find the point $P_{28} = (562, 174)$ or $(562, 576)$ on the curve E. It again takes two tries.

Of course we are using Mathematica to do this. Here is part of our Mathematica notebook. In versions of Mathematica from 2001, we needed to include the package

<< NumberTheory'NumberTheoryFunctions'

in order to take square roots mod n. Now we just use PowerMod[a,1/2,n].

For example, to do the Mathematica calculation for the second point on the curve mod 751, we define the following Mathematica functions f, g, h, k and perform the calculations.

```
f[x_]:=f[x]=Mod[PowerMod[x,3,751] - x + 188,751]
g[x_]:=g[x]=PowerMod[x,1/2,751]
h[x_]:=h[x]=Mod[x-376,751]
k[x_]:=k[x]=Mod[-x-376,751]
f[21]=416
g[416]=618
h[618]=242
k[618]=508
```

Exercise 8.5.15 *This exercise involves encrypting and decrypting messages using the curve just discussed.*

(a) Write the message

THEIR LANGUAGE IS THE LANGUAGE OF NUMBERS
AND THEY HAVE NO NEED TO SMILE

as a sequence of points on the curve

$y^2 + y \equiv x^3 - x \pmod{751}$.

using the method described above. This is a quote from the Dr. Who episode Logopolis.

(b) Translate the following sequence of points on the curve in part (a)

$$(421,013)(361,367)(621,220)(283,321)(421,013)(484,214)(461,283)$$
$$(324,360)(201,370)$$
$$(461,283)(261,663)(501,220)(543,314)(484,214)(562,174)(501,220)$$
$$(283,321)(543,314)$$

ElGamal Elliptic Curve Cryptosystem

The preceding exercise produces an encryption that is extremely easy to decrypt. We need to do better if we want our messages to stay out of the hands of the shadow creatures. So we must do some more work on these messages. In short, we must think about the ElGamal cryptosystem. Two other methods are also given in Koblitz [55].

Suppose that John on Babylon 5 wants to send a message to Delenn on Minbar – without the shadow creatures understanding the message. An elliptic curve $E(\mathbb{F}_q)$ is made public. A generator g of the group G of this elliptic curve is chosen, if possible. If this is not possible, John and Delenn want the subgroup $\langle g \rangle$ generated by g to be large (near the order of G).

Delenn picks a large integer $a \in G$ with $0 < a < |G|$. Then a is her secret deciphering key. She makes public the enciphering key ag.

John chooses an integer k. To send a message corresponding to a sequence of points P on the elliptic curve $E(\mathbb{F}_q)$, John sends a sequence whose entries have the form $(kg, P + kag)$, for k in a list of random integers. He does not need to know a to compute $k(ag)$, since he knows both k and the public key ag.

Then for Delenn to decipher an entry of the message, she computes $a(kg) = k(ag)$ and subtracts it from $P + kag$.

If the shadow creatures could find a from ag, they could decipher the message too. This is called the "discrete logarithm problem" for this elliptic curve situation. It is presumably too hard for the shadows to solve – assuming we choose large enough elliptic curves.

What is the advantage of elliptic curve cryptography over RSA described in Section 4.1? The public keys can be much smaller than those for RSA.

Exercise 8.5.16 *Suppose that you are Dumbledore. Use the preceding ElGamal cryptosystem and the elliptic curve in Exercise 8.5.15 to send the message WELL DONE SLYTHERIN to Professor Snape. Take $g=(0,0)$. You, Dumbledore, choose the following random list of integers to use as the integers k for your message:*

$$290, 437, 129, 484, 312, 206, 435, 508, 151, 335, 488, 501, 411, 429, 162, 535, 223.$$

Professor Snape chooses $a=361$. His public key is $361(0,0)$. What is the sequence of pairs of points $(kg, P+kag)$ that Dumbledore sends?

Exercise 8.5.17 *Show that, for odd prime p, the Legendre symbol $\left(\frac{n}{p}\right)$ from equation (8.34) defines a group homomorphism $f(n)=\left(\frac{n}{p}\right)$, mapping the multiplicative group $\mathbb{F}_p^*$ onto the group $\{\pm 1\}$ under multiplication. Show that the kernel has order $(p-1)/2$.*

Exercise 8.5.18 *Show that x^4+1 is always reducible in $\mathbb{F}_p[x]$, for all primes p.*

Hint. *For a prime $p>2$, there are three cases: either $-1, 2$, or -2 is a square* mod p. *In each case there will be a factorization as a product of two quadratic polynomials. For example, in the second case, if $2\equiv a^2 \pmod{p}$, $x^4+1=(x^2+ax+1)(x^2-ax+1)$.*

To end this section, which is really the end of this book, we include Figures 8.22 and 8.23 which are two pictures of level "curves" of $y^2-x^3-x+1 \pmod{29}$. Modern art?

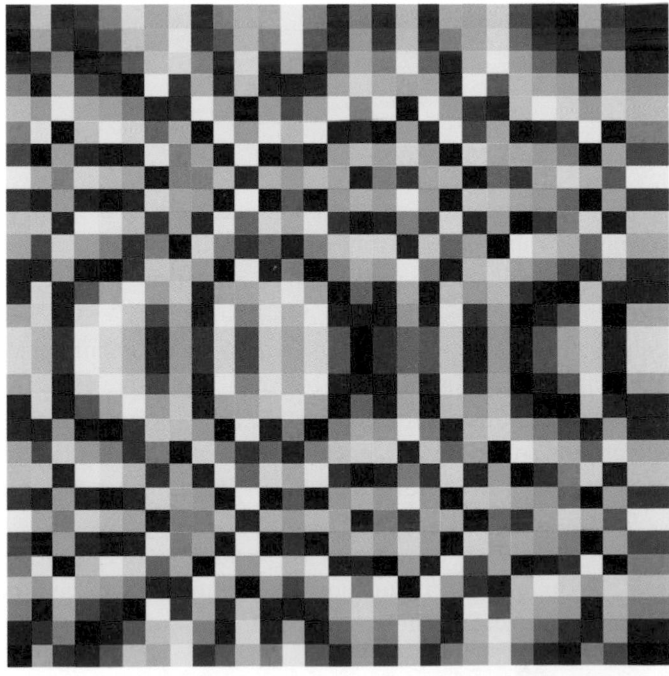

Figure 8.22 Level "curves" of $y^2-x^3-x+1 \pmod{29}$

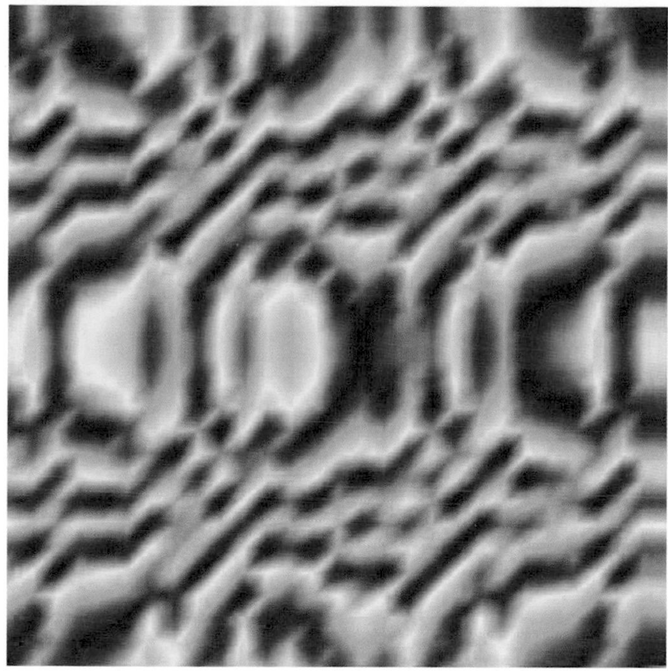

Figure 8.23 Smoothed level "curves" of $y^2 - x^3 - x + 1 \pmod{29}$

Figure 8.24 is a photoshopped version of the level curves of

$$(y + 2x)^4 + (x - 2y)^4 \pmod{101}.$$

The cover of this book involves another photoshopped version of Figure 8.22.

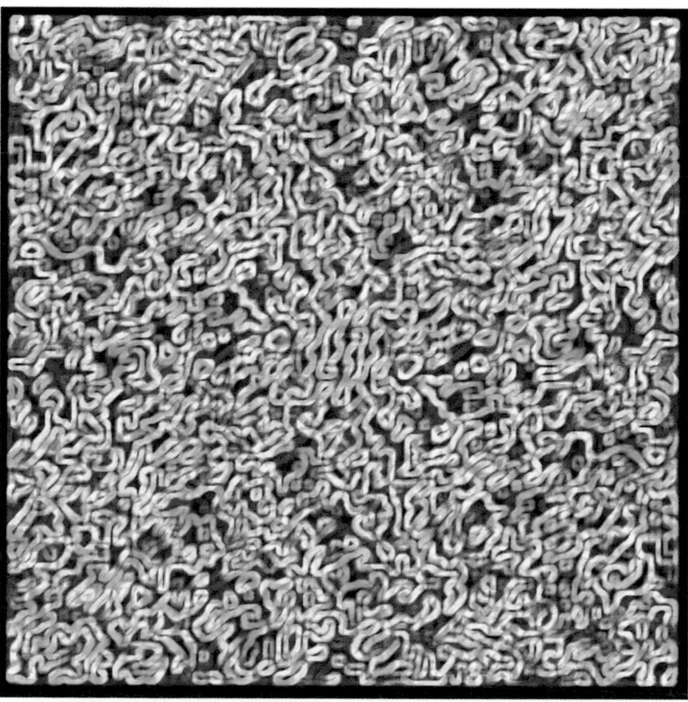

Figure 8.24 A photoshopped version of the level curves of $(y + 2x)^4 + (x - 2y)^4 \pmod{101}$.

References

[1] E. Arnold, R. Field, S. Lucas, and L. Taalman, Minimal complete Shidoku symmetry groups, *J. Combin. Math. Combin. Comput.*, 87 (2013), 209–228.

[2] M. Artin, *Algebra*, Prentice Hall, Englewood Cliffs, NJ, 1991.

[3] D. Austin, Random Numbers: Nothing Left to Chance, on the American Mathematical Society website under feature column (http://www.ams.org/samplings/feature-column/fcarc-random).

[4] F. Ayres, Jr., *Schaum's Outline on Theory and Problems of Matrices*, McGraw-Hill, New York, NY, 1962.

[5] A. Baker, *Transcendental Number Theory*, Cambridge University Press, Cambridge, 1975.

[6] T. Banchoff, *Beyond the Third Dimension*, Freeman, NY, 1990.

[7] F. Bardawil, T. Dodman, I. Jauslin, P. Marichalar, K. Oschema, and P. Redfield, Immigration, freedom, and the history of the Institute for Advanced Study, *Notices Amer. Math. Soc.*, 64, No. 10 (2017), 1160–1168.

[8] E.T. Bell, *Men of Mathematics*, Simon and Schuster, NY, 1937.

[9] G. Birkhoff and S. Maclane, *A Survey of Modern Algebra*, Macmillan, NY, 1953.

[10] A. Björner and F. Brenti, *Combinatorics of Coxeter Groups*, Springer, NY, 2005.

[11] R.P. Brent, Note on Marsaglia XOR Shift random number generators, *J. Stat. Software*, 11 (2004), 1–5.

[12] N.C. Carter, *Visual Group Theory*, on line at www.jstor.org.

[13] D. Charles, E. Goren, and K. Lauter, Cryptographic hash functions from expander graphs, *J. Cryptology*, 22, No. 1 (2009), 93–113.

[14] P. Chiu, Cubic Ramanujan graphs, *Combinatorica*, 12 (1992), 275–285.

[15] F. Chung and S. Sternberg, Mathematics and the buckyball, *American Scientist*, 81 (Jan.-Feb. 1993), 56–71.

[16] B. Cipra, The best of the 20th century: editors name top 10 algorithms, *SIAM News*, 33, No. 4 (2000).

[17] L.W. Cohen and G. Ehrlich, *The Structure of the Real Number System*, Van Nostrand, Princeton, NJ, 1963.

[18] R. Courant and D. Hilbert, *Methods of Mathematical Physics*, Vol. I, Wiley-Interscience, New York, NY, 1961.

[19] R. Courant and F. John, *Introduction to Calculus and Analysis*, Vol. II, Springer-Verlag, New York, NY, 1989.

[20] R. Courant and Robbins, *What is Mathematics? An Elementary Approach to Ideas and Methods*, Oxford University Press, New York, NY, 1979.

[21] J. T. Cushing, *Applied Analytical Mathematics for Physical Scientists*, Wiley, New York, NY, 1975.

[22] D. Cvetković, M. Doob, and H. Sachs, *Spectra of Graphs*, Academic, New York, NY, 1980.
[23] G. Davidoff, P. Sarnak, and A. Valette, *Elementary Number Theory, Group Theory and Ramanujan Graphs*, Cambridge University Press, Cambridge, 2003.
[24] P. Diaconis, *Group Representations in Probability and Statistics*, Institute of Mathematical Statistics, Hayward, CA, 1988.
[25] L. Dornhoff and F. Hohn, *Applied Modern Algebra*, Macmillan, New York, NY, 1978.
[26] A. Doxiadis and C. Papadimitriou, art by A. Papadatos and A. di Donna, *Logicomix*, Bloomsbury, New York, NY, 2009.
[27] G. Dresden, Small Rings (http://home.wlu.edu/~dresdeng/smallrings/), a research report of the work of 13 students and Professor Sieler at a Washington and Lee University class in abstract algebra (Math 322), 2005.
[28] D. Dummit and R. Foote, *Abstract Algebra*, Prentice-Hall, Englewood Cliffs, NJ, 1991.
[29] G. Ehrlich, *Fundamental Concepts of Algebra*, PWS-KENT, Boston, MA, 1991 (reprinted by Dover in 2011).
[30] F. Farris, *Creating Symmetry: The Artful Mathematics of Wallpaper Patterns*, Princeton University Press, Princeton, NJ, 2015.
[31] B. Fine, Classification of finite rings of order p^2, *Math. Magazine*, 66, 4 (1993), 248–252.
[32] J. B. Fraleigh, *A First Course in Abstract Algebra*, Addison-Wesley, Reading, MA, 1967.
[33] J. Gallian, *Contemporary Abstract Algebra*, Brooks/Cole, Belmont, CA, 2010.
[34] I. M. Gelfand and S. V. Fomin, *Calculus of Variations*, Prentice-Hall, Englewood Cliffs, NJ, 1963.
[35] W. J. Gilbert and W. K. Nicholson, *Modern Algebra with Applications*, 2nd edition, Wiley-Interscience, Hoboken, NJ, 2004.
[36] M. Goresky and A. Klapper, *Algebraic Shift Register Sequences*, Cambridge University Press, Cambridge, 2012.
[37] M. D. Greenberg, *Foundations of Applied Mathematics*, Prentice-Hall, Englewood Cliffs, NJ, 1978.
[38] M. Hall, *The Theory of Groups*, Macmillan, New York, NY, 1959.
[39] P. Halmos, *Finite-Dimensional Vector Spaces* (reprint of the 1958 second edition), Springer-Verlag, New York, NY, 1974.
[40] R. W. Hamming, *Digital Filters*, Prentice-Hall, Englewood Cliffs, NJ, 1977.
[41] G. Herschlag, R. Ravier, and J. C. Mattingly, Evaluating partisan gerrymandering, arXiv:1709.01596v1, September 5, 2017.
[42] I. Herstein, *Topics in Algebra*, Ginn, Waltham, MA, 1964.
[43] A. M. Herzberg and M. R. Murty, Sudoku squares and chromatic polynomials, *Notices Amer. Math. Soc.*, 54, No. 6 (2007), 708–717.
[44] J. Hoffstein, J. Pipher and J. H. Silverman, *An Introduction to Mathematical Cryptography*, 2nd edition, Springer, New York, NY, 2014.
[45] R. A. Horn and C. R. Johnson, *Matrix Analysis*, Cambridge University Press, Cambridge, 1990.
[46] T. E. Hull and A. R. Dobell, Random number generators, *SIAM Review*, 4, No. 3 (1962), 230–254.
[47] T. W. Hungerford, *Algebra*, Springer, New York, NY, 1974.
[48] A. F. Jarvis and E. Russell, Mathematics of Sudoku II, *Mathematical Spectrum*, 39 (2006), 54–58.

[49] D. Joyner, *Adventures in Group Theory: Rubik's Cube, Merlin's Machine, and other Mathematical Toys*, Johns Hopkins University Press, Baltimore, MD, 2002.

[50] T.W. Judson, *Abstract Algebra: Theory and Applications*, with Sage Exercises by R. A. Beezer, (http://mathbook.pugetsound.edu), 2015.

[51] M. Kac, Can you hear the shape of a drum?, *Amer. Math. Monthly*, 73 (1966), 1–23.

[52] R. Kanigel, *The Man Who Knew Infinity: A Life of the Genius Ramanujan*, Charles Scribner's Sons, New York, NY, 1991.

[53] J.C. Kemeny and J.L. Snell, *Finite Markov Chains* (reprint of the 1960 original), Springer-Verlag, New York, NY, 1976.

[54] D.E. Knuth, *The Art of Computer Programming, Vol. II: Semi-Numerical Algorithms*, Addison-Wesley, Reading, MA, 1981.

[55] N. Koblitz, *A Course in Number Theory and Cryptography*, Springer-Verlag, New York, NY, 1987.

[56] N. Koblitz, *Introduction to Elliptic Curves and Modular Forms*, Springer-Verlag, New York, NY, 1993.

[57] N. Koblitz, The uneasy relationship between mathematics and cryptography, *Notices Amer. Math. Soc.*, 54, No. 8 (2007), 972–979.

[58] N. Koblitz and A. Menezes, The brave new world of bodacious assumptions in cryptography, *Notices Amer. Math. Soc.*, 57, No. 3, (2016), 357–365.

[59] E.E. Kramer, *The Nature and Growth of Modern Mathematics*, Princeton University Press, Princeton, NJ, 1982.

[60] S. Krantz, *The Proof is in the Pudding: A Look at the Changing Nature of Mathematical Proof*, Springer, New York, NY, 2010.

[61] M. Krebs and A. Shaheen, *Expander Families and Cayley Graphs: A Beginner's Guide*, Oxford University Press, New York, NY, 2011.

[62] R. Kumanduri and C. Romero, *Number Theory with Computer Applications*, Prentice Hall, Upper Saddle River, NJ, 1998.

[63] S. Lang, *Undergraduate Analysis*, Springer-Verlag, New York, NY, 1997.

[64] S. Lang, *Elliptic Curves: Diophantine Analysis*, Springer-Verlag, New York, NY, 1978.

[65] S. Lang, *Algebra*, 3rd edition, Springer, NY, 2002.

[66] A.N. Langville and C.D. Meyer, *Google's Pagerank and Beyond: the Science of Search Engine Rankings*, Princeton University Press, Princeton, NJ, 2006.

[67] K. Lauter, The advantages of elliptic curve cryptography for wireless security, *IEEE Wireless Communications*, Feb. 2004, 2–7.

[68] L.M. Lederman and C.T. Hill, *Symmetry and the Beautiful Universe*, Prometheus Books, Amherst, NY, 2004.

[69] R. Lidl and H. Niederreiter, *Introduction to Finite Fields and their Applications*, Cambridge University Press, Cambridge, 1986.

[70] M. Livio, *The Golden Ratio: The Story of Phi, the World's Most Astonishing Number*, Broadway Books, New York, NY, 2003.

[71] C. Lorch and J. Lorch, Enumerating small sudoku puzzles in a first abstract algebra course, *Primus*, 18 (2008), 149–158.

[72] F.J. MacWilliams and N.J.A. Sloane, *The Theory of Error-Correcting Codes*, North Holland, Amsterdam, 1988.

[73] H.B. Mann (ed.), *Error Correcting Codes*, Proceedings of a Symposium conducted by the Mathematical Research Center, United States Army, at the University of Wisconsin, Madison, Wiley, New York, NY, 1968.

[74] G. Marsaglia, Random numbers fall mainly in the planes, *Proc. Natl. Acad. Sci. USA*, 61 (1968), 22–28.

[75] J.C. Mattingly and C. Vaughn, Redistricting and the will of the people, ArXiv:1410.8796v1, Oct. 29, 2014.

[76] G. McGuire, B. Tugemann, and G. Civario, There is no 16-clue Sudoku: solving the Sudoku minimum number of clues problem via hitting set enumeration, *Exp. Math.*, 23, No. 2 (2014), 190–217.

[77] C. D. Meyer, *Matrix Analysis and Applied Linear Algebra*, Society for Industrial and Applied Mathematics (SIAM), Philadelphia, PA, 2000.

[78] S. J. Miller and R. Takloo-Bighash, *An Invitation to Modern Number Theory*, Princeton University Press, Princeton, NJ, 2006.

[79] G. L. Mullen and D. Panario (eds.), *Handbook of Finite Fields*, CRC Press, Boca Raton, FL, 2013.

[80] D. Mumford, C. Series, and D. Wright, *Indra's Pearls*, Cambridge University Press, Cambridge, 2002.

[81] H. Niederreiter, *Random Number Generation and Quasi-Monte Carlo Methods, CBMS-NSF Regional Conference Series in Applied Mathematics, 63*, Society for Industrial and Applied Mathematics (SIAM), Philadelphia, PA, 1992.

[82] J. R. Norris, *Markov Chains* (reprint of the 1997 original edition), Cambridge Series in Statistical and Probabilistic Mathematics 2, Cambridge University Press, Cambridge, 1998.

[83] R. O'Donnell, *Analysis of Boolean Functions*, Cambridge University Press, Cambridge, 2014.

[84] O. Ore, *Number Theory and its History* (reprint of the 1948 original edition), Dover, New York, NY, 1988.

[85] V. Pless, *An Introduction to the Theory of Error-Correcting Codes*, Wiley, New York, NY, 1989.

[86] W. H. Press, B. P. Flannery, S. A. Teukolsky, and W. T. Vetterling, *Numerical Recipes: The Art of Computing*, Cambridge University Press, Cambridge, 1986.

[87] E. J. Purcell, *Calculus with Analytic Geometry*, Appleton-Century-Crofts, New York, NY, 1965.

[88] M. A. Rabin, Recursive unsolvability of group theoretic problems, *Ann. Math.*, 67 (1958), 172–194.

[89] P. Ribenboim, *Classical Theory of Algebraic Numbers*, Springer, New York, NY, 2001.

[90] I. Richards, Number theory, in L. A. Steen (ed.), *Mathematics Today*, Vintage, New York, NY, 1980, pp. 37–54.

[91] K. H. Rosen, *Elementary Number Theory and its Applications*, 3rd edition, Addison-Wesley, Reading, MA, 1993.

[92] K. H. Rosen (editor-in-chief), *Handbook of Discrete and Combinatorial Mathematics*, CRC Press, Boca Raton, FL, 2000.

[93] K. H. Rosen, *Discrete Mathematics and its Applications*, McGraw-Hill, Boston, MA, 2012.

[94] J. Rosenhouse and L. Taalman, *Taking Sudoku Seriously: The Math behind the World's Most Popular Pencil Puzzle*, Oxford University Press, Oxford, 2011.

[95] W. W. Rouse Ball and H. S. M. Coxeter, *Mathematical Recreations and Essays*, 12th edition, University of Toronto Press, Toronto, Canada, 1974.

[96] K. Rubin and A. Silverberg, Ranks of elliptic curves, *Bull. Amer. Math. Soc.*, 39 (2002), 455–474.

[97] P. Samuel (trans. A. J. Silberger) *Algebraic Theory of Numbers*, Houghton Mifflin, Boston, MA, 1970.

[98] P. Sarnak, *Some Applications of Modular Forms*, Cambridge University Press, Cambridge, 1990.

[99] P. Sarnak, What is an expander?, *Notices Amer. Math. Soc.*, 51 (2004), pp. 762–763.

[100] D. Schattschneider, *Visions of Symmetry*, Harry Abrams, New York, NY, 2002.

[101] O. Schreier and E. Sperner, *Introduction to Modern Algebra and Matrix Theory*, Chelsea, New York, NY, 1959.

[102] M. Senechal, The continuing silence of Bourbaki—An interview with Pierre Cartier, June 18, 1997, *The Mathematical Intelligencer*, No. 1 (1998), pp. 22–28.

[103] D. Shanks, *Solved and Unsolved Problems in Number Theory*, Chelsea, New York, NY, 1985.

[104] A. Silverberg, Introduction to elliptic curves, in *Arithmetic of L-functions*, IAS/Park City Mathematics Series, 18, American Mathematical Society, Providence, RI, 2011, pp. 155–167.

[105] J. H. Silverman, *A Friendly Introduction to Number Theory*, Prentice-Hall, Upper Saddle River, NJ, 1997.

[106] J. H. Silverman, *The Arithmetic of Elliptic Curves*, 2nd edition, Springer, Dordrecht, 2009.

[107] J. H. Silverman and J. Tate, *Rational Points on Elliptic Curves*, Springer, New York, NY, 1992.

[108] D. Singmaster and D. M. Bloom, *Amer. Math. Monthly*, 71, No. 8 (October, 1964), 918–920.

[109] D. Smith, *Variational Methods in Optimization*, Prentice-Hall, Englewood Cliffs, NJ, 1974.

[110] H. M. Stark, *An Introduction to Number Theory* (reprint of the original 1970 edition), MIT Press, Cambridge, MA, 1987.

[111] H. M. Stark, Modular forms and related objects, *Number Theory* (Montreal, Canada, 1985), *Canadian Mathematical Society Conference Proceedings*, 7, American Mathematical Society, Providence, RI, 1987, pp. 421–455.

[112] M. E. Starzak, *Mathematical Methods in Chemistry and Physics*, Plenum Press, New York, NY, 1989.

[113] C. Stein, S. Drysdale, and K. P. Bogart, *Discrete Mathematics for Computer Scientists*, Pearson/Addison-Wesley, Boston, MA, 2011.

[114] I. Stewart, *Why Beauty is Truth: A History of Symmetry*, Basic Books, New York, NY, 2007.

[115] G. Strang, *Linear Algebra and its Applications*, 2nd edition, Academic, New York, NY, 1976.

[116] A. Terras, *Fourier Analysis on Finite Groups and Applications*, Cambridge University Press, Cambridge, 1999.

[117] A. Terras, *Zeta Functions of Graphs: A Stroll through the Garden*, Cambridge University Press, Cambridge, 2011.

[118] A. Terras, *Harmonic Analysis on Symmetric Spaces—Euclidean Spaces, the Sphere, and the Poincaré Upper Half-Plane*, Springer, New York, NY, 2013.

[119] A. Terras, *Harmonic Analysis on Symmetric Spaces—Higher Rank Spaces, Positive Definite Matrix Space*, Springer, New York, NY, 2016.

[120] A. D. Thomas and G. V. Wood, *Group Tables*, Shiva, Orpington, 1980.

[121] N. D. Tyson, *Death by Black Hole*, Norton, New York, NY, 2007.

[122] N. Ya Vilenkin, *Stories about Sets*, Academic Press, New York, NY, 1968 (available as an ebook from Google).

[123] J. Walker, *Codes and Curves*, Student Mathematical Library, 7, IAS/Park City Mathematical Subseries, American Mathematical Society, Providence, RI; Institute for Advanced Study (IAS), Princeton, NJ, 2000.

[124] B. L. van der Waerden, *Algebra*, Volumes I, II, Springer-Verlag, New York, NY, 1991.

[125] H. Weyl, *Symmetry*, Princeton University Press, Princeton, NJ, 1952.

[126] H. Weyl, *Gesammelte Abhandlungen*, Volume III, Springer-Verlag, New York, NY, 1968.

[127] R. A. Wilson, *The Finite Simple Groups*, Springer-Verlag, New York, NY, 2009.

[128] R. M. Young, *Excursions in Calculus: An Interplay of the Continuous and the Discrete*, Mathematical Association of America, Washington, DC, 1992.

Index

Abelian, 43
action, 137
addition modulo n, 28, 58
adelic group, 152
adjacency matrix, 55
affine group, 55, 71, 99, 150
affine space, 284
algebra of quaternions, 222
algebra word origin, xvi
algebraic closure, 232
algebraic extension field, 230
algebraic topology, 152
algebraically closed field, 232, 239
alternating function, 224
alternating group, 87
antisymmetric, 34
Archimedean solids, 122
arithmetic–geometric mean inequality, 12
Artin conjecture on primitive roots, 79
associative, 10, 43
associative algebra, 223
autocorrelation, 255
automorphism, 92, 189, 238, 239
axiom of choice, 232

Banach–Tarski paradox, 233
basis, 215
BCH code, 262
benzene, xiii, 132, 133
Bézout's identity, 23, 184, 199
bijective, 37
binary operation, 42, 43
binomial theorem, 40
Boolean functions, 120, 264
Bourbaki, xviii, 218, 222
buckyball, 152
Burnside, 114, 117
Burnside's lemma, 117, 146

cancellation law, 10
Cantor, 6
Cartesian product, 7, 232
Cauchy, 13, 114, 118, 228, 243
Cauchy sequence, 13
Cauchy's theorem on elements of prime order in a group, 118
Cauchy–Binet formula, 228
Cayley, 28, 36, 43, 52
Cayley graph, 52, 55, 258
Cayley table, 28
Cayley's theorem, 91
Cayley–Hamilton theorem, 242
center, 70
centralizer, 118
character, 129
characteristic, 169
characteristic polynomial of a linear recurrent sequence, 251
characteristic polynomial of a matrix, 242
chemistry, xiii, 55, 114, 133, 273
Chinese remainder theorem, 110, 193
circular reasoning, 22
class equation, 118
classification of finite simple groups, 101, 152
Clay Mathematics Institute problems, 153
closure, 43
code, 257
combinations of n objects k at a time, 39
combinatorics, 38
commutative, 10, 43, 159
commutator subgroup, 101
companion matrix, 213, 253
complement, 6
complete graph, 153
complex conjugate, 222
complex conjugation, 192
complex numbers, xiii, 3, 5, 102, 104, 105, 129, 158, 168, 175, 192, 229, 232, 236, 272, 287
composition of functions, 35
congruence, 27
congruence class, 28

congruent, 27
congruent number problem, 287
conjugacy class, 92, 118
conjugate, 92, 93, 266
conjugation, 116, 118
conjugation in fields, 222, 238, 240, 254
conjugation in groups, 92
connected graph, 54
conservation of energy, 141
containment, 6
contrapositive, 4
converse, 4
convolution, 129
coset, 94, 99, 117, 174, 185, 200, 229, 232, 259
countable set, 38
Coxeter group, 145
Cramer's rule, 228
cryptanalysis, 124
cryptography, 124, 186, 269, 282, 293, 296
cryptology, 124
crystallography, 72, 134
cube, 8, 56
cycle, 82
cycle diagram, 80
cyclic code, 259
cyclic group, 47, 69, 73
cyclobutadiene, 133

Dedekind, 174
degree of a field extension, 230
degree of a polynomial, 163, 165
Delenn, 295
denumerable set, 38
derivative, 41, 104, 136, 236, 289
determinant, 61, 88, 104, 224
dicyclic group, 149
difference equation, 251
dihedral group, 46, 55, 66
dimension, 111, 215, 222
Diophantine equations, xvi
direct sum or product, 108, 179, 212
directed graph, 52
discrete logarithm problem, 75, 296
discriminant of a quadratic, 201
discriminant of an elliptic curve, 284
disjoint, 7
disjoint cycle notation, 82
distributive, 10, 159
divides, 19
division algebra, 223
division algorithm, 21, 182
divisor, 19, 181
DNA, 134
dodecahedron, 56, 122, 150, 269
double coset, 98
double helix, 134
Dr. Who, 295
dual group, 129

echelon form, 206
eigenvalue, 97, 133, 213, 233, 239, 240, 268, 272, 273
eigenvector, 272, 273
Eisenstein series, 288
elementary divisor, 211
elementary matrix, 209
elementary row operations, 206
elementary symmetric polynomials, 117, 204
ElGamal cryptosystem, 295
elliptic curve, 282
empty set, 6
encryption, 124
energy, 133, 137, 140, 141
equivalence class, 32
equivalence relation, 31
error-correcting code, 258
Euclid, xv
Euclid's lemma, 23
Euclidean algorithm, 22, 180, 184
Euclidean domain, 184
Euler, xvii, 26, 135
Euler phi-function, 60, 64
Euler's criterion, 202
Euler's identity, 104
Euler–Lagrange equation, 136
even permutation, 85
expander graph, 269
expansion by minors, 227, 229
exponent, 65
extension field, 171
external direct product, 113
extremal, 136

factor group, 99
factor ring, 174
feedback shift register, 185, 186, 251
Fermat's last theorem, xvi
Fermat's little theorem, 95
Fibonacci numbers, 17, 254
field, 167
field generated by an element of a larger field, 230
field of fractions or quotients, 203
field of rational functions, 204

finite-dimensional vector space, 215
finite field, xvi, xvii, 168, 175, 185, 221, 232, 233, 236
finite logarithm, 73
finite subgroup test, 69
finite upper half plane, 266
finitely generated group, 211
first principal of mathematical induction, 14
fixed point set, 116
floor, 20
formal power series, 164
Fourier, 106, 130
Fourier transform, 130, 263, 265
fractional linear transformation, 266
free group, 56, 153
free module, 217
Frobenius, 114, 223, 238, 280
Frobenius automorphism, 238
function, 35
functional, 136
fundamental group of a graph, 152
fundamental theorem of Abelian groups, 212
fundamental theorem of algebra, 232
fundamental theorem of arithmetic, 23, 180
fundamental theorem of Galois theory, 240
fundamental theorem on symmetric polynomials, 117, 204

Gödel incompleteness theorem, 3, 197
Galois, 243
Galois field, xvii, 222
Galois group, xv, 92, 101, 204, 239
Gauss, xiii, xvii, 5, 15, 24, 26, 106, 206, 222, 232
Gaussian elimination, 206, 210
Gaussian integers, 12, 161
general linear group, 61, 70, 97, 104, 116, 150, 210, 221, 229, 253, 266, 273
generator matrix, 258
generators of a group, 49, 52, 55, 69, 73, 79
geodesic, 135, 266
Golay code, 264
golden ratio, 254
golden rectangle, 254
Google bombing, 280
Google matrix, 278
Gram–Schmidt process, 273
graph, 55
graphs of Lubotzky, Phillips, and Sarnak, 270
greatest common divisor, 21, 23, 83, 184
group, 43
group action on a set, 114
group algebra, 131, 223
group of an elliptic curve, 285
groups of order less than or equal to 15, 148

Hadamard matrix, 262
Hamilton, 112, 222
Hamming, 257
Hamming code, 260
Hamming weight, 257
Hasse diagram, 34
Hasse's theorem on the number of points on an elliptic curve, 292
Heisenberg group, 151
Hermitian matrix, 273
Hilbert, 153, 174, 272
Hilbert's problems, 153
homogeneous linear recurrent sequence, 251
homomorphism, 102, 104, 189, 190

icosahedron, 56, 63
ideal, 173
ideal generated by a set, 174
identity, 10, 37, 43, 63, 159, 160, 162, 169, 221, 224, 227, 258, 285
identity matrix, 57, 61
image, 35, 91, 104, 105, 190, 220, 258
imaginary numbers, 5
impulse response sequence, 251
inclusion–exclusion principle, 41
indeterminate, 86, 115, 163, 164, 180, 199
induction hypothesis, 15
inductive definition, 16
infinite dimensional vector space, 215
infinite set, 37
injective, 36
inner automorphism, 92
inner product, 130
integers, xv, 10–12
integers modulo n, 28, 58, 98
integral domain, 166
internal direct product, 113
intersection, 6
invariance under a transformation, 138
inverse, 10, 43, 159
inverse function, 37
inverse image, 39, 192
inversion of Fourier transform, 132, 265
irreducible polynomial, 181
isomorphism, 89, 90, 104, 107, 170, 189, 190, 218, 222, 235, 238

Jacobi identity, 165
Jordan form, 213

kernel, 102, 103, 190, 220, 259
kinetic energy, 137
Klein, 50
Klein 4-group, 54, 60, 96, 108, 147
Krawtchouk polynomial, 265
Kronecker delta, 228

Lagrange's theorem, 95
Lagrangian, 137
Laplace expansion of determinant, 228
Latin square, 54
lattice in the plane, 288
leading coefficient of a polynomial, 181
least common multiple, 82, 109
left group action, 114
Legendre symbol, 291
Lehmer, 244, 250
Lie bracket, 165
Lie group, 152
line at infinity, 284
linear combination, 23, 133, 215
linear congruence, 29, 75, 77, 194
linear congruential random number generator, 245
linear function or mapping, 36, 62, 93, 103, 218
linear map, 218
linear recurrent sequence, 251
linearly independent vectors, 215
Lorentz group, 142
Lubotzky, Phillips, and Sarnak graphs, 271

majority rule, 265
mapping or map, 35
Markov chain, 274
Markov chain Monte Carlo methods, 244
Markov matrix, 274
mathematical induction, 14, 16
mathematics as a language, 5
matrix exponential, 107
matrix multiplication, 36
matrix of a linear mapping, 219
maximal ideal, 177, 183
Maxwell's equations, 138, 141
Mersenne prime, 4
methane, 57, 88
metric, 258
minimal polynomial, 231, 232
minor, 227
modular arithmetic, 26
modular form, 266, 271, 288
modular group, 51, 266
module, 217
momentum, 140
monic polynomial, 181
monomial, 165
monster group, 101, 152
Mordell's theorem on elliptic curves, 287
multilinear function, 224
multiple, 19
multiplication modulo n, 28, 59
multiplicative group of finite field is cyclic, 79, 202, 238
multiplicity of a root of a polynomial, 183

n factorial, 16
natural numbers, 9, 12, 24
natural projection, 105
Newton's law, 137
Noether, xvii, 104, 131, 135, 174
Noether's theorem, 139
non-Euclidean geometry, 5, 266
non-singular elliptic curve, 284
non-singular square matrix, 62, 107, 207, 229
norm, 129, 134, 222, 266
normal subgroup, 95
normalizer, 119
nullity, 220
nullspace, 220
number of elements in a finite set, 37

octahedron, 56, 122, 151, 267
odd permutation, 85
one-parameter group, 107, 141
one-step subgroup test, 68
one-to-one, 36
onto, 36
orbit, 83, 116
orbit/stabilizer theorem, 117
order, 11
order of a finite group, 65
order of an element in a group, 66, 77
orientation of the coordinates, 227
orthogonal group, 57
orthogonality, 131
outer automorphism, 92

p-group, 119
Pólya enumeration theory, 120
page rank, 279
parallelepiped, 227
parallelotope, 227
parity check matrix, 259
partial order, 34
partition, 33
Pascal's triangle, 40

period of a linear recurrent sequence, 252
permutation, 44, 81, 83
permutation matrix, 92, 209
permutation notation, 44
Perron theorem, 280
photo number 51, 134
physics, 50, 135, 152, 273
pigeonhole principle, 39
pivot, 206, 217
Platonic solids, 56
Poincaré, 266, 270
point at infinity, 284
polynomial in two indeterminates, 165
polynomial in n indeterminates, 86, 115
polynomial ring, 163, 175, 180
poset, 34
poset diagram, 34
potential energy, 137
power method to find dominant eigenvector, 279
power series, 18, 164
powers of group elements, 64
presentation, 55, 153
prime, 20
prime ideal, 176
primitive polynomial, 185, 186, 238
primitive root, 79
primitive root of unity, 233, 261, 264
principal ideal, 174, 210, 260
principal ideal domain, 210
principle of least action, 137
product of ideals, 179
projection, 105, 112
projective general linear group, 270
projective space, 284
projective special linear group, 152
proof by contradiction, 4
proper ideal, 173
proper subgroup, 67
proper symmetry, 56
pseudo random numbers, 244
public-key cryptography, 124, 125, 293, 295
Pythagoreans, 5, 24

quadratic formula, 101, 201
quadratic reciprocity law, 311, 312
quadratic residue code, 264
quantum physics, 50
quaternion, xvi, 112, 222
quaternion group, 112, 147
quintic equation, 101, 241
quotient group, 99
quotient ring, 174

Ramanujan, 272
Ramanujan graph, 269, 270
random number generator, 245
range, 220
rank of a free module, 217
rank of a linear map, 220
rank of an Abelian group, 287
rational canonical form, 213
rational functions, 203, 204, 230
rational numbers, xvi, 5, 12, 15, 158, 167, 202, 231, 241, 254, 286
real numbers, xiii, xvi, 5, 8, 12, 102, 105, 141, 158, 168, 226, 231, 247, 273
reciprocal polynomial, 241
reducible polynomial, 181
Reed–Muller code, 262
Reed–Solomon code, 262
reflexive, 31, 34
relation, 30
relation between group elements, 49, 55
relation between rank and nullity, 220
relatively prime, 21
restriction of a function, 38
resultant, 285
right group action, 114
ring, 159
ring generated by an element of a larger ring, 187, 230
ring of polynomials in several indeterminates, 204
roots of a polynomial, xv, 101, 164
rotation group, 141
row operations, 206
row rank of a matrix, 217, 258
row-reduced echelon form, 206
RSA cryptography, 124, 125
ruler and compass constructions, 241
Russell's paradox, 6

scalar, 214
Schreier graph, 96
Schroedinger equation, 138
Schur decomposition, 273
second principle of mathematical induction, 16
semi-direct product, 148
shidoku or junior sudoku, 143
sign of a permutation, 86
similar matrices, 93, 213, 214, 221, 273
simple algebra, 223

simple field extension, 239
simple group, 100
Smith normal form of a matrix, 211
solvable group, 241
space group, 72, 152
span of set of vectors, 215
spanning tree, 153
special linear group, 152
special orthogonal group, 56, 57
special relativity, 142
spectral theorem, 133, 273
spectroscopy, 134, 273
spectrum, 133, 134, 142, 268, 272
splat, 129
splitting field of a polynomial, 234
sporadic group, 101
stabilizer, 116
state vector of linear recurrent sequence, 253
statistical tests, 255
subfield, 171
subfield test, 171
subgroup, 67
subring, 161
subring test, 161
subset, 6
subspace, 216
sudoku, 143
sum of ideals, 179
surjective, 36
switching functions, 120, 264
Sylow p-subgroup, 119
Sylow theorems, 119
symmetric, 31
symmetric group, 81
symmetric matrix, 55
symmetric polynomial, 117
symmetry, xiii, 47, 48, 50, 54, 152, 154
system of Euler–Lagrange equations, 137

tessellation, 270
tesseract, 9, 108
tetrahedral group, 87
tetrahedron, 56, 87, 153
torsion subgroup, 287
torus, 111, 195, 288
transcendental extension field, 230
transitive, 11, 31, 34
translation-invariant metric, 258
transpose of a matrix, 57, 97, 141, 143, 219, 226, 272, 281
transposition, 85
triangle inequality, 12, 258
trichotomy, 11
truncated icosahedron, 152
two-step subgroup test, 68

undirected graph, 55
union, 6
unique factorization into primes, 23
unit, 20, 59, 162, 180
unitary matrix, 273

Vandermonde determinant, 88
vector, 214
vector space, xiii, 61, 108, 130, 131, 185, 214
very useful polynomial, 86
vibrating system, 133, 142
viruses, 58
volume function, 227
voting, 244, 264

webpages, 276
websites, 276
Wedderburn theory, 224
Wedderburn's theorem on finite division rings, 173
Wedderburn's theorem on simple algebras, 223
Weierstrass function, 287
well defined, 28, 35
well-ordering axiom, 12
Wilson's theorem, 64
word problem, 153

X-ray diffraction spectroscopy, 134

zero divisor, 10
Zorn's lemma, 232